GENERALIZED RECURSION THEORY II

STUDIES IN LOGIC

AND

THE FOUNDATIONS OF MATHEMATICS

VOLUME 94

NORTH-HOLLAND PUBLISHING COMPANY
AMSTERDAM • NEW YORK • OXFORD

GENERALIZED RECURSION THEORY II

Proceedings of the 1977 Oslo Symposium

Edited by

J. E. FENSTAD
University of Oslo, Norway

R. O. GANDY
University of Oxford, England

G. E. SACKS
Harvard University and M.I.T.
Cambridge, Mass., U.S.A.

N·H
P C

1978

NORTH-HOLLAND PUBLISHING COMPANY
AMSTERDAM · NEW YORK · OXFORD

ISBN: 0 444 85163 1

Published by:

North-Holland Publishing Company — Amsterdam • New York • Oxford

Sole distributors for the U.S.A. and Canada:

Elsevier North-Holland, Inc.
52 Vanderbilt Avenue
New York, N.Y. 10017

Library of Congress Cataloging in Publication Data

Symposium on Generalized Recursion Theory, 2d, University of Oslo, 1977.
Generalized recursion theory II.

(Studies in logic and the foundation of mathematics ; v. 94)
1. Recursion theory--Congresses. I. Fenstad, Jens Erik. II. Gandy, R. O. III. Sacks, Gerald E. IV. Title. V. Series.
QA9.6.S95 1977 511'.3 78-5366
ISBN 0-444-85163-1

PRINTED IN THE NETHERLANDS

PREFACE

The Second Symposium on Generalized Recursion Theory was held at the University of Oslo, June 13-17, 1977. The Symposium received generous financial support from the Norwegian Research Council, the International Union of History and Philosophy of Science, North-Holland Publishing Company, the Norwegian Mathematical Association, and from the University of Oslo. About 40 people attended the meeting.

The program consisted partly of short courses and survey lectures on topics of current interest. In addition there were a number of more specialized invited lectures. The participants were invited to submit papers for the proceedings, and the Editors are happy to present this collection of 19 papers covering almost all areas of generalized recursion theory.

The Editors are especially proud to include a new and important paper by S.C. Kleene, who gave the opening address of the Symposium. It was above all the fundamental work of S.C. Kleene in the 1950's that opened up the field of generalized recursion theory. It is our hope that the present volume as well as its predecessor testify of the richness and vitality of this branch of mathematics.

The Editors.

TABLE OF CONTENTS

PREFACE v

TABLE OF CONTENTS vii

J. BARWISE
Monotone quantifiers and admissible sets 1

J. BERGSTRA
The continuous functionals and ${}^{2}E$ 39

S. FEFERMAN
Recursion theory and set theory: A marriage of convenience 55

J.E. FENSTAD
On the foundation of general recursion theory: Computations versus inductive definability 99

S.D. FRIEDMAN
An introduction to β-recursion theory 111

S.D. FRIEDMAN
Negative solutions to Post's problem, I 127

J.M.E. HYLAND
The intrinsic recursion theory on the countable or continuous functionals 135

A.S. KECHRIS
Spector second order classes and reflection 147

S.C. KLEENE
Recursive functionals and quantifiers of finite types revisited I 185

M. LERMAN
Lattices of α-recursively enumerable sets 223

W. MAASS
High α-recursively enumerable degrees 239

W. MAREK and A.M. NYBERG
Extendability of ZF models in the von Neuman hierarchy to models of KM theory of classes 271

J. MOLDESTAD
On the role of the successor function in recursion theory 283

D. NORMANN
Set recursion 303

D. NORMANN and V. STOLTENBERG-HANSEN
A non-adequate admissible set with a good degree-structure 321

R.A. SHORE
On the $\forall\exists$-sentences of α-recursion theory 331

S.G. SIMPSON
Short course on admissible recursion theory 355

V. STOLTENBERG-HANSEN
Weakly inadmissible recursion theory 391

S.S. WAINER
The 1-section of a non-normal type-2 objects 407

J.E. Fenstad, R.O. Gandy, G.E. Sacks (Eds.)
GENERALIZED RECURSION THEORY II

MONOTONE QUANTIFIERS AND ADMISSIBLE SETS

Jon Barwise
University of Wisconsin
Madison, Wisconsin

1. Informal introduction and poll on "most" and "quite a few"
2. Monotone and bounded quantifiers
3. The axiomatic theory $\underset{\sim}{Q}^{\#}$-KPU
4. Examples of $\underset{\sim}{Q}^{\#}$-admissible sets
5. $\underset{\sim}{Q}$-constructible sets
6. $\underset{\sim}{Q}^{\#}$-$\mathbb{HYP}(\mathfrak{m}, \underset{\sim}{q})$ and the Represententability Theorem
7. $\underset{\sim}{Q}$-positive inductive definitions
8. $\underset{\sim}{Q}^{\#}$-deterministic inductive definitions
9. Results of the poll
10. A gap in Admissible Sets and Structures
References

Mostowski wanted to title one of his papers "Kleene's theories as I see them". We should give this paper a similar title, with Kleene replaced by Aczel and Moschovakis. We show how their work on induction in monotone quantifiers can be carried out in the context of our book Admissible Sets and Structures. All references of the form II 6.8 refer to the appropriate item in this book, which we naturally assume the reader has always close at hand.

Most of the material in this paper was presented in my course at UCLA in the spring of 1976. We would like to thank the people attending this class especially Y. N. Moschovakis, J. Schlipf and H. Enderton, for their suggestions

The preparation of this paper was partially supported by Grant MCS76-06541.

1. Informal introduction and poll

The study of generalized quatifiers, quantifiers other than "for all" and "there exist" has been taken up in recent years by two branches of logic, model theory and recursion theory. It would seem that admissible set theory would be one area where the two approaches could work together, and it is this topic which we take up here.

In model theory it has been traditional to study quatifiers based on cardinality considerations. It was in recursion theory, especially in the paper Aczel [1970] and the monograph Moschovakis [1974], that the importance of just plain monotonicity emerged, and it is with these quantifiers that we will concern ourselves. Everything in this paper was inspired by these two references, plus results from the folklore of the model theory of generalized quantifiers.

A quantifier Qx is monotone (increasing) if it satisfies the condition: for all unary predicates A, B with $A \subseteq B$,

$$QxA(x) \quad \text{implies} \quad QxB(x).$$

Examples of monotone quantifiers.

a) (English) "Most x", "Many x", "Quite a few x" are all monotone. One can think of most of the precise mathematical quantifiers below as attempts to make one of these quatifiers precise. It is interesting to speculate, however, whether any of these are in fact precise enough already to list some commonly accepted axioms about them. At the end of this section the reader will find a true/false questionnaire that I passed out in my department. The results of the poll are given in §9.

b) (Model theory) "for infinitely many", "for all but a finite number", "for uncountably many", "for all but countably many" are all monotone quatifiers.

c) (Measure theory) "for all but a set of measure 0".

d) (Topology) "for all x in some neighborhood of y" is a monotone quatifier which binds the variable x but not y. It is thus like the bounded quatifiers $\forall x \in y$ and $\exists x \in y$. Both are examples of what we call an indexical quatifier $Q^y x$, the y being the index.

e) (English, again) Some of Montague's work in the model theory of English suggests that noun phrases often act more as indexical monotone quatifiers than as simple constant symbols. The sentences:

1) Everyone walks to work;
2) Most people walk to work;
3) John walks to work

assert, respectively, that the set W of people who walk to work contains

1) the set A of all people under discussion;
2) a fairly good sized subset B of A, say $\frac{|B|}{|A|} > 1/2$;
3) $\{\text{John}\}$.

In this last, we can think of "John" as an indexical quatifier $Q^y x$ where y is "John".

f) (Recursion Theory) Here we assume that we can regard every finite sequence $x_1, \ldots, x_k$ of objects in our domain as (coded by) another object $\langle x_1, \ldots, x_k \rangle$ in our domain. The two most important monotone quatifiers here are the Souslin quantifier $SxA(x)$ defined by

$$\forall y_1 \ldots \forall y_n \ldots \bigvee_{n<\omega} A(\langle y_1 \ldots y_n \rangle)$$

and the game quantifier $GxA(x)$ defined by

$$\forall y_1 \; \exists z_1 \forall y_2 \; \exists z_2 \ldots \bigvee_{n<\omega} A(\langle y_1 z_1 \ldots y_n z_n \rangle).$$

It won't be important for what follows to understand these examples.

Given an ordinary first-order structure $\mathfrak{M}$, one may ask what sets in the universe V_M of all sets built on the urelements of $\mathfrak{M}$ does one really understand on the basis of quantification over $\mathfrak{M}$. If by "quantification" one means ordinary $\forall$ and $\exists$, then one way to answer this is by means of the admissible set $HYP(\mathfrak{M})$ studied in our book. But what if one wants to also allow some other monotone quantifiers over $\mathfrak{m}$ of the kind mentioned above? To answer this question, we introduce two new admissible sets HYP$(\mathfrak{m}, \underset{\sim}{q})$ and HYP$^{\#}(\mathfrak{m}, \underset{\sim}{q})$ where $\underset{\sim}{q}$ is a sequence of monotone quantification of $\mathfrak{m}$. By using ideas from Aczel [1970], Moschovakis [1974] and our book, we obtain several interesting characterizations of the sets in these admissible sets.

There is one important difference between the treatment here and that in the references just given. It has to do with the treatment of dual quantifiers. The <u>dual</u> $\breve{Q}$ of a quantifier Q is defined by

$$\breve{Q}x\ A(x) \quad \text{if and only if} \quad \neg Qx\ \neg A(x).$$

If Q is monotone so is $\breve{Q}$. Following the lead of Aczel and Moschovakis, recursion-theorists have always treated a quantifier and its dual as being on a par - if you can use one effectively, you can use the other, at least in any positive context. Model-theoretically, there are certain objections to this.

The dual of "for infinitely many x" is "for all but finitely many x" and the dual of "for uncountably many x" is "for all but countably many x". Notice that in both cases the original quantifier is <u>persistent</u>, that is, satisfies the condition

$$a \subseteq b \wedge Qx \in a\ A(x) \rightarrow Qx \in b\ A(x)$$

so that the dual satisfies the dual condition:

$$a \subseteq b \wedge \breve{Q}x \in b\ A(x) \rightarrow \breve{Q}x \in a\ A(x).$$

(This will be made precise in §2.) Model-theoretically, the original quantifiers persist from one structure $\mathfrak{m}$ to any $\mathfrak{n}$ containing $\mathfrak{m}$, the duals having the dual property.

There is an analogous but more complicated situation in English. The poll mentioned above and follow-up questioning suggests that well-determined English quantifiers do not always have well-determined duals. An English speaker uses "most" quite confidently in positive contexts, but in contexts where you would expect the use of a dual, he may use one of the several non-equivalent quantifiers, among which are "quite a few", "many", "a significant number" and occasionally even the non-monotonic quantifier "an unexpectedly large number of". (To see that this is non-monotonic, notice that "An unexpectedly large number of people voted for Witherall" does not imply "An unexpectedly large number of people of people voted".)

Thus, for model-theoretic and linguistic reasons, we treat a quantifier and its dual separately. Nothing is lost, though. For example, Moschovakis' notion of Q-admissible will be equivalent to our $Q, \check{Q}$-admissible. It is with the other approach that something is lost, namely the ability to study, say, induction in a quantifier in the absence of its dual.

To conclude this discussion of duals, we point out that the dual of "for all x in some neighborhood of y" is "every neighborhood of y contains an x such that" and the dual of a noun phrase like "John", thought of as a quantifier, is the same quantifier.

In many places the proofs of the results presented below will only be sketched. We will go into detail only where major changes are required. This seems to us one of the interesting aspects of the current project, that is, to see which proofs in the book are not the "right" proofs, in that they need to be drastically overhauled to get the stronger results presented here.

Here is the true/false poll mentioned above. We invite you to choose your own answers before turning to section 9. You may be undecided on some questions.

_____ 1. Most real numbers are not rational.

_____ 2. Most integers are not prime.

_____ 3. There are quite a few prime numbers.

_____ 4. If G is a free group generated by an infinite set I then most x in G are not in I.

_____ 5. If $\forall x[A(x)$ implies $B(x)]$ and Most x $A(x)$ then Most x $B(x)$.

_____ 6. If $\forall x[A(x)$ implies $B(x)]$ and Quite a few $x\ A(x)$ then Quite a few $x B(x)$.

_____ 7. $MxA(x)$ and $MxB(x)$ implies $\exists x[A(x)$ and $B(x)]$.

_____ 8. $MxA(x)$ and $MxB(x)$ implies $Qx[A(x)$ and $B(x)]$.

_____ 9. $MxA(x)$ and $MxB(x)$ implies $Mx[A(x)$ and $B(x)]$.

_____ 10. If not $MxA(x)$ then $Qx(\text{not } A(x))$.

_____ 11. If $Qx(\text{not } A(x))$ then not $MxA(x)$.

_____ 12. $MxA(x)$ and $QxB(x)$ implies $\exists x[A(x)$ and $B(x)]$.

In 7-12 we used M and Q for "Most" and "Quite a few".

2. Monotone and bounded quantifiers

Let L be a first-order language (a set of relation, function and constant symbols) and let $\underset{\sim}{Q}$ be a sequence of new quantifier symbols $Q_1, \breve{Q}_1, Q_2, \breve{Q}_2, \ldots, Q_k, \breve{Q}_k$. The logic $L(\underset{\sim}{Q})$ is formed like first-order logic but with a new formation rule added: if φ is a formula of $L(\underset{\sim}{Q})$ and x is a variable then $Qx\,\varphi$ is a formula of $L(\underset{\sim}{Q})$, for each Q in the sequence $\underset{\sim}{Q}$. The variable x is not free in $Qx\,\varphi$.

A <u>weak model</u> for $L(\underset{\sim}{Q})$, where $\underset{\sim}{Q} = Q_1, \breve{Q}_1, \ldots, Q_k, \breve{Q}_k$, consists of a pair $(\mathfrak{m}, \underset{\sim}{q})$ where:

i) $\mathfrak{m} = \langle M, \ldots \rangle$ is an ordinary structure for L

ii) $\underset{\sim}{q} = q_1, \ldots, q_k$ is a sequence of subsets of $P(M)$, the power set of M. These q_i are the <u>quantifiers</u> on $\mathfrak{m}$.

These are called weak models since they are not assumed to be monotone. Given $q \subseteq P(M)$ let $\breve{q} = \{X \subseteq M \mid (M - X) \notin q\}$. Satisfaction of formulas is defined as usual with the additional clauses:

i) $(\mathfrak{m}, \underset{\sim}{q}) \models Q_i x\, \varphi(x)$ iff $\{a \mid (\mathfrak{m}, q) \models \varphi[a]\} \in q_i$,

ii) $(\mathfrak{m}, \underset{\sim}{q}) \models \breve{Q}_i x\, \varphi(x)$ iff $\{a \mid (\mathfrak{m}, q) \models \varphi[a]\} \in \breve{q}_i$,

i.e. iff $\{a \mid (\mathfrak{m}, q) \not\models \varphi[a]\} \notin q_i$.

There is a useful completeness theorem for weak models that is implicit in the literature, see e.g., Keisler [1970]. The <u>axioms for</u> $L(\underset{\sim}{Q})$ are the usual ones for L plus, for each Q in the list $\underset{\sim}{Q}$, all the following:

A1) $\forall x(\varphi(x) \leftrightarrow \psi(x)) \rightarrow (Qx\,\varphi(x) \leftrightarrow Qx\,\psi(x))$

A2) $Qx\,\varphi(x) \leftrightarrow Qy\varphi(y)$

A3) $\check{Q}x\,\varphi(x) \leftrightarrow \neg Qx\neg\varphi(x)$.

In A2, y is not free in $Qx\varphi(x)$. In A3, Q is one of the Q_i, not $\check{Q}_i$. The rules for $L(\underset{\sim}{Q})$ are modus ponens and the usual one for universal generalization.

<u>2.1 Weak Completeness Theorem</u> (Folklore). A set T of sentences of $L(\underset{\sim}{Q})$ is consistent with the above axioms and rules iff T has a weak model $(\mathfrak{M}, \underset{\sim}{q})$. Moreover, if T is consistent, it has a model $(\mathfrak{M}, \underset{\sim}{q})$ with $\mathrm{Card}(M) \leq \mathrm{Card}(L) + \aleph_0$. This will be generalized in Theorem 2.6.

<u>2.2 Corollary.</u> If every finite subset of a set T of $L(\underset{\sim}{Q})$-sentences has a model, then T has a model.

Let $(\mathfrak{M}, \underset{\sim}{q})$ be a weak model, $\underset{\sim}{q} = q_1, \ldots, q_k$. The quantifier q_i is <u>monotone</u> if for all $X \subseteq Y \subseteq M$, if $X \in q_i$ then $Y \in q_i$. We can derive a completeness theorem for monotone quantifiers from Theorem 2.1 by a standard method.

<u>2.3 Completeness Theorem for Monotone Quantifiers.</u> A theory T of $L(\underset{\sim}{Q})$ has a model $(\mathfrak{M}, \underset{\sim}{q})$ where $q_{i_0}, \ldots, q_{i_n}$ are monotone iff T is consistent in $L(\underset{\sim}{Q})$ with the set of sentences

(M) $$\forall x(\varphi(x) \rightarrow \psi(x)) \rightarrow (Qx\varphi(x) \rightarrow Qx\psi(x))$$

for $Q = Q_{i_0}, \ldots, Q_{i_n}$.

<u>Proof.</u> Let $(\mathfrak{M}, \underset{\sim}{q})$ be a weak model of T plus all instances of (M). Let us show how to make a single q ($= q_1$ say) monotone. Given the original q_1, let q_1^d be the set of definable members of q_1. That is q_1^d is the set of sets of the form $\{a \mid \mathfrak{M}, \underset{\sim}{q}) \models \varphi[a, \underline{b}]\}$ where $(\mathfrak{M}, \underset{\sim}{q}) \models Q_1 x\varphi(x)[\underline{b}]$. An easy proof by induction shows that $(\mathfrak{M}, q_1, q_2, \ldots, q_k) \models \varphi[a_1 \ldots a_n]$ iff $(\mathfrak{M}, q_1^d, q_2, \ldots, q_k) \models \varphi[a_1, \ldots, a_n]$, for all $\underline{a} \in M$. Now let q_1' be

the set of all $X \subseteq M$ such that $Y \subseteq X$ for some $Y \in q_1^d$. We claim that $(\mathfrak{m}, q_1^d, q_2, \ldots, q_k) \models \varphi[\underline{a}]$, iff $(\mathfrak{m}, q_1^d, q_2, \ldots, q_k) \models \varphi[a_1, \ldots, a_n]$, for all $a \in M$. This is by induction on φ, the only interesting case being $(\mathfrak{m}, q_1', \ldots) \models Qx\varphi$ implies $(\mathfrak{m}, q_1^d \ldots) \models Qx\varphi$. Let $X = \{a \mid (\mathfrak{m}, q_1', \ldots) \models \varphi[a]\} \in q_1'$, so that $Y \subseteq X$ for some $Y \in q_1^d$. Suppose Y is defined in $(\mathfrak{m}, \underset{\sim}{q})$ by ψ. Thus, $(\mathfrak{m}, q_1^d, \ldots) \models Qx\psi(x)$ and $(\mathfrak{m}, q_1^d) \models \forall x(\psi(x) \rightarrow \varphi(x))$. An instance of (M) gives $(\mathfrak{m}, q_1^d, \ldots) \models Qx\psi(x)$ as desired. □

2.4 Corollary. The downward Löwenheim-Skolem Theorem as in 2.1 and Compactness Theorem as in 2.2 carry over to monotone quantifiers.

A monotone quantifier q on $\mathfrak{m}$ lives on $M_o \subseteq M$ if for all $X \subseteq M$, $X \in q$ iff $(X \cap M_o) \in q$. Since q is monotone, only the $(\Rightarrow)$ half of this says anything.

2.5 Lemma.

i) If q is monotone on $\mathfrak{m}$ and lives on M_o then the same is true of $\check{q}$.

ii) Any monotone quantifier q_o on $M_o \subseteq M$ has a trivial extension to a monotone quantifier q' on M that lives on M_o, namely $X \in q$ iff $(X \cap M_o) \in q_o$.

iii) If q is a quantifier with the property that $X \in q$ implies $(X \cap M_o) \in q$ then the monotone quantifier $q' = \{X \mid \exists Y \in q(Y \subseteq X)\}$ generated by q lives on M_o.

Proof. We only check (i). Let $X \in \check{q}$ but suppose that $(X \cap M_o) \notin \check{q}$. Then $(M - (X \cap M_o)) \in q$ but this set is $(M - X) \cup (M - M_o)$. Since q lives on M_o, $((M - X) \cap M_o) \in q$ so, by monotonicity, $(M - X) \in q$ which says that $X \notin \check{q}$. □

We now turn to indexical quantifiers. The language $L(\underset{\sim}{Q}^i)$ is like before, except the new formation rule reads: if φ is a formula and x, y are variables then $Q^y x\varphi$ is a formula for each Q in the list $\underset{\sim}{Q}$. The variable x is not free in $Q^y x\varphi$ but y is free in it. Thus, the meaning of $Q^y x\varphi$ will depend on an interpretation of y. A weak model is a pair $(\mathfrak{m}, \underset{\sim}{q})$ where $\underset{\sim}{q} = q_1, \ldots, q_k$ and each q_i is a function, $q_i = \{\langle a, q_i^a\rangle \mid a \in M\}$ which assigns a quantifier q_i^a to each $a \in M$. Satisfaction is defined as expected:

$$(\mathfrak{m}, \underset{\sim}{q}) \models Q^a x\, \varphi(x, a, \underline{b}) \quad \text{iff} \quad \{c \mid (\mathfrak{m}, \underset{\sim}{q}) \models \varphi(c, a, \underline{b})\} \in q^a$$

$$(\mathfrak{m}, \underset{\sim}{q}) \models \check{Q}^a x\varphi(x, a, \underline{b}) \quad \text{iff} \quad \{c \mid (\mathfrak{m}, \underset{\sim}{q}) \models \varphi(c, a, b)\} \in \check{q}^a.$$

A simple elaboration on the usual proof of Theorem 2.1 allows us to prove the completeness of the following axioms for $L(\underset{\sim}{Q}^i)$: all the usual axioms plus (A1)' - (A3)'.

(A1)': $\forall x(\varphi(x, y) \leftrightarrow \psi(x, y)) \rightarrow (Q^y x\varphi(x, y) \leftrightarrow Q^y x\psi(x, y))$

(A2)': $Q^y x\, \varphi(x) \leftrightarrow Q^y z\, \varphi(z)$

(A3)': $\check{Q}^y x\, \varphi(x) \leftrightarrow \neg Q^y z \neg\varphi(z)$.

The rules are, as before, modus ponens and generalization. The conditions on variables are similar to (A1) - (A3). Note that y is free in (A1)' - (A3)' so that, e.g., $\forall y(Q^y x\, \varphi(x) \rightarrow Q^y z\, \varphi(z))$ is a theorem.

<u>2.6 Weak Completeness for $L(\underset{\sim}{Q}^i)$.</u> A set T of sentences of $L(\underset{\sim}{Q}^i)$ is consistent with the above axioms and rules iff T has a weak model. Moreover, if T is consistent, it has a model $(\mathfrak{m}, \underset{\sim}{q})$ where $\text{Card}(M) \leq \text{Card}(L) + \aleph_0$.

<u>Proof sketch.</u> Let us treat the case where $\underset{\sim}{Q} = Q, \check{Q}$, just one indexed quantifier and its dual. Let C be a set of $\text{Card}(L) + \aleph_0$ new constant symbols. By the Henkin construction we may assume that T is maximal consistent such that $\exists x\, \varphi(x) \in T$ implies $\varphi(c) \in T$ for some $c \in C$. Let $\mathfrak{m}$ be the first-order Henkin model for T on the set $C(\text{mod} =)$. For each $b \in C$ let q^b consist of those subsets X of M of the form $X = \{c \mid \varphi(b, c)\}$ where the sentence $Q^b x\, \varphi(b, x)$ is in T. One proves that $\varphi \in T$ iff $(\mathfrak{m}, q) \models \varphi$, by induction on φ. The only interesting case is $Q^b x\, \varphi(b, x)$. So suppose $Q^b x\, \varphi(b, x)$ is in T. The induction hypothesis gives us

$$\{c \mid \varphi(b, c) \in T\} = \{c \mid (\mathfrak{m}, q) \models \varphi(c)\}.$$

The first set is in q^b by definition. Thus, $(\mathfrak{m}, q) \models Q^b x \varphi(b, x)$. Now suppose $(\mathfrak{m}, q) \models Q^b x \varphi(b, x)$. Then the set $X = \{c \mid (\mathfrak{m}, q) \models \varphi(b, c)\}$ $= \{c \mid \varphi(b, c) \in T\}$ is in q^b so there is some other $\psi(y, b)$ such that $Q^b y \psi(b, y) \in T$ and $X = \{c \mid \psi(b, c) \in T\}$. Using (A2), we can let y be x. Then for each c, $\varphi(b, c) \in T$ iff $\psi(b, c) \in T$. But then, by the witnessing property.

$$\forall x[\varphi(b, x) \leftrightarrow \psi(b, x)]$$

is in T. This, with (A1), gives $Qx\varphi(x) \in T$, as desired. □

Remarks.

1. To get a proof of Theorem 2.1, one need only erase all superscripts in the above proof. Alternately, one can derive 2.1 from 2.6. Of course, one could also prove a version of 2.6 that allowed both indexical and regular quantifiers.

2. After proving 2.6, H. Enderton pointed out an interesting fact. Namely, if instead of erasing all the superscripts you erase all the bound variables x from $Q^y x$, one obtains essentially Enderton's proof of a Completeness Theorem for Chang's modal model theory. One interprets $Q^y \varphi$ as some modal operator like "y finds it possible that φ", or "y knows that φ". This sort of formula can be viewed as a special case of our indexical quantifiers by forming $Q^y x \varphi(y)$ where x is not free in $\varphi(y)$. Here, though, one could also imagine applications to sentences like "John thinks that most x have A(x)" ($\text{Most}^{\text{John}}\, xA(x)$), thus taking into account the fact that people have different uses of quantifiers like most.

3. Unaware of either Enderton's results or ours, Sgro [to appear] proved a similar result in connection with the topological example mentioned in §1.

2.7 Corollary. The compactness theorem holds for $L(\underset{\sim}{Q}^i)$.

We now turn our attention to a special kind of indexical monotone quantifier, so-called bounded quantifiers.

Blanket assumption. For the rest of this paper, L^* refers to the language $L(\epsilon, \ldots)$ defined in I. 2.1 of Admissible Sets and Structures. $\mathfrak{A}$, $\mathfrak{B}$ range over structures for L^*, say $\mathfrak{A} = (\mathfrak{M}; A, E, \ldots)$. Recall that $a_E = \{b \in M \cup A \mid b\, E\, a\}$. If $p \in M$ then $p_E = 0$.

Definition. An indexical quantifier q on $\mathfrak{A}$ is a bounded quantifier if for each a, q^a is monotone and lives on a_E. Note that for $p \in M$, q^p must live on $p_E = 0$ so must be trivial, so we ignore them. Also note that, by Lemma 2.5, the dual of a bounded quantifier is a bounded quantifier.

Examples.

a) The usual bounded quantifiers $\forall x \in a$, $\exists x \in a$ are special cases. Let $\forall^a = \{X \subseteq M \cup A \mid a_E \subseteq X\}$, $\exists^a = \{X \subseteq M \cup A \mid X \cap a_E \neq 0\}$. Then $\forall^a x$ is $\forall x \in a$ and $\exists^a x$ is $\exists x \in a$.

b) Let $Q^a_\kappa = \{X \subseteq M \cup A \mid \mathrm{Card}(X \cap a_E) \geq \kappa\}$. Then $Q^a_\kappa x$ means "there exist $\geq \kappa\, x$ in a". Thus

$$Q^a_\omega\, x \quad \text{means} \quad \exists^{\geq \aleph_0} x \in a$$

$$\breve{Q}^a_\omega\, x \quad \text{means} \quad \text{"for all but a finite number of } x \text{ in } a\text{"}$$

To axiomatize bounded quantifiers we need, in addition to (A1)' - (A3)', the following:

(B1) $\forall x(\varphi(x,y) \rightarrow \psi(x,y)) \rightarrow (Q^y x \varphi(x,y) \rightarrow Q^y x \psi(x,y))$

(B2) $Q^y x \varphi(x,y) \rightarrow Q^y x(y \in x \wedge \varphi(x,y))$.

2.8 Completeness Theorem for bounded quantifiers. A set of T of $L(\mathcal{Q}^i)$ sentence has a model $(\mathfrak{A}, q)$ where each of $q_{i_1}, \ldots, q_{i_n}$ are bounded quantifiers iff T is consistent with all of (B1), (B2) for Q one of $Q_{i_1}, \ldots, Q_{i_n}$.

Proof. Almost like the proof of 2.3 from 2.1. □

2.9 Corollary. The compactness and Lowenheim-Skolem Theorem hold for bounded quantifiers.

Given a bounded quantifier Q on $(\mathfrak{A}, q)$, we say that Q is persistent on $(\mathfrak{A}, q)$ if for all a, b, X

$$X \in q^a \quad \text{and} \quad a \subseteq b \quad \text{implies} \quad X \in q^b .$$

We can axiomatize persistent bounded quantifiers by:

$$a \subseteq b \wedge Q^a x \varphi \rightarrow Q^b x \varphi .$$

Note that in this case $\check{Q}$ satisfies:

$$a \subseteq b \wedge \check{Q}^b x \varphi \rightarrow \check{Q}^a x \varphi .$$

We gave two examples of this in §1.

3. The axiomatic theory $\underset{\sim}{Q}^{\#}$-KPU.

The point of axiomatizing bounded quantifiers in §2 was to allow us to formulate the axioms for $\underset{\sim}{Q}^{\#}$-admissible sets.

Let $L^*(\underset{\sim}{Q})$ be the logic of bounded quantifiers developed in the previous section. Thus, we drop the superscript i in $L^*(\underset{\sim}{Q}^i)$. We assume axioms (A1)' - (A3)', (B1), (B2) throughout this section. We write Q for a typical member of the list $\underset{\sim}{Q} = Q_1, \check{Q}_1, \ldots, Q_k, \check{Q}_k$ and write $Qx \in y\varphi(x)$ for $Q^y x\varphi(x)$.

We want to consider the case where some of the quantifiers in the list $Q_1, \check{Q}_1, \ldots, Q_k, \check{Q}_k$ are sharper than others, sharper in a sense that comes from recursion theory. Consider, for example, some predicate $A(x)$ which is "r.e.-like" in that if $A(x)$ is true, we can eventually realize this fact but if $A(x)$ is false, we may not be able to realize the fact. Intuitively, Q is sharp if we can realize that $QxA(x)$ holds without knowing of each x, whether or not $A(x)$ holds. For example, "More than half x, $A(x)$" might be recognized to be true because we know that 101 of the 200 elements x in a model M have $A(x)$, even though we cannot figure out whether $A(x)$ holds for some of the other 99.

Thus, given $\underset{\sim}{Q} = Q_1, \check{Q}_1, \ldots, Q_k, \check{Q}_k$, we let $\underset{\sim}{Q}^{\#}$ denote the result of putting a sharp (#) superscript on some of the quantifiers in the list. The sharped members of $\underset{\sim}{Q}^{\#}$ are called sharped quantifiers. We write ${}^{\#}\underset{\sim}{Q}$ for the completely sharped list $Q_1^{\#}, \check{Q}_1^{\#}, \ldots, Q_k^{\#}, \check{Q}_k^{\#}$ and $\underset{\sim}{Q}$ if none are sharped.

3.1 Definitions

i) The class of $\Delta_0(\underset{\sim}{Q})$ formulas of $L^*(\underset{\sim}{Q})$ is the smallest collection containing the atomic formulas and negated atomic formulas, closed under $\wedge, \vee, \forall x \in a, \exists x \in a, Qx \in a$, for each Q in $\underset{\sim}{Q}$.

ii) The $\Sigma_1(\underset{\sim}{Q})$ formulas are those of the form $\exists y \varphi$ where φ is a $\Delta_0(\underset{\sim}{Q})$ formula.

iii) The $\Sigma(\underset{\sim}{Q}^{\#})$-formulas form the smallest collection containing the $\Delta_0(\underset{\sim}{Q})$ formulas and closed under $\wedge, \vee, \forall x \in a, \exists x \in a, \exists x$ and $Qx \in a$ for each sharped Q in $\underset{\sim}{Q}^{\#}$.

iv) The $\Pi(\underset{\sim}{Q})$ formulas are defined as in (iii) with $\exists x$ replaced by $\forall x$ and with $Qx \in a$ replaced by $\check{Q}x \in a$, where $\check{\check{Q}} = Q$.

Notice that the classes of $\Delta_0(\underset{\sim}{Q})$ and $\Sigma_1(\underset{\sim}{Q})$ formulas do not depend on which quantifiers are sharped, so we have left off the #. Also, a quantifier $Qx \in a$ never appears within the scope of a negation in the $\Sigma(\underset{\sim}{Q}^{\#})$-formulas. Thus, not even the $\Delta_0(\underset{\sim}{Q})$-formulas are closed under negation. However, since

$$\neg Qx \in a \varphi \longleftrightarrow \check{Q}x \in a \neg \varphi$$

is provable, we see that up to logical equivalence, the $\Delta_0(\underset{\sim}{Q})$-formulas are closed under $\neg$, and that the negation of a $\Sigma(\underset{\sim}{Q}^{\#})$ formulas is a $\Pi(\underset{\sim}{Q}^{\#})$-formula.

We now turn to the axioms of $\underset{\sim}{Q}^{\#}$-KPU.

3.2 Definition. $\underset{\sim}{Q}^{\#}$-KPU consists of the axioms of extensionality, pairing, union, the scheme of foundation for $L(\underset{\sim}{Q})$-formulas, the obvious schemes of $\Delta_0(\underset{\sim}{Q})$-separation and $\Delta_0(\underset{\sim}{Q})$-collection plus, for each sharped Q in $\underset{\sim}{Q}^{\#}$, the following scheme of Q-collection:

$$Qx \in a\, \exists y \varphi(x, y, a) \longrightarrow \exists b Qx \in a\, \exists y \in b \varphi(x, y, b),$$

where φ is in $\Delta_0(\underset{\sim}{Q})$.

We will see that $\check{Q}$-collection does not follow from Q-collection. Thus, if Q is sharped, it does not follow that we can treat $\check{Q}$ as sharped.

We will write $\underset{\sim}{Q}$-KPU for the case where none of the quantifiers are sharped and

${}^{\#}\underset{\sim}{Q}$-KPU where all are sharped, i.e., for $Q_1^{\#}, \check{Q}_1^{\#}, \ldots, Q_k^{\#}, \check{Q}_k^{\#}$-KPU.

The sharp (#)-notation comes from the theory of recursion in higher types. There is no real change in $\underset{\sim}{Q}$-KPU over KPU. It just amounts to have a lot of new basic relations, namely all the $\Delta_0(\underset{\sim}{Q})$-relations can now be thought of as atomic formulas. Thus, we will have little to say about it as an axiomatic theory until §8. We should point out, though, that Lemmas I.5.2 and I.5.4 will not hold for this theory. That is, we cannot treat $\Delta(\underset{\sim}{Q})$ relations and $\Sigma(\underset{\sim}{Q})$ functions as atomic. We are only able to do this in ${}^{\#}\underset{\sim}{Q}$-KPU. This turns out to be less serious than one would suspect.

We now list generalizations to $\underset{\sim}{Q}^{\#}$-KPU of some important facts from §I.4.

3.3 Lemma (see I.4.2). For each $\Sigma({}^{\#}\underset{\sim}{Q})$ formula φ the following are logically valid, that is, follow from (A1)' - (A3)' and (B1), (B2):

i) $\varphi^{(u)} \wedge u \subseteq v \rightarrow \varphi^{(v)}$,

ii) $\varphi^{(u)} \rightarrow \varphi$.

Proof. By induction on $\Sigma({}^{\#}\underset{\sim}{Q})$ formulas. Suppose $(Qx \in a\varphi)^u$ and $u \subseteq v$. Then $Qx \in a(\varphi^{(u)})$. By induction, $\forall x[\varphi^{(u)}(x) \rightarrow \varphi^{(v)}(x)]$ so, by monotonicity, $Qx \in a\varphi^{(u)}(x) \rightarrow Qx \in a\varphi^{(v)}(x)$. The proof of (ii) is similar. □

3.4 $\Sigma(\underset{\sim}{Q}^{\#})$-Reflection Principle. (See I.4.3) Let φ be a $\Sigma(\underset{\sim}{Q}^{\#})$ formula. Then $\underset{\sim}{Q}^{\#}$-KPU logically implies the universal closure of $\varphi \rightarrow \exists a\varphi^{(a)}$. In particular, every $\Sigma(\underset{\sim}{Q}^{\#})$ formula is $\underset{\sim}{Q}^{\#}$-KPU equivalent to a $\Sigma_1(\underset{\sim}{Q})$ formula.

Proof. One new case is added to the proof of I.4.3. Suppose $\psi(x) \leftrightarrow \exists a\psi^{(a)}(x)$. We need to prove that $Qx \in y\psi(x) \leftrightarrow \exists aQx \in y\psi^{(a)}(x)$, where Q is sharp. By 3.3ii, we need only check that $Qx \in y\psi(x) \rightarrow \exists aQx \in y\psi^{(a)}(x)$. Assume $Qx \in y\psi(x)$ so that $Qx \in y\psi^{(a)}(x)$. By Q-collection, $\exists bQx \in y\exists a \in b\psi^{(a)}(x)$. Let $a' = \cup b$. By 4.2i, $\exists a \in b\psi(x)^{(a)}$ implies $\psi^{(a')}(x)$ so, by monotonicity $Qx \in y\psi^{(a')}(x)$. □

There are two versions of the Σ-collection principle that are useful.

3.5 $\Sigma(\underset{\sim}{Q}^{\#})$-Collection Principle (See I.4.4). For every $\Sigma(\underset{\sim}{Q}^{\#})$ formula φ the following are theorems of $\underset{\sim}{Q}^{\#}$-KPU.

i) $\forall x \in a\ \exists y\ \varphi(x, y) \rightarrow \exists b[\forall x \in a\ \exists y \in b\ \varphi \wedge \forall y \in b\ \exists x \in a\ \varphi]$

ii) $Qx \in a\ \exists y\ \varphi(x, y) \rightarrow \exists b[Qx \in a\ \exists y \in b\ \varphi \wedge \forall y \in b\ \exists x \in a\varphi]$, for sharp Q.

Proof of (ii). Assume $Qx \in a\ \exists y\ \varphi(x, y)$. By $\Sigma(\underset{\sim}{Q}^{\#})$ reflection, there is a c such that $Qx \in a\ \exists y \in c\ \varphi^{(c)}(x, y)$. I.e., $Q^a x(x \in a \wedge \exists y \in c\ \varphi^{(c)}(x, y))$. Let $b = \{y \in c \mid \exists x \in a\varphi^{(c)}(x, y)\}$ by $\Delta_0(\underset{\sim}{Q})$ separation. Then $[x \in a \wedge \exists y \in c\ \varphi^{(c)}(x, y)]$ implies $[x \in a \wedge \exists y \in b\ \varphi^{(c)}(x, y)]$ which in turn implies, by 3.3(ii), $[x \in a \wedge \exists y \in b\ \varphi(x, y)]$. Thus, by monotonicity, $Q^a x(x \in a \wedge \exists y \in b\ \varphi(x, y))$, i.e., $Qx \in a\ \exists y \in b\ \varphi(x, y)$. □

The statements and proofs of $\Delta(\underset{\sim}{Q}^{\#})$-Separation, $\Sigma(\underset{\sim}{Q}^{\#})$-Replacement and strong $\Sigma(\underset{\sim}{Q}^{\#})$-Replacement for $\underset{\sim}{Q}^{\#}$-KPU are just as in I.4.5 - I.4.7. Similarly, definition by $\Sigma(\underset{\sim}{Q}^{\#})$-Recursion (as in I.6.4) can be verified in $\underset{\sim}{Q}^{\#}$-KPU as before.

The Truncation Lemma is one of the most useful technical results about KPU. We conclude this section by checking that a version of it holds for $\underset{\sim}{Q}^{\#}$-KPU.

Definition. $(\mathfrak{A}, \underset{\sim}{q}) \subseteq_{end} (\mathfrak{B}, \underset{\sim}{r})$ iff $\mathfrak{A} \subseteq_{end} \mathfrak{B}$ and for all $a \in A$ and all $X \subseteq a_E$, $X \in q_i^a$ iff $X \in r_i^a$.

3.6 Lemma. Let $(\mathfrak{A}, \underset{\sim}{q}) \subseteq_{end} (\mathfrak{B}, \underset{\sim}{r})$.

i) For all $\varphi \in \Delta_0(\underset{\sim}{Q})$ and all $\underline{a} \in A$,

$(\mathfrak{A}, \underset{\sim}{q}) \models \varphi[\underline{a}]$ iff $(\mathfrak{B}, \underset{\sim}{r}) \models \varphi[\underline{a}]$.

ii) For all $\varphi \in \Sigma(\underset{\sim}{Q})$ and all $\underline{a} \in A$,

$(\mathfrak{A}, \underset{\sim}{q}) \models \varphi[\underline{a}]$ implies $(\mathfrak{B}, \underset{\sim}{r}) \models \varphi[a]$.

In view of 3.6i and the fact that $\underset{\sim}{Q}$-KPU is just KPU relativized to all $\Delta_0(\underset{\sim}{Q})$ formulas, the truncation lemma for $\underset{\sim}{Q}$-KPU follows from that for KPU. The more general version for $\underset{\sim}{Q}^{\#}$-KPU, though, needs extra argument.

3.7 Theorem (Truncation Lemma, see I.8.9 and II.8.4) Let $\mathfrak{A}_{\mathfrak{m}} \subseteq_{end} \mathfrak{B}_{\mathfrak{n}}$ be (as in I.8.9) such that the "ordinals" of $\mathfrak{A}_{\mathfrak{m}}$ have no sup in $\mathfrak{B}_{\mathfrak{n}}$. Assume further that $(\mathfrak{A}_{\mathfrak{m}}, \underset{\sim}{q}) \subseteq_{end} (\mathfrak{B}_{\mathfrak{n}}, r)$ and that $(\mathfrak{B}_{\mathfrak{n}}, \underset{\sim}{r}) \models \underset{\sim}{Q}^{\#}$-KPU, where $\underset{\sim}{q}, \underset{\sim}{r}$ are bounded quantifiers.

i) $(\mathfrak{A}_{\mathfrak{m}}, \underset{\sim}{q}) \models \underset{\sim}{Q}^{\#}$-KPU - Foundation.

ii) In particular, if $\mathfrak{A}_{\mathfrak{m}} = \mathfrak{wf}(\mathfrak{B}_{\mathfrak{m}})$ and $\underset{\sim}{q} = \underset{\sim}{r} \upharpoonright \mathfrak{A}_{\mathfrak{m}}$ then $(\mathfrak{A}_{\mathfrak{m}}, \underset{\sim}{q})$ is a well founded, transitive model of $\underset{\sim}{Q}^{\#}$-KPU.

<u>Proof.</u> We need only check Q-collection in $\mathfrak{A}_{\mathfrak{m}}$. Assume $(\mathfrak{A}_{\mathfrak{m}}, \underset{\sim}{q}) = \mathfrak{A}$ is a model of $Qx \in a\, \exists y\, \varphi(x, y)$. Let $X = \{x \in a \mid \mathfrak{A} \models \exists y\, \varphi(x, y)\}$. Let γ be any "ordinal" of $B - A$ and let

$$Y = \{x \in a \mid (\mathfrak{B}, \underset{\sim}{r}) \models \exists \alpha < \gamma\, \exists y[rk(y) = \alpha \wedge \varphi(x, y)]\}.$$

Clearly $X \subseteq Y \subseteq a_E$ and $X \in q^a$ so $Y \in q^a$. Thus, $\mathfrak{B}$ is a model of:

$(1)_\gamma$ $Qx \in a\, \exists \alpha < \gamma\, \exists y[rk(y) = \alpha \wedge \varphi(x, y)]$.

By foundation, pick a least γ such that $(1)_\gamma$ is true in $\mathfrak{B}$. Since $(1)_\gamma$ holds for all $\gamma \in B - A$, the least such must be in $\mathfrak{A}$. Now, with this $\gamma \in A$ fixed, apply $\Sigma(\underset{\sim}{Q})$-collection in $\mathfrak{B}$ to get a set $b \in B$ such that both

(2) $Qx \in a\, \exists y \in b[\varphi(x, y)]$

(3) $\forall y \in b[rk(y) < \gamma]$

hold in $\mathfrak{B}$. By (3), $b \in \mathfrak{A}$. But (2) is $\Delta_0(\underset{\sim}{Q})$ so, by 3.6 i, must hold in $\mathfrak{A}$. □

4. Examples of $\underset{\sim}{Q}^{\#}$-admissible sets.

The definition of $\underset{\sim}{Q}^{\#}$-admissible is just what you'd expect. Recall that $\underset{\sim}{Q}^{\#}$ is a list of quantifiers, some of which may be sharped.

<u>4.1 Definition.</u> Let $\mathbb{A}$ be admissible and let $\underset{\sim}{q}$ be a sequence of bounded quantifiers on $\mathbb{A}$. $(\mathbb{A}, \underset{\sim}{q})$ is $\underset{\sim}{Q}^{\#}$-admissible iff $(\mathbb{A}, q) \models \underset{\sim}{Q}^{\#}$-KPU.

Of course, $(\mathbb{A}, \underset{\sim}{q})$ is $\underset{\sim}{Q}$-admissible iff for every $\Delta_0(\underset{\sim}{Q})$ definable relation R, $\langle \mathbb{A}, R \rangle$ is admissible in the usual sense. The notion of $\underset{\sim}{Q}^{\#}$-admissible is less straightforward.

As a first example, we note that if κ is regular then $(H(\kappa)_{\mathfrak{m}}, \underset{\sim}{q})$ is ${}^{\#}\underset{\sim}{Q}$-admissible for all bounded quantifiers $\underset{\sim}{q}$. (This is easy to check once one notices that $\underset{\sim}{Q}$-KPU + $\Sigma_1(\underset{\sim}{Q})$-separation implies ${}^{\#}\underset{\sim}{Q}$-KPU. In particular $\langle H(\kappa), q_\kappa\rangle$ is ${}^{\#}\underset{\sim}{Q}$-admissible, for κ-regular. On the other hand, if κ is singular, then there is a bounded quantifier q on $H(\kappa)$ such that $(H(\kappa), q)$ is not even Q-admissible. (Namely, if $\kappa = \sup_{\alpha<\beta} \lambda_\alpha$ where $\beta < \kappa$ we let $q^a = \{a\}$ if a is one of the λ_α's, otherwise let $q^a = 0$. Then the $\Delta_0(\underset{\sim}{Q})$ formula $Qx \in a\ (x = x)$ defines the set of λ_α's.)

Next we present an example of a class of $Q^{\#}$, $\check{Q}$-admissible sets which are not $Q^{\#}$, $\check{Q}^{\#}$-admissible. It goes back, basically, to recursive pseudo-wellorderings of ω, that is, to recursive linear orderings with descending sequences but with no hyperarithmetic descending sequences.

<u>4.2 Proposition.</u> Let $\mathfrak{m} = \langle\alpha(1+\eta), <\rangle$ where α is admissible, η is the order type of the rationals. Let $q = \{X \subseteq M \mid \exists\delta \in WF(<) \forall x \in M(\delta < x \rightarrow x \in X)\}$.

This $q = q^M$. For other $a \in \mathbb{HYP}(\mathfrak{m})$, let $q^a = 0$. Then $(\mathbb{HYP}(\mathfrak{m}), q)$ is $Q^{\#}$, $\check{Q}$-admissible but not $Q^{\#}$, $\check{Q}^{\#}$-admissible.

<u>Proof.</u> The Δ_0-formulas are essentially closed under $Qx \in a$.

$$Qx \in M\varphi(x) \leftrightarrow \exists\delta \in WF(<) \forall x \in M(\delta < x \rightarrow \varphi(x))$$
$$\leftrightarrow \exists d \in M \forall x \in M(d < x \rightarrow \varphi(x)).$$

Since if there is such a d, the least such must be in $WF(<)$ by an automorphism argument. Now assume $Qx \in M\, \exists y\varphi(x, y)$, where we can take $\varphi \in \Delta_0$ by the above. Then $\exists\delta \in WF(<) \forall x \in M(\delta < x \rightarrow \exists y\, \varphi(x, y))$. Pick such a δ. Now apply Δ_0-collection in $\mathbb{HYP}(\mathfrak{m})$ to get a set $b \in \mathbb{HYP}(\mathfrak{m})$ such that $\forall x \in M(\delta < x \rightarrow \exists y \in b\, \varphi(x, y))$. Then we have $Qx \in M\, \exists y \in b\, \varphi(x, y)$, as desired.

Now, to see that $(\mathbb{HYP}(\mathfrak{m}), q)$ does not satisfy $\check{Q}$-collection, note that $\check{Q}x \in M\varphi(x)$ is equivalent to

$$\forall y \in WF(<)\, \exists x \in WF(<)\, [y < x \wedge \varphi(x)].$$

Thus

$$\check{Q}x \in M \exists \beta \exists f[\beta \text{ an ordinal } \wedge f: \beta \cong \langle \text{pred}(x), < \rangle]$$

gives a failure of $\check{Q}$-collection, for the ordinals β needed here are unbounded in $\alpha = o(\mathbb{HYP}(\mathfrak{m}))$. □

For another example of $\underset{\sim}{Q}^{\#}$-admissible sets from the ones already obtained, we state the following simple application of the Truncation Lemmas and the Compactness Theorem for bounded quantifiers.

4.3 Theorem. For any $\underset{\sim}{Q}^{\#}$-admissible set $(\mathbb{A}_{\mathfrak{m}}, \underset{\sim}{q})$ above $\mathfrak{m}$ and any set T of $\Pi(Q)$ sentences true in $(\mathbb{A}_{\mathfrak{m}}, \underset{\sim}{q})$ (allowing a constant for the set M) we can find a $\underset{\sim}{Q}^{\#}$-admissible set $(\mathbb{B}_{\mathfrak{n}}, r) \models T$ where $o(B) = \omega$. In particular, we can require $(\mathfrak{m}, q^M) \equiv (\mathfrak{n}, r^N)$ for the language $L(Q)$.

5. $\underset{\sim}{Q}$-constructible sets.

In this section we show how to carry out the development of the $\underset{\sim}{Q}$-constructible sets in $\underset{\sim}{Q}$-KPU so that we can prove the existence of the smallest $\underset{\sim}{Q}$-admissible and smallest $\underset{\sim}{Q}^{\#}$-admissible sets above a structure $(\mathfrak{m}, \underset{\sim}{q})$.

Fix a sequence $\underset{\sim}{q}$ of bounded quantifiers on $V_{\mathfrak{m}}$. The basic assumption to replace II. 5.2 is the following: $\mathfrak{F}_1, \ldots, \mathfrak{F}_N, \mathfrak{F}_{N+1}$ are $\Sigma_1(\underset{\sim}{Q})$ operations so that the old conditions (i) - (v) hold with Δ_0 replaced by $\Delta_0(\underset{\sim}{Q})$ and KPU replaced by $\underset{\sim}{Q}$-KPU. The operation $\mathfrak{D}$ and $L^{\underset{\sim}{q}}$ are defined just as before from $\mathfrak{F}_1, \ldots, \mathfrak{F}_{N+1}$. $L^{\underset{\sim}{q}}(a, \alpha)$ is called the set of sets $\underset{\sim}{q}$-constructible from a by staged α.

5.1 Lemma. The operation L^q is $\Sigma_1(\underset{\sim}{Q})$ definable in $\underset{\sim}{Q}$-KPU.

Proof. This follows from the recursion theorem for $\underset{\sim}{Q}$-KPU. □

5.2 Theorem.

For every axiom φ of $\underset{\sim}{Q}^{\#}$-KPU, $\underset{\sim}{Q}^{\#}$-KPU $\vdash \varphi^{L^{\underset{\sim}{q}}(a)}$.

Proof. This is just like the proof of II.5.5. □

Now we turn to fulfilling the assumptions on $\mathfrak{F}_1, \ldots, \mathfrak{F}_{N+1}$. We need only add to the old $\mathfrak{F}_1, \ldots, \mathfrak{F}_N$ one new $\Sigma_1(\underset{\sim}{Q})$ operation whose graph is $\Delta_0(\underset{\sim}{Q})$-definable:

$$\mathfrak{F}_Q(a, b) = \{ y \mid Qx \in a(\langle x, y \rangle \in b)\}.$$

Notice that if c is transitive and $b \in c$ then $\mathfrak{F}_Q(a, b) \subseteq c$ so this new $\mathfrak{F}_Q$ preserves condition (*) on p.68. Thus, all we need check is that lemma II.6.1 goes through. The only new case is: if $\varphi(x_1, \ldots, x_{n+1})$ is a t-formula, so is $Qx_{n+1} \in x_j \varphi(x_1 \ldots x_{n+1})$, which we call $\psi(x_1 \ldots x_n)$. By induction $\mathfrak{F}_\varphi(a_1, \ldots a_n, a_j) = \{\langle x_{n+1}, \ldots, x_1\rangle \in a_j \times a_n \times \ldots \times a_1 \mid \varphi(x_1 \ldots x_{n+1})\}$. Thus, we may let $\mathfrak{F}_\psi(a_1 \ldots a_n) = \mathfrak{F}_Q(a_j, \mathfrak{F}_\varphi(a_1 \ldots a_n, a_j))$. □

We conclude this section with some remarks on $\underset{\sim}{Q}$-substitutability. We restrict ourselves to the situation which will concern us from here on, namely where all bounded quantifiers quantifiers in the list $\underset{\sim}{q}$ behave trivially except on the universe M of $\mathfrak{M}$; that is, $q^a = 0$ for $a \neq M$.

An operation symbol F is $\underset{\sim}{Q}$-substitutable if for every $\Delta_0(Q)$ formula $\varphi(w, \underline{v})$ there is a $\Delta_0(\underset{\sim}{Q})$ formula $\psi(\underline{u}, \underline{v}, \overline{M})$ ($\overline{M}$ is here a set constant denoting M) such that $\underset{\sim}{Q}$-KPU implies

$\varphi(F(\underline{u})\underline{v}) \leftrightarrow \psi(\underline{u}, \underline{v}, M)$. Using techniques from II. 7.5, it is easy to check that $\mathfrak{F}_Q$, and hence all of $\mathfrak{F}_1, \ldots, \mathfrak{F}_N, \mathfrak{D}$ are $\underset{\sim}{Q}$-substitutable.

<u>5.3 Corollary.</u> The relations on $(\mathfrak{M}, \underset{\sim}{q})$ which are elements of $L^{\underset{\sim}{q}}(\mathfrak{M}, \omega)$ are exactly the relations on $(\mathfrak{M}, \underset{\sim}{q})$ which are definable in the language $L(\underset{\sim}{Q})$ (allowing parameters from M).

<u>Proof.</u> Just like II. 7.1. □

We now come to the first mild surprise, and the first hint of the difference between $\underset{\sim}{Q}$-admissibility and $\underset{\sim}{Q}^{\#}$-admissibility.

<u>5.4 Definition.</u>

i) A structure $(\mathfrak{M}, \underset{\sim}{q})$ for $L(\underset{\sim}{Q})$ is <u>recursively saturated</u> if for every recursive set $\Phi(x, y, \ldots, y_k)$ of $L(Q)$ formulas, the following holds in $(\mathfrak{M}, \underset{\sim}{q})$, for $\underline{y} \in M^k$:

$$\bigwedge_{\Phi_0 \in S_\omega(\Phi)} \exists x \bigwedge \Phi_0(x, \underline{y}) \quad \text{implies} \quad \exists x \bigwedge \Phi(x, \underline{y}).$$

ii) Now let Q be one of the quantifiers in the list $\underset{\sim}{Q}$, or the dual of some such. A structure $(\mathfrak{M}, \underset{\sim}{q})$ is <u>$Q^{\#}$-recursively saturated</u> if it is recursively saturated (as in (i)) and, in addition, for each recursive $\Phi(x, \underline{y})$, and all $\underline{y} \in M^k$:

$$\bigwedge_{\Phi_0 \in S_\omega(\Phi)} Qx \bigwedge \Phi_0(x, \underline{y}) \quad \text{implies} \quad Qx \bigwedge \Phi(x, \underline{y}).$$

<u>5.5 Theorem.</u>

i) $(\mathfrak{M}, \underset{\sim}{q})$ is recursively saturated iff $L^{\underset{\sim}{q}}(\mathfrak{M}, \omega)$ is $\underset{\sim}{Q}$-admissible.

ii) For any Q in $\underset{\sim}{Q}^{\#}$, $L^{\underset{\sim}{q}}(\mathfrak{M}, \omega)$ satisfies Q-collection iff $(\mathfrak{M}, \underset{\sim}{q})$ is $\breve{Q}^{\#}$-recursively saturated.

<u>Proof.</u> The proofs of these are like the proof of IV. 5.3. The fact that the dual quantifier gets called in is amusing, but not that surprising when you recall that it is the contrapositive of the saturation condition which is used, and taking contrapositives brings in duals. □

As an application we prove the following.

5.6 Proposition. Let $\mathfrak{m}$ be an uncountable recursively saturated structure for L and let Qx be the quantifier "there exists uncountably many", with the intended interpretation q. Then $(\mathfrak{m}, q)$ is $\check{Q}^{\#}$-recursively saturated so $L^{\underset{\sim}{q}}(\mathfrak{m}, \omega)$ is $Q^{\#}$, $\check{Q}$-admissible.

Proof. Suppose $\mathfrak{m} \models \check{Q}x \wedge \Phi_0(x, \underline{y})$ for all finite Φ_0 contained in some (recursive) set $\Phi(x, \underline{y})$ of $L(Q)$ formulas. That is, all but countably many x, say those in a set X_{Φ_0}, satisfy $\wedge \Phi_0(x)$. Let $X = \bigcup\{X_{\Phi_0} | \Phi_0 \in S_\omega(\Phi)\}$. Then X is countable and all $x \notin X$ satisfy $\Phi(x)$. Thus, $\mathfrak{m} \models \check{Q}x \wedge \Phi(x, \underline{y})$. □

6. $\underset{\sim}{Q}^{\#}$-$\mathbb{HYP}(\mathfrak{m}, \underset{\sim}{q})$ and the Representability Theorem.

Let $(\mathfrak{m}, \underset{\sim}{q})$ be a structure $\mathfrak{m}$ with a sequence $\underset{\sim}{q} = q_1, \ldots, q_k$ of monotone quantifiers. We are now in a position to prove the existence of smallest $\underset{\sim}{Q}^{\#}$-admissible set above $\mathfrak{m}$. To apply §5 we extend $\underset{\sim}{q}$ to indexed quantifiers on $\mathbf{V}_{\mathfrak{m}}$ by setting $q^a = 0$ for $a \notin M$.

6.1 Definition.

i) The $\underset{\sim}{Q}^{\#}$-admissible hull of $(\mathfrak{m}, \underset{\sim}{q})$, $\underset{\sim}{Q}^{\#}$-$\mathbb{HYP}(\mathfrak{m}, \underset{\sim}{q})$ is the structure $(\mathfrak{m}; A, \in, \underset{\sim}{q})$ where $A = \bigcap\{B | M \in B$ and $(\mathfrak{m}; B, \in, \underset{\sim}{q})$ is $\underset{\sim}{Q}^{\#}$-admissible.

ii) The $\underset{\sim}{Q}^{\#}$-ordinal of $(\mathfrak{m}, \underset{\sim}{q})$ is the least ordinal not in $\underset{\sim}{Q}^{\#}$-$\mathbb{HYP}(\mathfrak{m}, \underset{\sim}{q})$ and is denoted by $\underset{\sim}{Q}^{\#}$-ord$(\mathfrak{m}, \underset{\sim}{q})$.

We write $\mathbb{HYP}(\mathfrak{m}, q)$ for the unsharped case $\underset{\sim}{Q}$-$\mathbb{HYP}(\mathfrak{m}, \underset{\sim}{q})$ and $\mathbb{HYP}^{\#}(\mathfrak{m}, \underset{\sim}{q})$ for the completely sharped case $^{\#}\underset{\sim}{Q}$-$\mathbb{HYP}(\mathfrak{m}, \underset{\sim}{q})$. It follows from 5.2i that $\underset{\sim}{Q}^{\#}$-$\mathbb{HYP}(\mathfrak{m}, q)$ is the smallest $\underset{\sim}{Q}^{\#}$-admissible set above $\mathfrak{m}$. This paper is primarily concerned with $\underset{\sim}{Q}^{\#}$-$\mathbb{HYP}(\mathfrak{m}, \underset{\sim}{q})$, $\underset{\sim}{Q}^{\#}$-ord$(\mathfrak{m}, \underset{\sim}{q})$ and their relationships with inductive definability and representability. Probably the most important theorem is the Representability Theorem, Theorem 6.2 below. Once that is proved, everything else works out without too much difficulty. First some definitions.

$\underset{\sim}{Q}^{\#}$-logic for $(\mathfrak{m}, \underset{\sim}{q})$ is obtained from $\mathfrak{m}$-logic (see III. 3.3) by adding the weak Q-rule for each unsharped Q in the list $\underset{\sim}{Q}$: if for each $p \in M$, either $T \vdash \varphi(\overline{p})$ or $T \vdash \neg\varphi(\overline{p})$, and if $\{p : T \vdash \varphi(p)\} \in q$ then $T \vdash Qx\,\varphi(x)$.

On the other hand, if Q is sharp, we add the stronger $\underset{\sim}{Q}$-rule: if $\{p : T \vdash \varphi(\overline{p})\} \in q$ then $T \vdash Qx\,\varphi(x)$.

For $^{\#}\underset{\sim}{Q}$, we call this logic $(\mathfrak{m}, \underset{\sim}{q})$-logic.

For Q the Souslin quantifier on ω, the Q-rule was introduced by Enderton. The Q-rule for general monotone Q and the weak rules were introduced by Aczel [1970].

A $\Sigma_1(\underset{\sim}{Q})$ formula $\varphi(x)$ (with constants $\overline{p}$ for $p \in M$ and $\overline{M}$) is a good Σ_1-definition of $a \in \mathbb{H}\mathrm{YP}^{\#}(\mathfrak{m}, \underset{\sim}{q})$ relative to $\underset{\sim}{Q}^{\#}$-KPU$^+$ if:

1) $\underset{\sim}{Q}^{\#}$-$\mathbb{H}$YP$(\mathfrak{m}, q) \models \varphi(a)$
2) $\underset{\sim}{Q}^{\#}$-KPU$^+$ proves $\exists! x\varphi(x)$ in $\underset{\sim}{Q}^{\#}$-logic for $(\mathfrak{m}, q)$.

6.2 Representability Theorem for $\underset{\sim}{Q}^{\#}$-$\mathbb{H}$YP$(\mathfrak{m}, \underset{\sim}{q})$. To each a in $\underset{\sim}{Q}^{\#}$-$\mathbb{H}$YP$(\mathfrak{m}, \underset{\sim}{q})$ we can assign a formula $\overline{a}$ with one free variable so that the following three conditions hold, where we write $\varphi(\overline{a})$ for $\exists x(\overline{a}(x) \wedge \varphi(x))$ and write "$\vdash$" for "provable from $\underset{\sim}{Q}^{\#}$-KPU using $\underset{\sim}{Q}^{\#}$-logic for $(\mathfrak{m}, q)$":

1) $\overline{a}$ is a good $\Sigma_1(\underset{\sim}{Q})$ definition of a relative to $\underset{\sim}{Q}^{\#}$-KPU$^+$.
2) the $\in$a-rule is derivable; that is, for all formulas ψ, if $\vdash \psi(\overline{x})$ for all $x \in a$ then $\vdash \forall v \in \overline{a}\psi(v)$.
3) true $\Delta_0(\underset{\sim}{Q})$ facts are provable; that is, for all $\Delta_0(\underset{\sim}{Q})$ formulas $\psi(v_1 \ldots v_n)$ and all $a_1, \ldots, a_n \in \underset{\sim}{Q}^{\#}$-$\mathbb{H}YP(\mathfrak{m}, \underset{\sim}{q})$, if $\psi(a_1, \ldots, a_n)$

is true then $\vdash \psi(\bar{a}_1, \ldots, \bar{a}_n)$.

Proof. First note that if

$\vdash \varphi(\bar{a})$ for any true $\Delta_0(\underset{\sim}{Q})$ $\varphi(a)$, then we can prove $\vdash \neg\varphi(\bar{a})$ for any false $\Delta_0(\underset{\sim}{Q})$ $\varphi(a)$, by using duals. This remark will be used below. We prove by induction on $\alpha < \underset{\sim}{Q}^{\#}\text{-ord}(\mathfrak{m}, \underset{\sim}{q})$ that:

$1)_\alpha$ This is (1) above for $a \in L(M, \alpha) \cup \{\alpha, L(M, \alpha)\}$.

$2)_\alpha$ The $\in a$-rule is derivable for all $a \in L(M, \alpha) \cup \{\alpha, L(M, \alpha)\}$.

$3)_\alpha$ All true $\Delta_0(\underset{\sim}{Q})$ facts about $a_1 \ldots a_n \in L(M, \alpha) \cup \{\alpha, L(M, \alpha)\}$ are provable in the appropriate logic, as is $L(\bar{M}, \bar{\alpha}) = \overline{L(M, \alpha)}$.

CASE A. $\alpha = 0$.

$1)_0$ is trivial since $\bar{p}, \bar{M}$ are names for p, M.

$2)_0$ is just the M-rule.

$3)_0$ This amounts to proving that any true sentence of $(\mathfrak{m}, \underset{\sim}{q})$ is provable in $\underset{\sim}{Q}$-logic for $(\mathfrak{m}, q)$. This is proved by showing:
$(\mathfrak{m}, \underset{\sim}{q}) \models \psi(p) \Rightarrow \vdash \psi(\bar{p})$ and $(\mathfrak{m}, \underset{\sim}{q}) \models \neg\psi(p) \Rightarrow \vdash \neg\psi(\bar{p})$. This is by induction on ψ and is easy.

CASE B. $\alpha = \beta + 1$.

$1)_\alpha$ If $\alpha \in L(M, \alpha)$ then $a = \mathfrak{F}_i(x, y)$ for $x, y \in L(M,\beta) \cup \{L(M,\beta)\}$. By $(1)_\beta$ we have good Σ_1 definitions of x, y so $\mathfrak{F}_i(\overline{x}, \overline{y})$ is a good name for a. Similarly, since β has a good name, so does $\alpha = \beta + 1$. Hence, so does $L(M, \alpha)$.

$3)_\alpha$ A $\Delta_0(Q)$ formula $\varphi(a)$ involving a as above is just $\varphi(\mathfrak{F}_i(x, y))$ so, if true, is provable by $(3)_\beta$ and the substitutability of the $\mathfrak{F}_i$. Since $\overline{L(M, \beta)} = L(\overline{M}, \overline{\beta})$ is provable, so is $L(\overline{M}, \overline{\beta+1}) = \overline{L(M, \beta+1)}$.

$2)_\alpha$ Subcase a. The ϵa-rule for $a = \alpha = \beta + 1$ follows easily from the $\epsilon\beta$-rule which is derivable by $(2)_\beta$.

Subcase b. The ϵa-rule for $a = L(M, \alpha) = L(M,\beta) \cup \{F_i(x, y) \mid i \le i \le N,$ $x, y \in L(M,\beta) \cup \{L(M,\beta)\}\}$. It is a theorem of $\underset{\sim}{Q}^{\#}$-KPU$^+$ that $\forall x \in \overline{a}\psi(x)$ is equivalent to the conjunction of all the following, for $1 \le i \le N$.

$$\forall x \in L(M, \overline{\beta})\, \psi(y)$$
$$\forall x, y \in L(M, \overline{\beta})\, \psi(\mathfrak{F}_i(x, y))$$
$$\forall x \in L(M, \overline{\beta})\, \psi(\mathfrak{F}_i(x, L(M, \overline{\beta})))$$
$$\forall y \in L(M, \overline{\beta})\, \psi(\mathfrak{F}_i(L(M, \overline{\beta}), y))$$
$$\psi(\mathfrak{F}_i(L(M, \overline{\beta}), L(M, \overline{\beta}))).$$

If for all $x \in a$, $\vdash \psi(\overline{x})$, then $(2)_\beta$ gives all the above provable.

Subcase c. $a \in L(M, \alpha)$. This is similar to subcase b. Since $L(M, \alpha)$ is transitive, every $x \in a$ is of the form $\mathfrak{F}_i(y, z)$ for some $y, z \in L(M,\beta) \cup \{L(M,\beta)\}$. Thus we can prove that $\forall x \in \overline{a}\psi(x)$ is equivalent to a big conjunction.

$$\forall y, z \in L(M,\beta)\, [\mathfrak{F}_i(y, z) \in a \rightarrow \psi(\mathfrak{F}_i(y, z))]$$
$$\forall y \in L(M,\beta)\ [\mathfrak{F}_i(y, L(M, \beta)) \in a \rightarrow \psi(\mathfrak{F}_i(y, L(M,\beta)))]$$
$$\forall z \in L(M,\beta)\, [\mathfrak{F}_i(L(M, \beta) z)) \in a \rightarrow \psi(\mathfrak{F}_i(L(M,\beta) z))]$$
$$\mathfrak{F}_i(L(M, \beta), L(M, \beta)) \in a \rightarrow \psi(\mathfrak{F}_i(L(M, \beta), L(M, \beta))).$$

By $(3)_\alpha$ proved above, whenever $\bar{a}_1(y, z) \in a$, this fact is provable. Then, if for each $x \in a$, $\vdash \psi(\bar{x})$, then we can use $(2)_\beta$ to prove each of the above conjuncts. This finishes the proof of $(2)_\alpha$.

CASE C. α is a limit $<\underset{\sim}{Q}$-ord$(\mathfrak{m}, \underset{\sim}{q})$. $L(M, \alpha)$ must fail to satisfy $\Delta_0(\underset{\sim}{Q})$ collection so fix a failure:

$$L(M, \alpha) \models \forall x \in \bar{b}\ \exists y \varphi(x, y)$$

but no $L(M, \beta)$ for $\beta < \alpha$ satisfies this. Choose $\gamma < \alpha$ large enough so that b and all the parameters in φ are in $L(M, \gamma)$. The good Σ_1 definition $\bar{\alpha}(z)$ of α is:

"z is the least ordinal such that $\forall x \in b\, \exists y \in L(M, z)\varphi(x, y)$".

We must prove, in $\underset{\sim}{Q}$-KPU$^+$, using $\underset{\sim}{Q}$-logic for $(\mathfrak{m}, \mathfrak{q})$, that $\exists! z\, \varphi(z)$. This will prove the only non-trivial part of $(1)_\alpha$. For each $x \in b$, pick a $\delta < \alpha$ so that $\gamma \leq \delta$ and

$$\exists y \in L(M, \delta)\varphi(x, y)$$

is true. By $(3)_\delta$, we have $\vdash \exists y \in L(M, \bar{\delta})\ \varphi(\bar{x}, y)$, so $\vdash \exists \delta\, \exists y \in L(M, \delta)\varphi(\bar{x}, y)$. By the $\in b$-rule, of $(2)_\gamma$, we have

$$\forall x \in b\, \exists \delta\, \exists y \in L(M, \delta)\varphi(x, y).$$

Now by Σ-reflection in Q-KPU$^+$, $\vdash \exists! z\, \bar{\alpha}(z)$. This proves $(1)_\alpha$.

$(2)_\alpha$. The only non-trivial parts are the $\in\alpha$-rule and the $\in L(M, \alpha)$-rule. The second follows from the first. Suppose for all $\beta < \alpha$, $\vdash \psi(\bar{\beta})$. We must prove $\vdash \forall \beta < \bar{\alpha}\, \psi(\beta)$. That is, we must prove, in the correct logic, that:

$\forall \beta$ [if $\beta <$ the least z such $\forall x \in b\, \exists y \in L(M, z)\varphi(x, y)$, then $\psi(\beta)$].

Work in the theory and define $\beta^* =$ the least β s.t. $\neg\psi(\beta)$, if such exists, otherwise let $\beta^* = \bar{\alpha}$. We prove that $\beta^* \geq \bar{\alpha}$. The proof uses the $\in b$-rule As before, for all $x \in b$ we can find a $\delta < \alpha$ so that

$$\exists y \in L(M, \delta)\varphi(x, y)$$

is true. Hence

$$\vdash \exists y \in L(M, \bar{\delta})\ \varphi(\bar{x}, y).$$

But also, by hypothesis, $\vdash \psi(\bar{\delta})$ and $\vdash \psi(\bar{\beta})$ all $\beta < \delta$ so, by the

$\in\delta+1$ rule,

$$\vdash \forall\beta \leq \overline{\delta}\psi(\beta).$$

Thus we have

$$\vdash \exists y \in L(M,\overline{\delta})\varphi(\overline{x},y) \wedge \overline{\delta} < \beta^*.$$

Then we can apply $\exists$-rule and then $\in b$ rule to get

$$\vdash \forall x \in \overline{b}\exists\, \delta[\exists y \in L(M,\delta)\varphi(x,y) \wedge \delta < \beta^*]$$

which is the assertion that $\vdash \overline{\alpha} \leq \beta^*$. So much for $(2)_\alpha$.

$(3)_\alpha$. This is proved by induction on $\Delta_0(\mathcal{Q})$ formulas. As in $(3)_0$ you prove that $\psi(a_1 \ldots a_n)$ implies $\vdash \psi(\overline{a}_1 \ldots \overline{a}_n)$ and $\neg\psi(a_1 \ldots a_n)$ implies $\vdash \neg\psi(\overline{a}_1 \ldots \overline{a}_n)$, by induction on $\Delta_0(\mathcal{Q})$ formulas $\psi(x_1 \ldots x_n)$. You use $(2)_\alpha$ to get over the bounded quantifiers $\forall x \in b$.

CASE D. α a limit $< \mathcal{Q}^{\#}$-ord$(\mathfrak{m}, \mathfrak{q})$. If $L(M,\alpha) \not\models \Delta_0(Q)$-collection, use the proof of Case C. So suppose that $L(M,\alpha) \not\models Q$-collection where Q is sharp. Suppose

$$L(M,\alpha) \models Qx \in b\ \exists y\ \varphi(x,y)$$

but no $L(M,\beta)$ for $\beta < \alpha$ is a model of this. Let $\overline{\alpha}(z)$ be the formula:

"z is the least ordinal such that $Qx \in b\ \exists y \in L(M,z)\varphi(x,y)$".

We need the <u>strong</u> Q-rule to prove $\exists z\, \alpha(z)$, but from there on the proof is the same. □

<u>6.3 Theorem.</u> Let S be a relation on $\mathfrak{m}$.

i) If S is $\Sigma_1(\mathcal{Q})$ on $\mathcal{Q}^{\#}$-$\mathbb{HYP}(\mathfrak{m}, \mathfrak{q})$ then S is weakly representable in $\mathcal{Q}^{\#}$-KPU$^+$ using $\mathcal{Q}^{\#}$-logic for $(\mathfrak{m}, \mathfrak{q})$.

ii) If S is in $\mathcal{Q}^{\#}$-$\mathbb{HYP}(\mathfrak{m}, \mathfrak{q})$ then S is strongly representable in $\mathcal{Q}^{\#}$-KPU$^+$ in $\mathcal{Q}^{\#}$-logic for $(\mathfrak{m}, \mathfrak{q})$.

6.3 follows from 6.2 and $6.2^{\#}$ by the usual methods. The next two sections are devoted to proving converses to the two halves of 6.3.

7. Q-Positive Inductive definitions.

To prove the converses of the results of the previous section we need analogues of Gandy's Theorem (VI. 2.6) to the effect that Σ-inductive definitions have Σ_1 fixed points on admissible sets. For this, we need analogues of the Second Recursion Theorem for KPU. We treat the ${}^{\#}Q$-KPU case first. The $Q^{\#}$-KPU case is more subtle.

<u>7.1 Second Recursion Theorem for ${}^{\#}Q$-KPU.</u> Let $\varphi(x_1 \ldots x_n, R_+)$ be an R-positive $\Sigma({}^{\#}Q)$ formula with R n-ary. (There may be other free variables acting as parameters.) There is a $\Sigma({}^{\#}Q)$ formula $\psi(x_1 \ldots x_n)$ such that

$$ {}^{\#}Q\text{-KPU} \vdash \forall \vec{x}\, [\psi(\vec{x}) \leftrightarrow \varphi(\vec{x}, \{\langle z_1 \ldots z_n\rangle \mid \psi(z_1 \ldots z_n)\})] . $$

<u>Proof.</u> The proof of the corresponding result in the book does not go through here. For this reason we discovered a much simpler proof.

We only sketch the proof here. Suppose $n = 2$ for simplicity. The formula $\psi(x_1, x_2)$ is

$$\exists r[r \text{ is a set of ordered parts, } \varphi(x_1, x_2, r) \text{ and } \forall z \in r\, \varphi(\text{1st}(z), \text{2nd}(z), r)] .$$

Here $\varphi(x_1, x_2, r)$ is the result of replacing $R(t_1, t_2)$ in $\varphi(x_1, x_2, R)$ by $\exists y \in r(y = \langle t_1, t_2\rangle)$ and 1st(z), 2nd(z) are the y_0, y_1 so that $z = \langle y_0, y_1\rangle$. Note that ψ is a $\Sigma({}^{\#}Q)$ formula as is $\varphi(x_1, x_2, \psi(\cdot,\cdot))$, the result of replacing R by ψ in φ. It is tedious but not too difficult, using $\Sigma({}^{\#}Q)$-reflection and $\varphi(x_1, x_2, r_1) \wedge r_1 \subseteq r_2 \rightarrow \varphi(x_1, x_2, r_2)$, to verify that this ψ has the required property. □

Theorem 7.1 is not true for $Q^{\#}$-KPU even if we restrict $\varphi(x_1 \ldots x_n, R_+)$ to be $\Delta_0(Q)$. In the proof of 7.1 you see that a relation symbol R is replaced by a $\Sigma({}^{\#}Q)$-formula ψ and that may introduce unbounded $\exists$'s within the scope of a bounded $Qx \in a$ for some unsharped Q. $Q^{\#}$-KPU can't handle this. To get a reasonable version of 7.2 and the following for $\underset{\sim}{Q}^{\#}$-KPU, we will need a new kind of $\Sigma(\underset{\sim}{Q}^{\#})$-formula, called a <u>deterministic</u> $\Sigma(\underset{\sim}{Q}^{\#})$-formula.

<u>7.2 Corollary.</u> If $(\mathbb{A}, \underset{\sim}{q})$ is $^{\#}\underset{\sim}{Q}$-admissible and $\varphi(\underline{x}, R_+)$ is an R-positive $\Sigma(^{\#}\underset{\sim}{Q})$ inductive definition on $(\mathbb{A}, \underset{\sim}{q})$ then the smallest fixed point I_φ of φ is $\Sigma_1(\underset{\sim}{Q})$-definable on $(\mathbb{A}, q)$ and $\|\varphi\| \leq 0(\mathbb{A})$.

<u>Proof.</u> Given 7.1, the proof is just like the proof on pp. 208-210 of the book. □

<u>7.4 Definition.</u> A relation S on $(\mathfrak{m}, \underset{\sim}{q})$ is $\underset{\sim}{Q}$-<u>positively inductive</u> if S is Φ-inductive (see VI. 3.1) where Φ is the set of all formulas of $L(\underset{\sim}{Q})$ of the form $\varphi(R_+)$ where R is a new relation symbol. Similarly for the notions of $\underset{\sim}{Q}$-<u>positively hyperelementary</u> and the extended notions (see VI. 3.7).

This allows us to prove the following theorem characterizing the relations on $(\mathfrak{m}, \underset{\sim}{q})$ which are $\Sigma_1(\underset{\sim}{Q})$ definable on $\mathbb{HYP}^{\#}(\mathfrak{m}, \underset{\sim}{q})$.

<u>7.5 Theorem.</u> Let S be a relation on $(\mathfrak{m}, \underset{\sim}{q})$. The following are equivalent:

i) S is $\Sigma_1(\underset{\sim}{Q})$ on $\mathbb{HYP}^{\#}(\mathfrak{m}, \underset{\sim}{q})$.

ii) S is weakly representable in $^{\#}\underset{\sim}{Q}$-KPU$^+$ using $(\mathfrak{m}, \underset{\sim}{q})$-logic.

iii) S is $\underset{\sim}{Q}$-positive inductively* definable.

Proof. (i) ⇒ (ii) was proved in §6. (ii) ⇒ (iii). The definition of provable in $(\mathfrak{m}, \mathfrak{q})$-logic can be given as a $\underset{\sim}{Q}$-positive inductive definition. The crucial clause, corresponding to the Q-rule, is

$$Qv \in M(\ulcorner \psi(v) \urcorner \in R) \rightarrow \ulcorner Qv \in M\psi(v) \urcorner \in R,$$

which takes the form in the inductive definition Γ:

$$x \in \Gamma(R) \text{ iff } [\Theta(x_1, R_+) \vee x \text{ is of the form } \ulcorner Qv\psi(v) \urcorner$$

$$\text{and } Qy \in M\ R \ulcorner \psi(y) \urcorner)]$$

where $\Theta(x_1\ R_+)$ is the clause for $\mathfrak{m}$-logic and does not involve $\underset{\sim}{Q}$ at all. Thus the definition of Γ is $(\underset{\sim}{Q}, R)$-positive. □

If $(\mathfrak{m}, \mathfrak{q})$ has an inductive pairing function then we may delete the * in (iii) and in 7.6 .

7.6 Corollary. $\underset{\sim}{Q}^{\#}$-ord$(\mathfrak{m}, \mathfrak{q}) = \sup\{\|\varphi\| : \varphi$ is a $\underset{\sim}{Q}$-positive inductive * definition on $(\mathfrak{m}, \mathfrak{q})\}$.

8. $\underset{\sim}{Q}^{\#}$-deterministic inductive definitions.

In this section we give a treatment of the $\underset{\sim}{Q}^{\#}$-KPU case parallel to that for $^{\#}\underset{\sim}{Q}$-KPU in §7. Our first goal is to find a weakening of the notion of $\Sigma(\underset{\sim}{Q}^{\#})$-formula that will allow us to prove a second recursion theorem for $\underset{\sim}{Q}^{\#}$-KPU.

8.1 Definition.

i) The $\underset{\sim}{Q}^{\#}$-deterministic $\Sigma(\underset{\sim}{Q}^{\#})$ formulas of $L^{*}(\underset{\sim}{Q})$ form the smallest collection det-$\Sigma(\underset{\sim}{Q}^{\#})$ containing the atomic and negated atomic formulas, closed under $\wedge, \vee, \exists, \exists x \in a, \forall x \in a, Qx \in a$ for sharp Q, and the peculiar clause for unsharped Q: if φ, ψ are in det-$\Sigma(\underset{\sim}{Q}^{\#})$ so is the formula:

$$\exists x \in a(\varphi \wedge \psi) \vee [\forall x \in a(\varphi \vee \psi) \wedge Qx \in a\ \varphi]$$

written $Qx \in a(\varphi | \psi)$, and read, $Qx \in a\varphi$ given that φ is equivalent to $\neg\psi$ on a. Similar definitions apply to det-$\Delta_0(\underset{\sim}{Q})$ and det $\Sigma_1(\underset{\sim}{Q})$.

ii) Recall the first-order language $L(\underset{\sim}{Q})$. The corresponding notion for $L(\underset{\sim}{Q}^{\#})$ is the collection of $\underset{\sim}{Q}^{\#}$-deterministic formulas of $L(\underset{\sim}{Q}^{\#})$, the smallest collection containing all atomic and negated atomic formulas, closed under $\wedge, \vee, \exists, \forall$, Q (for sharp Q) and the formation rule: if φ, ψ are $\underset{\sim}{Q}^{\#}$-deterministic, so is

$$Qx(\varphi \mid \psi)$$

by which we mean the formula

$$\exists x(\varphi \wedge \psi) \vee [\forall x(\varphi \vee \psi) \wedge Qx\varphi]$$

iii) The deterministic extended formulas also allow $\exists a \in \mathbb{HF}(\mathfrak{m})$ and bounded set quantifiers, as usual.

The intuition behind these formulas comes from the Δ-separation principle in KPU. Given two Σ-formulas φ, ψ, if $\varphi \leftrightarrow \neg\psi$ then we may form $\{x \in a \mid \varphi\}$ because, basically, what φ does not tell us, ψ does.

There is a confusing point here. Every $L(\underset{\sim}{Q})$ formula φ is equivalent to a Q-deterministic formula φ', since $Qx\psi(x)$ can always be written as $Qx(\psi \mid \neg\psi)$. However, this does not preserve positive occurrences of R in ψ. Thus, not every R-<u>positive</u> formula $\varphi(R_+)$ will be equivalent to a Q-deterministic R-<u>positive</u> formula. Thus, inductive definitions cannot always be written deterministically.

<u>8.2 Lemma.</u> If φ, ψ are $\underset{\sim}{Q}^{\#}$-deterministic formulas of $L(\underset{\sim}{Q}^{\#})$ so is (up to logical equivalence) the formula.

$$\forall x(\varphi \vee \psi) \wedge [Qx\varphi \vee \exists x(\varphi \wedge \psi)] .$$

<u>Proof.</u> The formula in question is equivalent to

$$\forall x(\varphi \vee \psi) \wedge Qx(\varphi \mid \psi). \qquad \Box$$

Given $\underset{\sim}{Q}^{\#}$, we write $\check{\underset{\sim}{Q}}^{\#}$ for the list where $\check{Q}$ is sharped in $\check{\underset{\sim}{Q}}^{\#}$ iff Q is sharped in $\underset{\sim}{Q}^{\#}$, recalling that $\check{\check{Q}} = Q$.

8.3 Corollary. The negation of a $\underset{\sim}{Q}^{\#}$-deterministic formula is equivalent to a $\underset{\sim}{\breve{Q}}^{\#}$-deterministic formula. The proof is by induction on formulas the only cause for alarm being the formation rule $Qx(\varphi \mid \psi)$. The induction hypothesis assures us that $\neg\varphi \leftrightarrow \breve{\varphi}$, $\neg\psi \leftrightarrow \breve{\psi}$ where $\breve{\varphi}, \breve{\psi}$ are $\underset{\sim}{\breve{Q}}^{\#}$-deterministic.

$$\begin{aligned}\neg Qx(\varphi \mid \psi) &\leftrightarrow \neg[\exists x(\varphi \wedge \psi) \vee [\forall x(\varphi \vee \psi) \wedge Qx\varphi]] \\ &\leftrightarrow \forall x(\neg\varphi \vee \neg\psi) \wedge [\exists x(\neg\varphi \wedge \neg\psi) \vee \neg Qx\varphi] \\ &\leftrightarrow \forall x(\breve{\varphi} \vee \breve{\psi}) \wedge [\breve{Q}x\breve{\varphi} \vee \exists x(\breve{\varphi} \wedge \breve{\psi})].\end{aligned}$$

The last formula is $\underset{\sim}{\breve{Q}}^{\#}$-deterministic by 8.2. □

The best way to understand the deteministic $\Sigma(\underset{\sim}{Q}^{\#})$ formulas is to verify some cases of the following.

8.4 Lemma

i) Let $\varphi(R_+), \psi(x_1 \ldots x_n)$ be det $\Sigma(\underset{\sim}{Q}^{\#})$ formulas, where R is n-ary. Then the formula $\varphi(\psi/R)$ which results from the R-positive formula $\varphi(R_+)$ by replacing $R(t_1 \ldots t_n)$ by $\psi(t_1 \ldots t_n)$ is also det $\Sigma(\underset{\sim}{Q}^{\#})$, with the usual caveat about clashing variables.

ii) Now let $\underset{\sim}{Q}^{\#}$-KPU be formulated in a fixed language L^* and restrict attention to those formulas in this language. Then

a) Every $\Delta_0(\underset{\sim}{Q})$ formula is provably equivalent to a det-$\Delta_0(\underset{\sim}{Q}^{\#})$ formula in $\underset{\sim}{Q}^{\#}$-KPU, and vice versa.

b) If φ is a det $\Sigma(\underset{\sim}{Q}^{\#})$-formula then $\underset{\sim}{Q}^{\#}$-KPU $\vdash \varphi \to \exists a \varphi^{(a)}$.

So, in particular, every det $\Sigma(\underset{\sim}{Q}^{\#})$ formula is provably equivalent to a $\Sigma_1(\underset{\sim}{Q})$ formula.

Proof. (i) is proved by induction on R-positive formulas $\varphi(R_+)$ and is easy. (iia) follows from 8.2 since

$$Qx \in a\varphi \twoheadrightarrow Qx \in a(\varphi \mid \neg\varphi).$$

Let us now prove (iib). The crucial class is:

$Qx \in a(\varphi \mid \psi) \to \exists b[Qx \in a(\varphi \mid \psi)]^{(b)}$. Assume $Qx \in a(\varphi \mid \psi)$. I.e., assume $\exists x \in a(\varphi \wedge \psi) \vee [\forall x \in a(\varphi \vee \psi) \wedge Qx \in a\varphi]$.

Case a) $\exists x(\varphi \wedge \psi)$. By the induction hypothesis there is a ψ, b_1, b_2

such that $x \in a$, $\varphi^{(b_1)}(x)$, $\psi^{(b_1)}(x)$. Let $b = b_1 \cup b_2$, so that $\exists x \in a(\varphi^{(b)}/x) \wedge \psi^{(b)}/x))$.

Case b) $\neg\exists x \in a(\varphi \wedge \psi)$. Use $\veebar$ for exclusive disjunction ($A \veebar B$ means A or B but not both). Then we have

$$\forall x \in a(\varphi \veebar \psi) \wedge Qx \in a\varphi .$$

By induction we have $\forall x \in a\, \exists c[\varphi^{(c)}(x) \vee \psi^{(c)}(x)]$. We collect these c's together by $\Delta_0(Q)$-collection and take a union to get a b such that

$$\forall x \in a[\varphi^{(b)}(x) \vee \psi^{(b)}(x)] \quad \text{and hence:}$$

$$\forall x \in a[\varphi^{(b)}(x) \veebar \psi^{(b)}(x)] .$$

Thus, $\{x \in a \mid \varphi^{(b)}(x)\} = \{x \in a \mid \varphi(x)\}$ and $Qx \in a\varphi(x)$ so $Qx \in a\varphi^{(b)}(x)$. □

<u>8.5 Second Recursion Theorem for</u> $\underset{\sim}{Q}^{\#}$-KPU. Let $\varphi(x_1, \ldots, x_n, R_+)$ be an R-positive deterministic $\Sigma(\underset{\sim}{Q}^{\#})$-formula. There is a $\Sigma_1(\underset{\sim}{Q})$ formula $\psi(x_1 \ldots x_n)$ such that $\underset{\sim}{Q}^{\#}$-KPU $\vdash$

$$\forall x_1 \ldots \forall x_n[\psi(x_1 \ldots x_n) \leftrightarrow \varphi(x_1, \ldots, x_n, \{\langle y_1 \ldots y_n\rangle \mid \psi(y_1 \ldots y_n)\})].$$

<u>Proof.</u> The proof is just like the other, given 8.4. □

<u>8.6 Corollary.</u> Let $(\mathbb{A}, \underset{\sim}{q})$ be $\underset{\sim}{Q}^{\#}$-admissible and let $\varphi(x_1, \ldots, x_n, R_+)$ be a deterministic $\Sigma(\underset{\sim}{Q}^{\#})$ inductive definition. Then I_φ is definable by a $\Sigma_1(\underset{\sim}{Q})$ formula and $\|\varphi\| \le o(\mathbb{A})$. □

<u>8.7 Definition.</u> A relation S on $(\mathfrak{M}, \underset{\sim}{q})$ is $\underset{\sim}{Q}^{\#}$-<u>deterministically inductive</u> on $(\mathfrak{M}, \underset{\sim}{q})$ if S is Φ_0-inductive where Φ_0 is the set of the set of $\underset{\sim}{Q}^{\#}$-deterministic formulas $\varphi(R_+)$. Similarly for $\underset{\sim}{Q}^{\#}$-deterministic hyperelementary and the other notions.

<u>8.8 Theorem.</u> Let S be a relation on $(\mathfrak{M}, \underset{\sim}{q})$. The following are equivalent:

i) S is $\Sigma_1(\underset{\sim}{Q})$ on $\underset{\sim}{Q}^{\#}$-$\mathbb{HYP}(\mathfrak{M}, \underset{\sim}{q})$

ii) S is weakly representable in $\underset{\sim}{Q}^{\#}$-KPU$^+$ using $\underset{\sim}{Q}^{\#}$-logic for $(\mathfrak{M}, \underset{\sim}{q})$.

iii) S is $\mathfrak{Q}^{\#}$-deterministically inductive* on $(\mathfrak{M}, \mathfrak{q})$.

<u>Proof.</u> (ii) ⇒ (iii). "$Qx\psi(x)$ is provable from $\mathfrak{Q}^{\#}$-KPU^{+} by an application of the weak (Q)-rule" is given by a $Q^{\#}$-deterministic formula expressing:

$$\exists p[\ulcorner\psi(\bar{p})\urcorner \in R \wedge \ulcorner\neg\psi(\bar{p})\urcorner \in R]$$

or

$$\forall p[\ulcorner\psi(\bar{p})\urcorner \in R \vee \ulcorner\neg\psi(\bar{p})\urcorner \in R] \wedge Qp(\ulcorner\psi(\bar{p})\urcorner \in R).$$

The other parts follow from earlier results. □

<u>8.9 Corollary.</u> The $\mathfrak{Q}^{\#}$-ordinal of $(\mathfrak{M}, q)$ equals $\sup\{\|\varphi\| \mid \varphi$ is a $\mathfrak{Q}$-deterministic inductive* definition on $(\mathfrak{M}, \mathfrak{q})\}$. □

To see just how much larger $^{\#}\mathfrak{Q}$-ord$(\mathfrak{M}, \mathfrak{q})$ is in general than $\mathfrak{Q}$-ord$(\mathfrak{M}, \mathfrak{q})$ the reader is referred to Aczel [1970].

Special cases of the results in these last two sections are due to Aczel and Moschovakis. We again refer the reader to the references. In addition, some extremely interesting results about the $(Q^{\#}, \breve{Q})$-case have been found in the forthcoming dissertation of Phokion Kolaitis of UCLA.

9. Results of the poll.

Seventy-one people responded to the questionnaire. In tabulating the results I counted responses like "True??", "usually true" or "quite often false" as undecided.

1.	97% true	1% false	1% undecided
2.	87% true	10% false	3% undecided
3.	89% true	7% false	4% undecided
4.	76% true	13% false	11% undecided

(No number theorist or algebraist answered "false" to either 2 or 4, whereas most of those who did answer false were topologists.)

5. ("Most" is monotone.)

92% true	7% false	1% undecided

6. ("Quite a few" is monotone.)

92% true 5% false 3% undecided

Thus, it seems safe to say that they are monotone quantifiers, at least for mathematicians.

If most x have property A and most have property B, then how many x have both properties? The next three questions addressed this question.

7. Some x has both properties.

   76% true 20% false 4% undecided

8. Quite a few x have both properties.

   47% true 45% false 8% undecided

9. Most x have both properties.

   24% true 72% false 4% undecided

The interesting thing here is that most mathematically precise versions of "most" would give a "true" to the strongest conclusion 9, whereas this was rejected fairly conclusively.

The final three questions had to do with whether "most" and "quite a few" are duals. I split the equivalence

(*) $\neg Mx\, A(x) \leftrightarrow Qx \neg A(x)$

into two halves. I then choose a consequence of the more dubious half (11) and monotonicity as the final question.

10. The $\rightarrow$ half of (*).

    82% true 10% false 8% undecided

11. The $\leftarrow$ half of (*).

    30% true 62% false 8% undecided

12. If most x have A(x) and quite a few x have B(x) then some x has A and B.

    18% true 76% false 8% undecided

Every single person who voted true on 12 also voted true on both 2 and 3, thus exhibiting an inconsistency. 43% of those who voted true on 11 and 5 voted false for their consequence 12.

Lemma. If Q is any monotone quantifier then

$$Qx\, A(x) \wedge \check{Q}x\, B(x) \text{ implies } \exists x(A(x) \wedge B(x)).$$

Proof. Assume $Qx\, A(x)$ but $\forall x(A(x) \rightarrow \neg B(x))$. Then, by monotonicity, $Qx\, A(x) \rightarrow Qx \neg B(x)$ so $Qx \neg B(x)$. But $\check{Q}x\, B(x) \rightarrow \neg Qx \neg B(x)$ (all we need here is 11) so $\neg \check{Q}x\, B(x)$. □

10. A gap in Admissible Sets and Structures.

Our purpose here is to correct the one serious mathematical flaw in the book that has been called to our attention. It was discovered by the eagle eye of Professor S. C. Kleene when he used the book for a course. Since we are able to fix it, we appreciate his calling it to our attention.

The gap is in the proof of Lemma IV.4.4 (Main Lemma) on which the proof of Theorem 7.1 hinges.

Recall the aim of the section in question. We are given an infinite set M together with a pairing function $p: M \times M \rightarrow M$. Our aim is to define a notation system for HF_M and insure that the relations mirroring $\in, \notin, =, \neq$ are indirectively definable on the structure $\langle M, p \rangle$. The trouble with the proof of the old lemma comes with set terms of high notational rank that denote sets of low set theoretic rank. Thus, we redefine the set T_s of set terms (definition 4.6ii) and the functions $|\cdot|$ mapping T_s onto HF_M as follows. T_s is to be the union $\bigcup_{n<\omega} T_s^{(n)}$ where the $T_s^{(n)}$ and $|\cdot|$ are defined inductively as follows.

a) $\emptyset \in T_s^{(0)}$ and $|\emptyset| = 0$.

b) If $x \in T_s^{(n)}$ then $x \in T_s^{(n+1)}$.

c) If $x = u \,\sigma\, v$ where $u \in T_s^{(n)}$, $v \in T_u \cup T_s^{(n)}$ and if there is no $z \in T_s^{(n)}$ such that $|z| = |u| \cup \{|v|\}$, then $x \in T_s^{(n+1)}$ and $|x| = |u| \cup \{|v|\}$.

With this definition, if we put a notation x for a set b into $T_s^{(n)}$ we will usually put in several such, but we will then never put in any notations for b at any later stage. We now define $\mathcal{E}, \check{\mathcal{E}}, \approx, \check{\approx}$ as before:

$x \,\mathcal{E}\, y$ iff $x, y \in T$ and $|x| \in |y|$

$x \,\breve{\mathcal{E}}\, y$ iff $y \in T$ and if $x \in T$ then $|x| \notin |y|$

$x \approx y$ iff $x, y \in T$ and $|x| = |y|$

$x \,\breve{\approx}\, y$ iff $y \in T$ and if $x \in T$ then $|x| \neq |y|$.

The set N of notations for finite ordinals is defined as before. The relations, $R, \breve{R}$ used in the book are to be dropped entirely.

Main Lemma. T_u is definable on (M, p) and the following relations are inductive on (M, p):

i) $y \in N$ (a relation of y)

ii) $y \in N$ and $x \in T_s^{(y)}$ (a relation of y and x)

iii) $y \in N$ and $x \notin T_s^{(y)}$ (a relation of y and x)

iv) $x \in T_s$ (a relation of x)

v) $x \,\mathcal{E}\, y$ (a relation of x and y)

vi) $x \,\breve{\mathcal{E}}\, y$ "

vii) $x \approx y$ "

viii) $x \,\breve{\approx}\, y$ "

<u>Sketch of the proof:</u> Notice that $x \in T_s$ iff $\exists y(y \in N \wedge x \in T_s^y)$ so iv) follows from ii) and the closure properties of inductive relations. We define a simultaneous inductive definition of the remaining seven relations, leaving it to the reader to check, by induction on y in all cases, that the inductive definition does in fact give the desired relations, that is, that they are the smallest seven relations satisfying lines (1) - (7) below.

(1) $y \in N$ iff $y = \emptyset \vee \exists z[y = z \text{♂} z \wedge z \in N]$

(2) $y \in N \wedge x \in T_s^{(y)}$ iff $(y = \emptyset \wedge x = \emptyset)$ or $y = z \text{♂} z$ for some $z \in N$ and $x \in T_s^{(z)}$ or else:

$$x = u \text{♂} v \text{ for some } u \in T_s^{(z)},\ v \in T_s^{(z)} \cup T_u,$$

where for all w, either $w \notin T_s^{(z)}$ or $(w \in T_s^{(z)} \wedge (\exists t[t \,\mathcal{E}\, w \wedge t \,\breve{\mathcal{E}}\, u \wedge t \,\breve{\approx}\, v] \vee \exists t[(t \,\mathcal{E}\, u \vee t \approx v) \wedge t \,\breve{\mathcal{E}}\, w]))$.

(3) $y \in N$ and $x \notin T_s^{(y)}$ iff $y = \emptyset$ and $x \neq \emptyset$ or $y = z \text{♂} z$ for some $z \in N$ and either x is not of the form $u \text{♂} v$ or else $x = u \text{♂} v$ where one of the

following happens:

$$u \notin T_s^{(z)}$$

$$v \notin T_u \cup T_s^{(z)}$$

$$\exists w[w \in T_s^{(z)} \wedge \forall t((t \,\check{\varepsilon}\, u \wedge t \,\check{\approx}\, v) \vee t \,\varepsilon\, w) \wedge \forall t(t \,\check{\varepsilon}\, w \vee t \,\varepsilon\, u \vee t \approx v)]$$

(4) $x \,\varepsilon\, y$

(5) $x \,\check{\varepsilon}\, y$ — as in book, p.224

(6) $x \approx y$

(7) $x \,\check{\approx}\, y$ if $(y \in T_u \wedge x \neq y)$ or else $y \in T_s$ and one of the following:

(7.1) $\exists n[n \in N \wedge x \in T_s^{(n)} \wedge y \notin T_s^{(n)}]$

(7.2) $\exists n[n \in N \wedge x \notin T_s^{(n)} \wedge y \in T_s^{(n)}]$

(7.3) $\exists n[n \in N \wedge x \in T_s^{(n)} \wedge y \in T_s^{(n)} \wedge$

$$\exists z((z \,\varepsilon\, x \wedge z \,\check{\varepsilon}\, y) \vee (z \,\varepsilon\, y \wedge z \,\check{\varepsilon}\, x))] .$$

Notice that $\check{\approx}$ does not appear on the right side of (7), but does appear on the right side of (2), (3), (4), which are used in (7). If you think of these seven lines as gradually building up the seven relations, it is $\check{\approx}$ that gets built up the slowest in that we need ε, $\check{\varepsilon}$ restricted to $T_s^{(n)}$ before we define $\check{\approx}$ on $T_s^{(n)}$. The complete proof of the lemma is tedious in the extreme, but can be written out given the patience. Indeed, Prof. Kleene, using our above sketch, wrote out a complete proof for the course he taught from the book. □

|1| P. Aczel: Representability in some systems of second-order arithmetic, Israel J. Math. 8 (1970) 309-328.

|2| J. Barwise: Admissible sets and structures, Springer-Verlag, Heidelberg, 1975.

|3| H. J. Keisler: Logic with the quantifier "there exist uncountably many", Ann. Math. Logic, 1 (1970) 1-93.

|4| P. Kolaitis: Recursion and nonmonotone induction in a quantifier, (to appear).

|5| Y. N. Moschovakis: Elementary induction on abstract structures, North Holland Publ. Co. (1974).

|6| A. Mostowski: Representability of sets in formal systems, in Recursive function theory, A.M.S. Symposia in Pure Math. 5 (1961) 29-48.

|7| J. Sgro: The interior operator logic and product topologies, (to appear).

|8| J. Sgro: An application of topological Model Theory to Chang's Modal Logics, (to appear).

J.E. Fenstad, R.O. Gandy, G.E. Sacks (Eds.)
GENERALIZED RECURSION THEORY II

THE CONTINUOUS FUNCTIONALS AND ^2E

J.A. Bergstra
Institute of Applied Mathematics and Computer Science
University of Leiden
Wassenaarseweg 80, Leiden
The Netherlands

We give an alternative definition of C (the continuous functionals) which expresses the fact that C is the maximal type-structure which does not lead to discontinuities at type-2 (given some elementary means of relative definability). Then we solve a problem of Grilliot [3] using arguments from recursion on the continuous functionals.

§ 1. Preliminaries

1.1. We start with a definition of the continuous functionals, of natural types, using associates. $CTp(0) = \omega$.

$CTp(1) = \omega \to \omega$. $f \in CTp(1)$ has itself as an associate only.

$\alpha^F \in \omega \to \omega$ is an associate of $F^{n+1} : CTp(n) \to \omega$ if the following holds:

For every G^n and associate α^G of G^n the following conditions are satisfied:

i) $\exists n \ \alpha^F(\overline{\alpha^G}(n)) > 0$,

ii) $\forall n[\alpha^F(\overline{\alpha^G}(n)) > 0 \Rightarrow \alpha^F(\overline{\alpha^G}(n)) = F(G) + 1]$.

$CTp(n+1) = \{F^{n+1} : CTp(n) \to \omega | F^{n+1}$ has an associate$\}$.

We write $\leq$ for 'is recursive in' in the sense of [4], and $\leq_{RC}$ for 'is recursively countable in'. ($F \leq_{RC} G$ if there exists a recursive operator which transforms any associate of G into an associate of F).

REC is the class of recursive functionals, RC the class of recursively countable functionals.

1.2. *Definition* For all $n \geq 3$ the partial continuous function ψ^n is defined by:

$$\psi^n(F^{n-1},p) \cong q \quad \text{if} \quad \forall\alpha^F[\{p\}(\alpha^F) \cong q] .$$

In [1] we considered, in a different formulation, recursion relative to the ψ^n . The following theorems are easy to prove.

1.3. *Theorem* If $F^n \leq G^m, \psi^k$ then $F^n \leq_{RC} G^m$.

1.4. *Theorem* For all n : $F^{n+1} \in CTp(n+1)$ iff for no $\alpha \in CTp(1)$ ${}^2E \leq F^{n+1}, \psi^n, \alpha$.

1.5. *Theorem* F^n is recursively countable iff $F^n \leq \psi^{n-1}$.

Theorem 1.4. gives a connection between ${}^2E$ and the continuous functionals.

Remark In [1] we motivated the use of partial recursively continuous functionals like ψ^n by the possibility to define systems of recursion between REC and RC . As an example we gave: χ^3 , defined by $\chi(F,\alpha,\beta) = n$ if $\forall m > CM(F,\beta)[\alpha(m) = n]$.
Here, $CM(F,\beta)$ is the modulus of continuity of F along β . But it turns out that (for total ${}^nF$) :

1.6. *Theorem* ${}^nF \leq \chi^3 \Rightarrow {}^nF \in REC$

To see this one proves that there exists a partial recursive extension χ' of χ (Using a trick like the modulus of continuity functional in [2]). At first sight, however, it seems that to compute $\chi(F,\alpha,\beta)$ one needs an associate for F .

We will now return to theorem 1.4.. An unsatisfactory aspect from the computational point of view are the ψ^n . However, we can prove a similar theorem which avoids these functionals.

§ 2. An alternative definition of the continuous functionals

We will give a definition of the continuous functionals which uses as only topological notion the continuity of a functional of type 2. Further we use a notion of relative definability which is introduced by means of a typed

λ-calculus.

We consider type structures $A = \langle A_0, A_1 \ldots \rangle$ where $A_0 = \omega$ and for all $i > 0$ $A_i \subseteq A_{i-1} \to A_0$.

2.1. *Definition* of a typed λ-calculus (term system).

'Syntax' x^i_j are variables for objects of type i.

$\langle , \rangle$ is a recursive pairing function for numerals (i.e. $\langle \underline{n}, \underline{m} \rangle \geq \underline{\langle n,m \rangle}$ for $n, m \in \omega$).

$(\ )_0$ and $(\ )_1$ are the corresponding projection functions

$\lambda x^i \cdot \tau$, abstraction operator; $(,)$ brackets.

Terms – x^0_j are terms

– if τ, ρ are terms then so are:

$\langle \tau, \rho \rangle$,

$(\tau)_0, (\tau)_1$ and

$x^{n+2}_i (\lambda x^n_j \cdot \tau)$ for $n, i, j \in \omega$.

All terms have type 0. In a straigthforward manner one defines the interpretation $A, s \models \tau$ of a term τ for a given valuation $s : \text{var} \to A$ (provided that λ-abstraction does not lead outside A, otherwise $A, s \models \tau$ does not exist).

As A is extensional $(A_{i+1} \subseteq A_i \to A_0)$ there are no difficulties in the interpretation of abstraction-terms. If s assigns the functionals $\vec{F_i}$ to the free variables of τ then we write $\tau_A(\vec{F_i})$ instead of $A, s \models \tau$ and we delete the subscript A if no confusion arises.

2.2. *Definition* $F : A_n \to \omega$ is explicitly definable from $\vec{G}^{n_i}_i$ if for some term τ:

$$\forall H^n \in A_n [F(H) = \tau_A(H, \vec{G}^{n_i}_i)] .$$

$$(\text{i.e. } F = \lambda H \cdot \tau(H, \vec{G}^{n_i}_i)) .$$

Remark In this case F is primitive recursive in the G_i.

Notation We write $\leq_E$ for 'explicitly definable from'.

2.3. *Definition* $A = \langle A_0,\ldots,A_n\rangle$ is closed under explicit definition if every $F : A_i \to \omega (i<n)$ which is explicitly definable from a sequence $\vec{G}_i^{n_i}$ of elements of A is in fact an element of A_{i+1}.

2.4. *Lemma* For all i,j there are terms $\tau_{i,j}$ such that we may write

$$Tr_i^j(x^i) = \begin{cases} \tau_{i,j}(x^i) & \text{if } j = 0 \\ \lambda y^{j-1}\cdot\tau_{i,j}(x^i,y^{j-1}) & \text{if } j > 0 \end{cases}$$

and such that the Tr_i have the usual properties of lifting up and pushing down operators.

Proof straightforward. ⊠

2.5. *Lemma* Let $\langle B_0\ldots B_n\rangle$ be closed under $\leq_E$. Then there exists a unique maximal set B_{n+1} (denoted by $UME(\langle B_0,\ldots,B_n\rangle)$) such that $\langle B_0,\ldots,B_n,B_{n+1}\rangle$ is again closed under $\leq_E$.

Proof We take for B_{n+1} the set of all $F^{n+1} : B_n \to B_0$ which satisfy the following condition:

If $H^m \leq_E F$, $\vec{G}_i^{n_i}$ with $m, n_i \leq n$ then $H_m \in B_m$.

The maximality of B_{n+1} is obvious. Remains to show that $\langle B_0,\ldots,B_{n+1}\rangle$ is closed under $\leq_E$.

Now suppose that $F^{i+1} = \lambda x^i\cdot\tau(x^i,\vec{G}_i^{n_i})$ $(n_i\leq n+1)$. We must prove $F^{i+1} \in B_{i+1}$.

There are two cases:

i) <u>$i = n$</u>. Suppose $F^{n+1} \notin B_{n+1}$. Then there exist $j \leq n$, ρ, $\vec{H}_i^{n_i}$ $(n_i\leq n)$ such that $T^j = \lambda x^{j-1}\cdot\rho(x^{j-1},F^{n+1},\vec{H}_i^{m_i})$ is not in B_j. Using an easy normalisation argument one finds a term σ such that $T_j = \lambda x^{j-1}\cdot\sigma(x^{j-1},\vec{H}_i^{m_i},\vec{G}_i^{n_i})$.

This reduces the case to the next case.

ii) <u>$i < n$</u>. We write $\vec{G}_i^{n_i} = \vec{H}_i^{n+1}$, $\vec{L}_i^{m_i}$ $m_i < n$. The fact that we delete arguments of type n is justified by the possibility that we code them into an object of type $n+1$. $\langle J_1^j,\ldots,J_n^j\rangle$ denotes the functional $\lambda x^{j-1}\cdot\langle J_1^j(x),\ldots J_n^j(x)\rangle$. (Where $\langle a_1^0,\ldots,a_n^0\rangle =$

$<\ldots<<a_1^0,a_2^0>,a_3^0\ldots>\ldots>)$.

With induction on the structure of the terms we can prove the following.

Lemma Let ρ be a term which has its free variables in $x_i^{\vec{m}_i}$ and $y_i^{\vec{n}+1}$, then there exists a functional $U_\rho \in \mathcal{B}_n$ s.t. for all $K_i^{\vec{m}_i}$ $\rho(K_i^{\vec{m}_i},\vec{H}_i^{n+1}) = U_\rho(<\vec{Tr}^{n-1}_{m_i}(K_i^{m_i})>)$.

Proof - If $\rho = x_j^0$ then U_ρ exists by closure under $\leqslant_E$.

- Let $\rho = <\rho_1,\rho_2>$.

  Then $U_\rho(x) = <U_{\rho_1}(x),U_{\rho_2}(x)>$; similar with $(\ )_0$, $(\ )_1$.

- if $\rho = x^{j+1}(\lambda x^j.\sigma(x^j,\text{—}))$ then there are again two cases: $j < n$, $j = n$.

  i) <u>$j < n$</u> . Then $\rho(<\vec{Tr}^{n-1}_{m_i}(\vec{x}_i^{m_i})>)$ is found from U_σ and $<Tr\ldots>$ by means of explicit definition relative to functionals of type $\leq n$.

  ii) <u>$j = n$</u> . Then x^{j+1} must be one of the H_i^{n+1} . In this case $U_\rho(A)$ is found by explicit definition relative to:

  A of type $n - 1$, U_σ of type n and H_i^{n+1} . This is possible by the definition of $\mathcal{B}_{n+1}$ ($H_i^{n+1} \in \mathcal{B}_{n+1}$ of course) ⊠

Using this lemma we see that $\lambda x^{j-1}.\sigma(x^{j-1},H_i^{\vec{m}_i}, G_i^{\vec{n}_i}) = \lambda x^{j-1}.U^n(<Tr^{n-1}_{j-1}(x^{j-1}),Tr^{n-1}_{m_i}(H_i^{\vec{m}_i})>)$.

Hence it is in $\mathcal{B}_j$ as $<\mathcal{B}_0,\ldots,\mathcal{B}_n>$ is closed under $\leqslant_E$. ⊠

2.6. *Definition* $A_0 = \omega = CTp(0)$

$A_1 = \omega \to \omega = CTp(1)$

$A_2 = \{F;A_1 \to A_0 | F \text{ is continuous}\} = CTp(2)$.

A_{n+3} $UME(<A_0,\ldots,A_{n+2}>)$ (= the unique maximal extension of $<A_0,\ldots A_{n+2}>$ which is closed under explicit definition and exists as a consequence of the previous lemma).

2.7. *Theorem* for all n $A_n = CTp(n)$.

Proof With induction on n . For $n = 0,1,2$ the situation is obvious. So we assume that

$$A_0 = CTp(0) \wedge \ldots \wedge A_{n+2} = CTp(n+2) .$$

$A_{n+3} \subseteq CTp(n+3)$: Given any $F^{n+3} \notin CTp(n+3)$ we show that $\langle A_0, \ldots A_{n+2}, B_{n+3} \rangle$ is not closed under explicit definition as soon as B_{n+3} contains F^{n+3} . To prove this we try to define an associate α^F of F .

$$\alpha^F(\sigma) = \begin{cases} k+1 & \text{if for all } G \in CTp(n+2)(=A_{n+2}) \text{ and all } \beta^G \text{ (associate of } G\text{) the following implication holds:} \\ & \beta^G \text{ extends } \sigma \Rightarrow F(G) = k \\ 0 & \text{otherwise} \end{cases}$$

As F is not in $CTp(n+3)$ it has no associate. Therefore for some G^{n+2} with associate β^G we have

$$\forall m[\alpha^F(\overline{\beta^G}(m)) = 0] \quad \text{or}$$

$$\exists m[\alpha^F(\overline{\beta^G}(m)) > 0 \wedge \alpha^F(\overline{\beta^G}(m)) \neq F(G) + 1]$$

The second situation, however, is impossible therefore there exist associates β^{G_n} for functionals $G_n \in CTp(n+2)$ such that:

i) $\forall m[\overline{\beta^{G_m}}(m) = \overline{\beta^G}(m)]$ and

ii) $F(G_m) \neq F(G)$.

Now we will define a continuous D^{n+2} and a function α such that ${}^2E$ is explicitly definable from F, G, D, α . (Now ${}^2E \notin CTp(2)$ and we are done).

$$\alpha(x) = \begin{cases} 1 & \text{if } x = F(G) \\ 0 & \text{otherwise} \end{cases}$$

$$E(\delta) = \alpha(F(\lambda H^{n+1} \cdot \tau(\delta, D, H))) .$$

Here τ and D must be chosen such that

$$\forall m[\delta(m) \neq 0] \Rightarrow \lambda H^{n+1} \cdot \tau(\delta, D, H) = G \quad \text{and}$$

$$\exists m[\delta(m) = 0] \Rightarrow \exists m\ \lambda H^{n+1} \cdot \tau(\delta, D, H) = G_m .$$

Definition: $\langle\langle \delta, H^{n+1} \rangle\rangle =$

$\lambda x^n \cdot \langle \delta((x^n(\lambda y^{n-2} \cdot 0))_0), H^{n+1}(\lambda y^{n-1} \cdot (x^n(y^{n-1}))_1) \rangle$

(if $n < 2$ we need a slightly different definition).

For each L^{n+1} there exist δ_L and H_L^{n+1} such that

$L^{n+1} = \langle\langle \delta_L, H_L^{n+1} \rangle\rangle$

Take $\delta_L = \lambda m \cdot (L(\lambda z^{n-1} \cdot m))_0$ and

$H_L^{n+1} = \lambda U^A \cdot (L(\lambda y^{n-1} \cdot \langle 0, U^n(y^{n-1}) \rangle))$.

We put $\tau(\delta, D, H) = D(\langle\langle \delta, H \rangle\rangle)$.

Now we must find a countable D such that

*) $\forall m[\delta(m) \neq 0] \Rightarrow D(\langle\langle \delta, H \rangle\rangle) = G$ and

**) $\exists m[\delta(m) = 0] \Rightarrow D(\langle\langle \delta, H \rangle\rangle) = G_m$ for some m .

We show how to compute $D(\langle\langle \delta, H \rangle\rangle)$ from δ and an associate β^H of H .

As an oracle we need a function γ which encodes β and the $\beta_i (= \beta^{G_i})$

Let $\beta_{[\delta]}(x) = \begin{cases} \beta(x) & \text{if } \forall i \leq x\ [\delta(i) \neq 0] \\ \beta_i(x) & \text{if } i = \mu j[\delta(j) = 0] \text{ otherwise} \end{cases}$

Now $D(\langle\langle \delta, H \rangle\rangle) = CAP(\beta_{[\delta]}, \beta^H) = \beta_{[\delta]}(\overline{\beta}^H(k)) - 1$ where $k = \mu l[\beta_{[\delta]}(\overline{\beta}^H(l)) > 0]$.

It is easy to see that *) and **) are satisfied.

This shows that we can find a satisfactory countable D in $CTp(n+2)$.

$A_{n+3} \supseteq CTp(n+3)$:

To show this we note that $\langle CTp(0), \ldots, CTp(n+3) \rangle$ is closed under explicit definition as it is closed under relative $S_1, \ldots, S_9$ recursion. ■

2.8. *Corollary* Another definition of C is as follows:

$CTp(0) = \omega$, $CTp(1) = \omega \to \omega$, $CTp(2)$ is the set of continuous elements of $Tp(2)$.

$CTp(n+1)$ = the set of all F^{n+1} such that ^2E is not

$S_1,\ldots,S_9$ recursive in F^{n+1} and any $G^n \in CTp(n)$.

Proof Immediate. ⊠

§ 3. Some properties of semirecursive sets of functionals

3.1. *Introduction* We consider a type structure A :

$$A = \langle A_0, A_1, \ldots \rangle \text{ with}$$

$$A_0 = \omega$$

$$A_{n+1} \subseteq A_n \to A_0 .$$

A subset V of A_n is recursive in functionals $\vec{F}$ from A if for some $p \in \omega$:

$^nG \in V \iff \{p\}(^nG,\vec{F}) \cong 0$ and

$^nG \notin V \iff \{p\}(^nG,\vec{F}) \cong 1$

V is semirecursive (in $\vec{F}$) if for some $p \in \omega$

$^nG \in V \iff \{p\}(^nG,\vec{F})\downarrow$

Here $\{\cdot\}(\text{——})\cong$ denotes Kleene's computation relation as defined in [4].

In [3] Grilliot states the following three problems concerning semirecursive sets:

- *The union problem*: Is the union of two s.r. sets again s.r.?
- *The negation problem*: If V and $\overline{V}$ are s.r. must V be recursive?
- *The reduction problem*: If V , W are s.r. do there exist s.r. V_1 , W_1 such that $V_1 \cap W_1 = \emptyset$, $V_1 \cup W_1 = V \cup W$, $V_1 \subseteq V$ and $W_1 \subseteq W$?

We add to these problems the following one:

- *The density problem*: If V , W are s.r. such that $V \subsetneq W$ does there exist an s.r. U such that $V \subsetneq U \subsetneq W$?

Of course these problems have relativised versions and can be posed in all kinds of type structures at every type.

3.2. *Semirecursive subsets of Tp(2) (unrelativised case)*

Platek [5] (page 131) solves the negation problem negatively.

Grilliot [3] (page 16) solves the union problem negatively. He leaves open the reduction problem which we solve negatively below. We leave the density question unanswered.

3.2.1. *Theorem* There are s.r. sets $A, B \subseteq Tp(1) \times Tp(1) \times Tp(2)$ such that for no part. rec. ψ_A , ψ_B the following holds:

i) $Dom(\psi_A) \subseteq A$, $Dom(\psi_B) \subseteq B$

ii) $Dom(\psi_A) \cup Dom(\psi_B) = A \cup B$

iii) $Dom(\psi_A) \cap Dom(\psi_B) = \emptyset$

Proof We begin with some preliminaries concerning trees of computation. The presentation will be rather informal. To any computation $\{p\}(\vec{\alpha},F)$ (which possibly diverges) we can associate a tree $Tr(p,\vec{\alpha},F)$ which is a prefix-closed set of sequence numbers. (We take only arguments $\vec{\alpha}$ and F for the sake of easy presentation, obviously one may define $Tr(p,-)$ for all sequences of arguments – .)

$Tr(p,\vec{\alpha},F)$ has the following properties:

i) if $\{p\}(\vec{\alpha},F)\downarrow$ then it is (uniformly) recursive in $p, \vec{\alpha}, F$.

ii) if $\{p\}(\vec{\alpha},F)\uparrow$ then it is (uniformly) semi-recursive in $p, \vec{\alpha}, F$.

iii) $\{p\}(\vec{\alpha},F)\downarrow$ iff $Tr(p,\vec{\alpha},F)$ is well-founded.

iv) the σ in $Tr(\beta,\vec{\alpha},F)$ label subcomputations:

notation: σ labels the computation.

$$\{p_\sigma\}(\vec{m}_\sigma,\vec{\alpha},F)$$

It is not necessary that this computation converges for σ to be in $Tr(p,-)$.

P_σ and $\vec{m}_\sigma$ are coded in σ somehow.

v) in a computation $\{p\}(\vec{\alpha},F)$ F is only evaluated at arguments β if for some σ P_σ is an S_8-index (application) and $\{p_\sigma\}(\vec{m}_\sigma\vec{\alpha},F) = F(\beta)$

where $\beta = \lambda x\cdot\{q\}(x,\vec{m}_\sigma,\alpha,F)$. Here $q = p_{\sigma^x}$ where σ^x is the immediate successor of σ corresponding to the subcomputation with argument x .

notation: for $\beta : \theta_p(\sigma,\vec{\alpha},F)$.

In fact we will only consider computations with F_0 as an argument where

$$F_0 = \lambda\ \alpha\cdot 0\ .$$

For computations of this kind we have the following continuity property:

Suppose 1) $\{p\}(\vec{\alpha},F_0)\downarrow$

2) $\sigma \in Tr(p,\vec{\alpha},F_0)$

3) p_σ is an S_8-index

4) $n \in \omega$

Then for some $l \in \omega$ the following holds:

<u>If</u> a) the α_i' extend $\overline{\alpha}_i(l)$ for $i \leq lth(\vec{\alpha})$ and

b) $\sigma \in Tr(p,\vec{\alpha}',F_0)$ and

c) $\theta(\sigma,\vec{\alpha}',F_0)$ is total

<u>Then</u> for all $j \leq n$

$\theta_p(\sigma,\vec{\alpha},F_0)(j) = \theta_p(\sigma,\vec{\alpha}',F_0)(j)$

(Note that we do not require $\{p\}(\vec{\alpha}',F_0)\downarrow$. The meaning of this property is that we can fix initial segments of functions occurring in $\{p\}(\vec{\alpha}',F_0)$ by fixing initial segments of $\vec{\alpha}$. The proof of it is easy and only uses the continuity of F_0).

Now we can prove the theorem.

For $\alpha \in Tp(1)$ we define $\phi(\alpha) \in \omega \xrightarrow{\text{part}} \omega$ by

$$\phi(\alpha)(x) = \mu y[\alpha(\langle x,y\rangle) \neq 0]\ .$$

We extend ϕ to sequence numbers by

$$\phi(\sigma)(x) = \mu y[\langle x,y\rangle < lth(\sigma) \text{ and } \sigma_{\langle x,y\rangle} \neq 0]\ .$$

Now we define A and B by:

$$(\alpha,\beta,F) \in A \text{ iff } F(\phi(\alpha)) = 0 \text{ and}$$

$(\alpha,\beta,F) \in B$ iff $F(\phi(\beta)) = 0$.

Suppose that ψ_A and ψ_B satisfy the properties i) and ii) (in the formulation of the theorem).

We will construct α and β such that

$\psi_A(\alpha,\beta,F_0)\downarrow$ and $\psi_B(\alpha,\beta,F_0)\downarrow$.

This will contradict requirement iii).

Let a, b be indices of ψ_A, ψ_B .

3.2.2. *Lemma* There are α, β such that:

i) $\phi(\alpha)$ and $\phi(\beta)$ are total and unequal,

ii) $\phi(\alpha)$ does not occur in the comp. $\{b\}(\alpha,\beta,F_0)$,

iii) $\phi(\beta)$ does not occur in the comp. $\{a\}(\alpha,\beta,F_0)$.

The lemma is sufficient to prove the theorem. To see this consider $\psi_A(\alpha,\beta,F_0)$ where α, β are as in the lemma. If $\psi_A(\alpha,\beta,F_0)\uparrow$ then $\{a\}(\alpha,\beta,F_0)\uparrow$. Define F_0^γ by

$$F_0^\gamma(\delta) = \begin{cases} 1 & \text{if } \gamma = \delta \\ 0 & \text{otherwise} \end{cases}$$

As $\phi(\beta)$ does not occur in the comp. $\{a\}(\alpha,\beta,F_0)$ we know $\{a\}(\alpha,\beta,F_0^{\phi(\beta)})\uparrow$. But $(\alpha,\beta,F_0^{\phi(\beta)}) \notin B$. Therefore $\psi_A(\alpha,\beta,F_0^{\phi(\beta)})$ must converge as $F_0^{\phi(\beta)}(\alpha) = 0$ (by the difference of $\phi(\alpha)$ and $\phi(\beta)$. This however implies $\psi_A(\alpha,\beta,F_0)\downarrow$. Similarly one proves $\psi_B(\alpha,\beta,F_0)\downarrow$.

Proof of the lemma We use a spoiling construction to find extending sequences σ_n^α and σ_n^β of sequence numbers and take $\alpha = \lim_n \sigma_n^\alpha$ and $\beta = \lim_n \sigma_n^\beta$. We want to ensure that:

i) $\phi(\alpha)$ and $\phi(\beta)$ are total. (This is possible by the fact that each σ has an extension γ such that $\sigma(\gamma)$ is total; in fact we want $\phi(\sigma_{2n}^\alpha)(n)$ and $\phi(\sigma_{2n}^\beta)(n)$ to be defined for all n .)

ii) $\phi(\alpha) \neq \phi(\beta)$. (This is possible by choosing σ_0^α and σ_0^β such that $\phi(\sigma_0^\alpha)$ and $\phi(\sigma_0^\beta)$ are incompatible.)

iii) Let $|\sigma|$ be a code for σ in ω (In fact we may take $|\sigma| = \sigma$ but this seems notationally confusing). Then we want:

1. <u>if</u> $\tau \in Tr(a,\alpha',\beta',F_0)$ for some α',β' which extend $\sigma^{\alpha}_{2|\tau|+1}$ (and $\sigma^{\beta}_{2|\tau|+1}$) and $\theta_a(\tau,\alpha',\beta',F_0)$ is total <u>then</u> $\phi(\beta') \neq \theta_a(\tau,\alpha',\beta',F_0)$
2. a similar condition concerning $\phi(\alpha')$ and $\theta_b(\tau,\alpha',\beta',F_0)$.

Claim If the sequences σ^{α}_n and σ^{β}_n satisfy i), ii) and iii) they provide the example required by the lemma.

Proof The only non-trivial point is to verify conditions ii) and iii) of the lemma. Consider condition ii).

Suppose $\phi(\alpha)$ occurs in $\{b\}(\alpha,\beta,F_0)$. Let τ be in $Tr(b,\alpha,\beta,F_0)$ such that $\phi(\alpha) = \theta_b(\sigma,\alpha,\beta,F_0)$. Now α,β extend $\sigma^{\alpha}_{2|\tau|+1}, \sigma^{\beta}_{2|\tau|+1}$. It is easy to see that this leads to a contradiction.

Finally we describe the definition of σ^{α}_n and σ^{β}_n . The steps to satisfy conditions i) and ii) above are obvious (and done in the easiest way).

We describe how to satisfy the first part of condition iii). The second part is handled similarly.

Let $n = 2|\tau| + 1$. Suppose $\sigma^{\alpha} = \sigma^{\alpha}_{n-1}$ and $\sigma^{\beta} = \sigma^{\beta}_{n-1}$ are defined.

We define two functions $D_0(\sigma^{\beta})$ and $D_1(\sigma^{\alpha})$ by

$$D_0(\sigma)(x) = \begin{cases} 0 & \text{if } \sigma(x) \text{ is undefined} \\ \sigma(x) & \text{otherwise.} \end{cases}$$

$$D_1(\sigma)(x) = \begin{cases} 1 & \text{if } \sigma(x) \text{ is undefined} \\ \sigma(x) & \text{otherwise} \end{cases}$$

$(\sigma(x) = \sigma_x)$.

Clearly $D_0(\sigma^{\beta})$ is non-total and $D_1(\sigma^{\alpha})$ is total.

Therefore $(D_1(\sigma^\alpha), D_0(\sigma^\beta), F_0) \in A\backslash B$ and hence

$\{a\}(D_1(\sigma^\alpha), D_0(\sigma^\beta), F_0)\downarrow$

We consider two cases.

I) $\tau \notin Tr(a, D_1(\sigma^\alpha), D_0(\sigma^\beta), F_0)$.

In this case we choose extensions ρ^α and ρ^β of σ^α, σ^β , such that $\rho^\alpha < D_1(\sigma^\alpha)$ and $\rho^\beta < D_0(\sigma^\beta)$, for which the following holds:

if $\alpha' > \rho^\alpha$ and $\beta' > \rho^\beta$ then $\tau \notin Tr(a, \alpha', \beta', F_0)$.

To see that this is possible note the following:

if $\tau \in Tr(a, \alpha', \beta', F.)$ this is caused by the fact that some values computed as $\{\rho_{\tau'}\}(\vec{m}_{\tau'}, \alpha', \beta', F_0)$ for τ' "before" τ lead to the computation belonging to τ . Now we can fix all values computed in subcomputations which can "lead to τ " by fixing initial segments of $D_1(\sigma^\alpha)$, $D_0(\alpha^\beta)$. Then we are sure that τ will not occur in computations with extensions of these initial segments. (However, this non-occurrence may as well have the reason that some subparts of the computation are divergent.) Now we take $\sigma_n^\alpha = \rho^\sigma * <0>$ (to ensure that its length increases). Further we choose σ_n^β as an extension of ρ^β . It is now obvious that the first part of condition iii) in the lemma is satisfied.

II) $\tau \in Tr(a, D_1(\sigma^\alpha), D_0(\sigma^\beta), F_0)$

Now choose $l > lth(\sigma^\beta)$.

Clearly $\phi(D_0(\sigma^\beta))(l)$ is undefined.

Now choose ρ^α and ρ^β between σ^α and $D_1(\sigma^\alpha)$ (resp. between σ^β and $D_0(\sigma^\beta)$) such that the following holds:

whenever $\alpha' > \rho^\alpha$, $\beta' > \rho^\beta$, $\tau \in Tr(b, \alpha', \beta', F_0)$

and $\theta_b(\tau, \alpha', \beta', F_0)$ is total

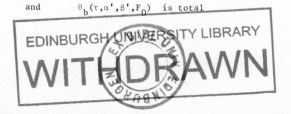

then $\theta_b(\tau,\alpha',\beta',F_0)(1) \cong h(1) \underset{\text{def.}}{\cong} \theta_b(\tau,D_1(\sigma^\alpha),D_0(\sigma^\beta),F_0)(1)$

(This again uses the continuity property of computations with continuous arguments.)

Now we want to take σ^α_n and σ^β_n as extensions of ρ^α and ρ^β such that $\phi(\sigma^\alpha_n) \neq h$. (This gives $\phi(\alpha) \neq \phi_b(\tau,\alpha',\beta',F)$ when even both functions are total and $\alpha' > \sigma^\alpha_n$, $\beta' > \sigma^\beta_n$. To reach this we take σ^α_n to be an extension of ρ^α for which $\phi(\sigma^\alpha_n)(1)\downarrow$ and $\phi(\sigma^\alpha_n)(1) > h(1)$. This is easily possible as $\phi(\rho^\alpha)(1)\uparrow$.

This ends the description of a construction of functions α and β satisfying the conditions of the lemma. ◘

3.3. *Semi-recursive subsets of CTp(3)*

In [1] we proved that in CTp(3) the predicate

$\{p\}(^3F)\downarrow$ has the quantifier form $\forall\alpha\exists n\ P(p,\alpha,n,h^F)$ with recursive P.

Here h^F is the graph of F on the primitive recursive functionals. On the other hand all predicates of this form are semi-recursive. From these facts it follows that we have the reduction and union property in this case. As Platek's counterexample to the negation property already works in CTp(2) the negation property does not hold now.

3.4. *Subsets of Tp(2) which are semi-recursive in $^3 0, {}^2F$*

3.4.1. *Theorem* The negation problem in the type-2 case relativised to $^3 0$ and E_1 has a negative solution.

Proof Take $V = \{(p,{}^2F) \mid \{p\}({}^2F,{}^2E_1)\downarrow\}$

V is obviously not recursive in E_1.

But: i) V is semi-recursive in $^3 0$ and E_1.

and ii) $\overline{V}$ is also semi-recursive $^3 0$ and E_1.

To prove this we must do some work.

Let $R_1(p,\alpha,F)$ denote:

" α codes a locally correct computation tree for $\{p\}(F,E_1)$".

R_1 is a predicate recursive in E_1.

Then we have:

$\{p\}(F,E_1)\uparrow \Leftrightarrow \forall\alpha[R_1(p,\alpha,F) \Rightarrow IB(\alpha)]$

(here $IB(\alpha)$ means α has an infinite branch, and α is considered as a tree, IB is recursive in E_1)

$\Leftrightarrow 0^3(\lambda\alpha\cdot\mu n[R_1(p,\alpha,F) \Rightarrow IB(\alpha)])\downarrow$. ⊠

3.4.2. *Theorem* The union property holds in the case relativised to any ^2F and $^3 0$.

Proof The predicate $\{p\}(^2G,{}^2F,{}^3 0)\downarrow$ has the quantifier form $\forall\alpha\exists n\, R(\alpha,n,{}^2G,{}^2F)$ with recursive R.

To see this note that $\{p\}(^2G,{}^2F,{}^3 0)\uparrow$ if it has an infinite branch (in its computation tree) which can be coded in a function. As local correctness is a matter of (defined) values only one does not need actual applications of $^3 0$ to check it. Therefore $\{p\}(^2G,{}^2F,{}^3 0)\uparrow$ is Σ^1_1.

From the quantifier form it follows that the disjunction of two s.r. (Π^1_1) predicates is again Π^1_1 and hence s.r. (in $^2F,{}^3 0$). ⊠

References

[1] Bergstra J.A. (1976) Computability and Continuity in Finite Types (Ph.D. Thesis Utrecht).

[2] Gandy R.O. (1967) Computable Functionals of Finite Type I In: Sets Models and Recursion Theory, Ed. J.N. Crossley (North-Holland).

[3] Grilliot T.J. (1967) Recursive Functions of Finite Higher Types (Ph.D. Thesis Duke Univ.).

[4] Kleene S.C. (1959) Recursive Functionals and Quantifiers of Finite Type I T.A.M.S. 91 1 - 52.

[5] Platek R. (1966) Foundations of Recursion Theory (Ph.D. Thesis Standford University).

J.E. Fenstad, R.O. Gandy, G.E. Sacks (Eds.)
GENERALIZED RECURSION THEORY II

Recursion theory and set theory: a marriage of convenience [1]

by

Solomon Feferman

1. Introduction. We expand here on a program which was initiated in [F1] and elaborated in one direction in [F2]. The aim of the program is to provide an abstract axiomatic framework to explain the success of various analogues to classical (set-theoretical) mathematics which have been formulated in operationally explicit terms. These analogue developments fall roughly into two groups: (a) recursive and/or constructive mathematics, and (b) hyperarithmetic and/or predicative mathematics.

The framework proposed in [F1] was given by two theories T_o and T_1 with the following features:

(i) they are theories whose universe of discourse includes operations and classes as elements;

(ii) the notions in (i) are not interreducible, operations being given by rules of computation (in some sense or other) and classes by predicates (from a fairly rich language).

(iii) operations may be applied to any elements, including operations and classes;

(iv) the theories are non-extensional;

(v) T_1 is obtained from T_o by adjunction of a single axiom for an operation e_N which gives quantification over N;

---

[1] Text of a talk presented at the conference: Generalized Recursion Theory II, Oslo June 13-17, 1977. Research and preparation supported by NSF grant No. MCS 76-07163.

(vi) T_0 restricted to intuitionistic logic is constructively justified;

(vii) T_1 minus its theory of generalized inductive definitions is predicatively justified.

(viii) T_0 has a model in which the elements of $N \to N$ represent all the recursive functions;

(ix) T_1 has a model in which the elements of $N \to N$ represent all the hyperarithmetic functions.

(x) T_1 has a model in which the elements of $N \to N$ represent all set-functions of natural numbers.

The plan of the program is to explain cases in which analogues have been successful, e.g. in recursive mathematics as follows. Say one has a theorem $\phi^{(set)}$ of set-theoretical mathematics which has a positive recursive analogue $\phi^{(rec)}$. Then one tries to find a theorem ϕ of T_0 such that on the one hand ϕ specializes to $\phi^{(rec)}$ in the model (viii) and on the other hand to $\phi^{(set)}$ in the model (x). Similarly for the other analogues, using (vi)-(ix).

This plan was carried out in some detail for a portion of model theory in [F2], using an extension $T_1^{(\Omega)}$ of T_1; that theory had the same features as T_1, but also axioms for a class Ω of ordinals were adjoined. We explained thereby the success of Cutland's analogue development [C] in which: hyperarithmetic models ~ countable models, and Π^1_1 chains of hyperarithmetic models ~ models of cardinality $\leq \aleph_1$.

In this paper we expand the systems T_0, T_1 to new theories $T_0(S)$, $T_1(S)$ so as to increase their flexibility and range of applicability. S is a class which acts like the class of all sets in set-theory, and the new axioms (in § 2 below) provide strong, natural closure conditions on S. Otherwise the principal features of $T_0(S)$ and $T_1(S)$ are the same as for T_0 and T_1. These now constitute our proposed marriage of recursion theory and set theory for the "convenience" of achieving the program explained above. It seems that any such framework must give up some significant features or principles of ordinary set theory. Our choice is to give up the identification of functions with graphs and to give up extensionality. As to the latter, the principle of extensionality has no essential mathematical use; its standard purpose is to map an equivalence relation $\equiv_A$ in a class A onto the equality relation by passing to $A/\equiv_A$. Instead, one simply works with the structure $(A, \equiv_A)$ accompanied by the new "equality" $\equiv_A$. However, it is possible that extensional, non-classical systems can also be used for our purposes, (as has been suggested by H. Friedman). In any case, the choice of axioms should be based on pragmatic considerations (not necessarily in conflict with constructive principles) and, as such, is still subject to experimentation.

§3 goes into some detail about how a variety of models of $T_0(S)$ and $T_1(S)$ can be constructed directly. There are two steps to be considered. First is the choice of an applicative model, either using familiar recursion theories or by generation. Examples of the former are denoted $\mathrm{Rec}(\omega)$ (ordinary recursion theory) and $\exists^N\text{-}\mathrm{Rec}(\omega)$ (Π_1^1 recursion theory). Examples of the latter are given over any set-theoretical model $\mathfrak{M}$, resulting in three applicative structures $\mathrm{Rec}(\mathfrak{M})$, $\exists^N\text{-}\mathrm{Rec}(\mathfrak{M})$ and $\text{Set-Fun}(\mathfrak{M})$; in the first two $\mathrm{Rec}(\omega)$ and $\exists^N\text{-}\mathrm{Rec}(\omega)$ are lifted to $\mathfrak{M}$, and in the third all set functions of $\mathfrak{M}$ are fed into a generalized recursion theory. Next, given an applicative structure $\mathfrak{a}$ it is shown how to build a model $\mathfrak{a}^*$ of $T_0(S)$ in which any given collection of sets is represented. This finally leads to models such as $(\mathrm{Rec}(\omega))^*$, $\mathfrak{M}^*_{\mathrm{Rec}}$, $\exists^N\text{-}\mathrm{Rec}(\omega))^*$, $\mathfrak{M}^*_{\exists^N\text{-}\mathrm{Rec}}$ and $\mathfrak{M}^*_{\text{Set-Fun}}$; the last three of these are also models of $T_1(S)$.

§4 outlines how the abstract constructive measure theory of Bishop-Cheng [Bi,C] can be formalized in $T_0(S)$. That involved prima-facie use of a power-class operation which had been an obstacle in T_0 and other approaches. It is now handled essentially via $\mathcal{P}_S(X)=\{a \mid a \in S \wedge a \subseteq X\}$. A possible application of interest is given using the models $\mathrm{Rec}(\omega)^*$ or $\mathfrak{M}^*_{\mathrm{Rec}}$ of $T_0(S)$: *if a recursive Borel set* A *is measurable in the sense of* [Bi,C] *and* $\mu(A) > 0$ *then* A *contains a recursive member* (4.6). Some suggestions about how $T_0(S)$ might further be used to generalize classical and recursive mathematics are given in 4.7.

In §5 a theory of *accessible ordinals* $\mathcal{O}_S$ and (*regular*) *number classes* ($\Omega_x^{(r)}$ and) Ω_x for $x \in \mathcal{O}_S$ is developed within $T_0(S)$. In any model of $T_0(S)$ there are associated ordinals $|\mathcal{O}_S| = \sup\{|x| : x \in \mathcal{O}_S\}$ and $|\Omega_x|$ (defined similarly). Under the interpretation by $\mathfrak{M}^*_{\text{Set-Fun}}$ we have $|\mathcal{O}_S|$ = *least inaccessible ordinal* and $|\Omega_x| = \omega_x$. On the other hand in both $\mathrm{Rec}(\omega)^*$ and $\Xi^N\text{-}\mathrm{Rec}(\omega)^*$ we have $|\Omega_1| = \omega_1^c$ = *least nonrecursive ordinal*. It is conjectured that $|\mathcal{O}_S|$ = *least recursively inaccessible ordinal* and $\forall x \in \mathcal{O}_S[\,|x| = \alpha \Rightarrow |\Omega_x^{(r)}| = \omega_\alpha^c (= \tau_\alpha)]$ in these latter models. If so, this theory provides an approach to recursively accessible ordinals which is conceptually superior to that of Richter [R].

The paper concludes in 5.4 with a discussion of some further axioms which may be added to $T_1(S)$ and which are true in $\mathfrak{M}^*_{\Xi^N\text{-}\mathrm{Rec}}$, such as the *selection principle* Sel_{Ω_1} for Ω_1. $T_1(S) + (\mathrm{Sel}_{\Omega_1})$ can be used for all the purposes in *model theory* which had been provided by $T_1^{(\Omega)}$ in [F2]. Now one can look for further applications in model theory by use of the development of higher number classes in $T_1(S)$. Another possible application is to "long" *hierarchies of normal* (*critical*) *functions* (originally due to Bachmann), which make use of higher number classes to define large countable ordinals. In certain specific cases these have been verified to be recursive by tedious calculations. The idea would be to obtain such results instead as a consequence of a treatment of these hierarchies within the framework of $T_1(S)$, using the fact that $|\Omega_1| = \omega_1^c$ in $\mathfrak{M}^*_{\Xi^N\text{-}\mathrm{Rec}}$.

2. *The theories* $T_0(S)$ *and* $T_1(S)$. Knowledge of [F1], [F2] is not presumed here.

2.1 *Syntax of the theories*. The basic language is described as follows.

(Expansions of this syntax will consist simply in the adjunction of further constant symbols.)

Individual (general) variables: $a,b,c,\ldots,f,g,h,\ldots,x,y,z$

Class variables: $A,B,C,\ldots,X,Y,Z$

Individual constants: $\underline{0},\underline{k},\underline{s},\underline{d},\underline{p},\underline{p}_1,\underline{p}_2,\underline{j},\underline{i}_n (n < \omega)$

Class constant: S

Basic terms: variables or constants of either sort.

Individual terms are denoted $t, t_1, t_2, \ldots$

Class terms are denoted $T, T_1, T_2, \ldots$

Atomic formulas:

(i) Equations between terms of either sort

(ii) $\mathrm{App}(t_1,t_2,t_3)$, also written $t_1 t_2 \simeq t_3$

(iii) $t \in T$

Formulas are generated by $\neg, \wedge, \vee, \Rightarrow$, and the quantifiers $\exists$ and $\forall$ applied to either sort of variable.

$\phi, \psi, \theta, \ldots$ range over formulas. We may write ϕ with a distinguished free variable as $\phi(x,\ldots)$ or $\phi(x)$. Then $\phi(t,\ldots)$, $\phi(t)$ resp., denotes $\mathrm{Sub}(t/x)\phi$; similarly for class variables. The Gödel-number of a formula ϕ is denoted $\ulcorner\phi\urcorner$.

We write $C\ell(a)$ for $\exists A(a = A)$ and $x \in a$ for $\exists A(a = A \wedge x \in A)$.

2.2 Stratified and elementary formulas. By a stratified formula we mean one which contains equations only between individual terms. If $\phi(X)$ is stratified and $\psi(x)$ is any formula then $\phi(\hat{x}\,\psi(x))$ or $\phi(\hat{\psi})$ is defined to be the result of substituting $\psi(t)$ for each occurrence of $(t \in X)$ in ϕ. This is assumed to avoid collision of variables. Also for stratified formulas it makes sense to write $\phi(X^+)$ for a formula with only positive occurrences of subformulas $(t \in X)$.

By an elementary formula is meant a stratified formula without bound class variables and without the constant S.

Note that the formulas $C\ell(a)$, $x \in a$ are not stratified.

2.3 Application terms. These are generated in an extension of the basic language as follows:

(i) every basic term of either sort is an application term;

(ii) if τ_1, τ_2 are application terms so also is $\tau_1\tau_2$.

In the following, $\tau, \tau_1, \tau_2, \ldots$ range over application terms. $\tau_1\tau_2 \ldots \tau_n$ is written for $(\ldots(\tau_1\tau_2)\ldots)\tau_n$ (association to the left). Certain formulas involving application terms are translated into the basic language as follows:

$\tau \simeq x$ is $\tau = x$ when τ is a basic term

$\tau_1\tau_2 \simeq x$ is $\exists y_1 \exists y_2 [\tau_1 \simeq y_1 \wedge \tau_2 \simeq y_2 \wedge y_1 y_2 \simeq x]$

$\tau_1 \simeq \tau_2$ is $\forall x[\tau_1 \simeq x \Leftrightarrow \tau_2 \simeq x]$

$\tau\downarrow$ is $\exists x(\tau \simeq x)$ ("τ is defined")

$\phi(\tau)$ is $\exists x[\tau \simeq x \wedge \phi(x)]$.

$\tau_1 = \tau_2$ is written for $\tau_1 \simeq \tau_2$ when $\tau_1\downarrow$ and $\tau_2\downarrow$ is known or assumed. $\tau_1 \neq \tau_2$ is written for $\neg\ (\tau_1 \simeq \tau_2)$ under the same conditions.

The constant $\underline{p}$ will act as a pairing operator. We write

$$(\tau_1, \tau_2) = \underline{p}\ \tau_1\tau_2 \quad \text{and}$$

$$(\tau_1, \ldots, \tau_n, \tau_{n+1}) = ((\tau_1, \ldots, \tau_n), \tau_{n+1})$$

Tuples are indicated by bars: $\bar{\tau} = (\tau_1, \ldots, \tau_n)$.

2.4 Class terms. Consider any stratified formula $\phi(x, X^+ ; \bar{y} ; \bar{A})$, for which we also write $\phi(x,X)$. We write $\mathrm{Clos}_\phi(X)$ for $\forall x[\phi(x,X) \Rightarrow x \in X]$. We write $\phi_c(x, - ;\bar{y}; \bar{A})$ or $\phi_c(x, -)$ for $[\mathrm{Clos}_\phi(X) \Rightarrow x \in X]$.

Then we shall use $\underline{i}_{\ulcorner\phi\urcorner}(\bar{y},\bar{A})$ to denote the smallest class X satisfying $Clos_\phi$, i.e. the class inductively defined by ϕ. We thus write

$$\{x|\phi_c(x, -; \bar{y}; \bar{A})\} \text{ or } \cap X[Clos_\phi(X)] \text{ for } \underline{i}_{\ulcorner\phi\urcorner}(\bar{y},\bar{A}).$$

Note that this is given as an operation $\underline{i}_k$ applied to the tuples of individual and class parameters of ϕ (for $k = \ulcorner\phi\urcorner$). As a special case of this, given $\phi(x; \bar{y}; \bar{A})$ which does not contain X we write $\{x|\phi(x; \bar{y}; \bar{A})\}$ or $\{x|\phi(x)\}$ for $\{x|\phi_c(x; \bar{y}; \bar{A})\}$. As another special case we write

$$\{x|x \in B \wedge \phi_c(x, -; \bar{y}, \bar{A})\} \text{ for } \{x|\phi^*_c(x, -; \bar{y}; \bar{A}, B\}$$

where $\phi^*(x,X; \bar{y}, \bar{A},B)$ is $x \in B \wedge \phi(x, X; \bar{y}, \bar{A})$. The axioms will guarantee that all these operations lead to classes.

We write

$$A \subseteq B \text{ for } \forall x(x \in A \Rightarrow x \in B), \text{ and}$$

$$A \equiv B \text{ for } A \subseteq B \wedge B \subseteq A.$$

Further we write

$$f: A \to B \text{ for } \forall x \in A\, \exists y \in B(fx \simeq y) \text{ (or } \forall x \in A(fx \in B)).$$

2.5 The axioms of $T_o(S)$.

I. Applicative axioms

(i) (Unicity) $xy \simeq z_1 \wedge xy \simeq z_2 \Rightarrow z_1 = z_2$

(ii) (Constants) $(\underline{k}\, xy\downarrow) \wedge \underline{k}\, xy = x$

(iii) (Substitution) $(\underline{s}\, xy\downarrow) \wedge \underline{s}\, xy\, z \simeq xz(yz)$

(iv) (Definition by cases) $(\underline{d}\, a\, b\, x\, y\downarrow) \wedge (x = y \Rightarrow \underline{d}\, a\, b\, x\, y = a)$
$\wedge (x \neq y \Rightarrow \underline{d}\, a\, b\, x\, y = b)$

(v) (pairing, projections) $(\underline{p}x_1x_2\downarrow) \wedge (\underline{p}_i z\downarrow) \wedge \underline{p}_i(\underline{p}x_1x_2) = x_i$

(vi) (zero) $\underline{p}\,xy \neq 0$

II. Special axioms

(i) (Classes are elements) $\forall X\ \exists x(X = x)$

(ii) (Totality of class operations on elements) $(\underline{i}_n z\downarrow) \wedge (\underline{j}\, z\downarrow)$.

III. Elementary inductive definitions. For each elementary $\phi(x, X^+)$ and any $\psi(x)$,

$$\exists C\{x \mid \phi_c(x, -)\} \simeq C \wedge \mathrm{Clos}_\phi(C) \wedge [\mathrm{Clos}_\phi(\hat{\psi}) \Rightarrow \forall x \in C.\psi(x)]\}.$$

IV. Join

$$\forall x \in A.C\ell(fx) \Rightarrow \exists C\{j(A,f) \simeq C \wedge$$
$$\forall z[z \in C \Leftrightarrow \exists x \in A\ \exists y\ (y \in fx \wedge z = (x,y))]\}.$$

V. S-axioms.

These will be explained after drawing consequences of I-IV, and introducing more notation. (Not all of that will be needed to state V, but serves later purposes as well).

Remarks. (1) The axioms I-IV are slightly stronger than the system T_0 introduced in [F1]. The axioms of elementary comprehension and inductive generation in T_0 are subsumed under the present IV. Also the logic is not restricted to be intuitionistic as it was in [F1]. (2) By II(i), operations applied to classes are special cases of operations applied to elements. II(ii) is taken for convenience. The operation $\underline{j}$ applied to any element z always give some element. It is only assumed to give a class when z is of the form (A,f) where $\forall x \in A.C\ell(fx)$. Similarly $\underline{i}_k z$ is always defined but it is only assumed we get a class when z is of the form $(\bar{y},\bar{A})$ where $\bar{y} = (y_1,\ldots,y_n)$, $\bar{A} = (A_1,\ldots,A_m)$ and $k = \ulcorner\phi(x,X^+; y_1,\ldots,y_n, A_1,\ldots,A_m)\urcorner$ with ϕ elementary.

2.6 Consequences of the applicative axioms (Refer to [Fl]3.3 for more details.)

(1) (Explicit definition). With each applicative term $\tau(x)$ is associated a term $\lambda x.\tau(x)$ such that

$$(\lambda x.\tau(x))\downarrow \wedge \forall y\,((\lambda x\,\tau(x))y \simeq \tau(y)).$$

Informally, $\lambda x.\tau(x)$ "exists as a rule" whether or not $\tau(x)\downarrow$ for any given x.

(2) (Zero, successor). Define $x' = (x,0)$. By I(v), (vi) we conclude $x' \neq 0$, $x' = y' \Rightarrow x = y$, $x = y' \Rightarrow y = \underline{p}_1 x$.

(3) (Recursion theorem). By the usual diagonalization we can define r such that

$$\forall f\{(r\,f\downarrow) \wedge \forall x[(r\,f)x \simeq f(r\,f)x\,]\}.$$

(4) (Non-extensionality) ([Fl]3.4). We can disprove $\forall f,g[\forall x(fx \simeq gx) \Rightarrow f=g]$. The idea is to associate with each f an f^* with the same domain and which is identically 0 on that domain (use defn. by cases). Then f is total $\Leftrightarrow f^* = \lambda x.0$, if extensionality holds. Diagonalizing gives a contradiction.

2.7 Consequences of axiom III.

(1) (Elementary inductive definitions). Given elementary $\phi(x,X^+;\bar{y};\bar{A})$, let $\bigcap X[\mathrm{Clos}_\phi(X)] \simeq C$. Then

$$\forall x[x \in C \Leftrightarrow \forall X(\mathrm{Clos}_\phi(X) \Rightarrow x \in X)].$$

For the proof of $\Rightarrow$, consider any X, apply $\mathrm{Clos}_\phi(\hat{\psi}) \Rightarrow \psi(x)$ to $\psi(x) = (x \in X)$. For the proof of $\Leftarrow$, apply $\mathrm{Clos}_\phi(C)$.

(2) (Elementary comprehension). Given elementary $\phi(x;\bar{y};\bar{A})$ we have defined $\{x|\phi(x)\}$ as $\{x|\phi_c(x)\}$; call this C. Then $\mathrm{Clos}_\phi(C)$ shows

$\forall x[\phi(x) \Rightarrow x \in C]$ and $\mathrm{Clos}_{\phi}(\hat{\phi})$ shows $\forall x \in C.\phi(x)$. Hence $\{x|\phi(x)\}$ is a class and

$$y \in \{x|\phi(x)\} \Leftrightarrow \phi(y).$$

(3) (Class constructions). The following are obtained directly as special cases of (2):

$$V = \{x|x = x\}, \quad \Lambda = \{x|x \neq x\}$$

$$\{y_1,\ldots,y_n\} = \{x|x = y_1 \vee \ldots \vee x = y_n\}$$

$$A \cap B = \{x|x \in A \vee x \in B\}, \quad A \cup B = \{x|x \in A \vee x \in B\}$$

$$- A = \{x|x \notin A\}$$

$$A \times B = \{x|\exists y \in A\, \exists z \in B\;\; x = (y,z)\}$$

$$(A \to B) = \{f|\; f : A \to B\}$$

$$f[A] = \{y|\exists x \in A\,(fx \simeq y)\}$$

$$\mathfrak{D} f = \{x|(fx\downarrow)\}.$$

(4) (The natural numbers). We introduce these by

$$N = \bigcap X[0 \in X \wedge \forall x(x \in X \Rightarrow x' \in X)]$$

i.e. as $\bigcap X[\mathrm{Clos}_{\phi}(X)]$ for $\phi(x,X) = [x = 0 \vee \exists y \in X\,(x = y')]$. Then we have:

$$0 \in N, \quad x \in N \Rightarrow x' \in N, \text{ and}$$

$$\psi(0) \wedge \forall x(\psi(x) \Rightarrow \psi(x')) \Rightarrow \forall x \in N.\,\psi(x), \text{ for any } \psi.$$

(5) (Primitive recursion on N). Using the recursion theorem and definition by cases we obtain existence of r_N satisfying

$$r_N(0,a,f) \simeq a, \quad r_N(x',a,f) \simeq f(x,r_N(x,a,f)).$$

Note that for any A, $r_N\colon N \times A \times A^{N \times A} \to A$. With explicit definition we can now generate all primitive recursive operators, in particular bounded minimum and bounded quantification.

(6) (Partial recursion on N). The unbounded minimum μf is defined as $g(f,0)$ where $g(f,x) \simeq (\mu y \leq x)(fy \simeq 0)$ if $\exists y \leq x(fy \simeq 0 \wedge \forall z < y(fz\downarrow))$ and $g(f,x) \simeq g(f,x')$ if $\forall y \leq x(fy\downarrow \wedge fy \neq 0)$ (g obtained by recursion theorem and def. by cases). Then we can get for each $k \in N$ existence of f_k with $f_k x \simeq \{k\}(x)$ for all $x \in N$. Also $\lambda(z,x).\{x\}(x)$ is obtained.

(7) (Non-extensionality for classes). Similarly to 2.6(4) we can disprove $\forall A, B(A \equiv B \Rightarrow A = B)$; cf. [Fl]3.4.

2.8 Consequences of the join axiom IV. We write $\Sigma_{x \in A} fx$ for $j(A,f)$ so that

$$z \in \Sigma_{x \in A} fx \Leftrightarrow \exists x \in A\ \exists y(y \in fx \wedge z = (x,y))$$

whenever $\forall x \in A\, \exists X(fx \simeq X)$. Note that the defining property of $\Sigma_{x \in A} fx$ is not stratified.

(1) (Product). Suppose $\forall x \in A.C\ell(fx)$. The class

$$\Pi_{x \in A}\, fx =_{def} \{g \mid \forall x \in A((x,gx) \in C)\}$$
$$\text{where } C = \Sigma_{x \in A}\, fx$$

exists by join and elementary comprehension, and satisfies

$$g \in \Pi_{x \in A}\, fx \Leftrightarrow \forall x \in A\,(gx \in fx)\,.$$

In other words, once we have Σ, the unstratified definition of Π can be reduced to the stratified (indeed elementary) definition given above. Note that if $fx = B$ for each $x \in A$ then $(\Sigma_{x \in A}\, fx) \equiv A \times B$ and $(\Pi_{x \in A}\, fx) \equiv (A \rightarrow B)$.

(2) (Union and intersection). Similarly we can infer existence of $\bigcup_{x \in A} fx$ and $\bigcap_{x \in A} fx$.

(3) (Membership on classes of classes). Suppose $\forall x \in A.\,C\ell(x)$, i.e. A is a class of classes. Then the class

$$E_A = \Sigma_{x \in A} x$$

represents the membership relation on A:

$$z \in E_A \Leftrightarrow \exists x \in A\, \exists y(z = (x,y) \wedge y \in x).$$

(4) (Non-existence of a class of all classes). Suppose there exists A such that $\forall x[x \in A \Leftrightarrow C\ell(x)]$. Using E_A we can form $C = \{x | x \in A \wedge x \notin x\}$, from which we get a contradiction. Thus we cannot in general introduce $\mathcal{P}(B)$, a class of all sub-classes of B.

(5) (Relative power class). Given any class A of classes, we can form

$$\mathcal{P}_A(B) = \{a | a \in A \wedge a \subseteq B\}, \text{ i.e. } \{a | a \in A \wedge \forall x((a,x) \in E_A \Rightarrow x \in B\}.$$

We shall make particular use of a generalization of $\mathcal{P}_S(B)$ later in the paper.

2.9 Classes with "equality" relations. A class A with equality I on it is a pair $a = (A,I)$ where $A^2 \subseteq I$ and $A^2 \cap I$ is an equivalence relation on A; if this holds we write $C\ell\text{-}Eq(a)$. Note that $A = p_1 a$ and $I = p_2 a$ in this case. Classes with equality relations arise naturally in the practice of explicit mathematics (cf. e.g. § 4 below) and are in any case essential for a non-extensional development. S will satisfy $\forall a \in S[C\ell\text{-}Eq(a)]$. Furthermore in the set-theoretic models of $T_0(S)$ and $T_1(S)$ (in §3) we shall show that $(A,I) \in S$ implies A/I is a set of the model.

(1) Notation. Given $C\ell\text{-Eq}(a)$, $a = (A, I)$, we write $\equiv_a$ or $\equiv_A$ for I and $x \in a$ for $x \in p_1 a$. By $(A, =)$ we mean A with the relation $I = \{(x,y) \mid x = y\}$; $a = (A, =)$ is called a discrete class in this case. The natural numbers will be dealt with as a discrete class, for example.

(2) (The subclass relation in $C\ell$-Eq). We write $a \subseteq_h b$ or $h : a \subseteq b$ for $h : A \to B \wedge \forall x,y \in A[x \equiv_A y \Leftrightarrow hx \equiv_B hy]$ when $a = (A, \equiv_A)$, $b = (B, \equiv_B)$. (Thus in the set-theoretic interpretation, h induces an injection of $A/\equiv_A$ into $B/\equiv_B$.) $a \subseteq b$ is written for $\exists h(a \subseteq_h b)$. a, b are isomorphic when there exist h_1, h_2 inverse to each other (on A, B resp.) such that $a \subseteq_{h_1} b \wedge b \subseteq_{h_1} a$.

(3) (Finitary operations on $C\ell$-Eq). Let $a = (A, \equiv_A)$, $b = (B, \equiv_B)$. We put

$$\text{(i)} \quad \begin{cases} a \times b = (A \times B, \equiv_{A \times B}), \quad \text{where} \\ (x_1, x_2) \equiv_{A \times B} (y_1, y_2) \Leftrightarrow x_1 \equiv_A y_1 \wedge x_2 \equiv_B y_2 . \end{cases}$$

We put

$$\text{(ii)} \quad \begin{cases} b^a = (B^A, \equiv_{B^A}), \quad \text{where} \\ f \in B^A \Leftrightarrow f : A \to B \wedge \forall x, y \in A[x \equiv_A y \Rightarrow fx \equiv_B fy], \quad \text{and} \\ f \equiv_{B^A} g \Leftrightarrow \forall x \in A[fx \equiv_B gx]. \end{cases}$$

The operation (i) is distinguished from Cartesian product on classes (2.7(3)) by the context.

(4) Infinitary operations on $C\ell$-Eq. Suppose A is a discrete class and that for each $x \in A$, fx is in $C\ell$-Eq, say $fx = b_x = (B_x, \equiv_x)$. Then we put

$$\text{(i)} \quad \begin{cases} \Sigma_{x \in A} fx = (\Sigma_{x \in A} B_x, \equiv_\Sigma) \quad \text{where} \\ (x_1, y_1) \equiv_\Sigma (x_2, y_2) \Leftrightarrow x_1 = x_2 \wedge y_1 \equiv_{x_1} y_2 . \end{cases}$$

Note that Bx is $p_1(fx)$ and $\equiv_x$ is $p_2(fx)$. Under the same conditions on A and f we define

$$\text{(ii)} \quad \begin{cases} \Pi_{x \in A} fx = (\Pi_{x \in A} Bx, \equiv_\Pi) \text{ where} \\ g \equiv_\Pi h \Leftrightarrow \forall x \in A[gx \equiv_x hx]. \end{cases}$$

We rely on the context to distinguish Σ and Π as operations on sequences of classes (2.8) or on sequences of classes with equality relations, as here.

Remark. The present operations can be generalized still further to define $\Sigma_{x \in a} fx$ and $\Pi_{x \in a} fx$ for $a = (A, \equiv_A)$ and $fx = (B_x, \equiv_x)$ under the following circumstances. Namely we must be provided with a system of maps $h_{x,y} : fx \subseteq fy$ for $x,y \in A$ with $x \equiv_A y$ such that $h_{x,y}$ and $h_{y,x}$ are inverses and $h_{x,z} = h_{y,x} \circ h_{x,y}$ when $x \equiv_A y \equiv_A z$. For full generality, closure under these extended operations could and should be included in the S-axioms; however, only the operations with discrete index classes will be used in the applications and, for simplicity, closure will be assumed only for these.

(5) (Inductive separation). Given $a_i = (A_i, \equiv_i)$ where $\equiv_i$ is I_i $(1 \leq i \leq m)$ we write $\bar{a}$ for $(a_1, \ldots, a_m)$ and $\phi(\ldots, \bar{a})$ for a formula which includes among its class parameters A_i and I_i $(1 \leq i \leq m)$. Given $b = (B, \equiv_B)$ and elementary $\phi(x, X^+ ; \bar{y}, \bar{a})$ we shall consider the process of separation applied to b, yielding

$$\text{(i)} \qquad (\{x | x \in B \wedge \phi_c(x, - ; \bar{y}, \bar{a})\}, \equiv_B)$$

when we make no change in the equivalence relation.

(6) (Coarsening). Suppose given a class with equality $a = (A, I)$. By a coarsening of a we mean a structure $a' = (A, I')$ where $A^2 \cap I \subseteq I'$. Only explicitly definable coarsenings will be considered below. (In the set-theoretic interpretation there is a natural map of A/I onto A/I'.)

We now formulate the remaining axioms of $T_o(S)$.

2.10 The S-axioms - (group V of $T_o(S)$).

(i) $a \in S \Rightarrow C\ell\text{-}Eq(a)$

(ii) $(N, =) \in S$

(iii) $a, b \in S \Rightarrow a \times b \in S \wedge b^a \in S$

(iv) $(A, =) \in S \wedge f : A \to S \Rightarrow \Sigma_{x \in A} fx \in S \wedge \Pi_{x \in A} fx \in S$.

(v) For each elementary $\phi(x, X^+ ; \bar{y}, \bar{a})$ with $\bar{a} = (a_1, \ldots, a_m)$ and $a_1, \ldots, a_m \in S$ and for each $b = (B, \equiv_B) \in S$ we have: $(\{x | x \in B \wedge \phi_c(x, - ; \bar{y}, \bar{a})\}, \equiv_B) \in S$.

(vi) Under the same hypothesis as (v) we have: if $I' = \{z | \phi_c(z, - ; \bar{y}, \bar{a})\}$ and $\forall x, y \in B [x \equiv_B y \Rightarrow (x,y) \in I]$ then $(B, I') \in S$.

Remark. These axioms are related to the ones for "bounded classes" given in [F1] 7.3. An essential difference is that the predicate Bd is replaced here by the class constant S.

2.11 The system $T_1(S)$. This has only one additional axiom, which is really an expansion of the applicative axioms I. It involves a new constant $\underline{e}_N$ for the operation of existential quantification over N.

I(vii) ($\exists^N$ - axiom)

$$[\underline{e}_N f \simeq 0 \Leftrightarrow \exists x \in N(fx \simeq 0)] \wedge [\underline{e}_N f \simeq 1 \Leftrightarrow \forall x \in N(fx \simeq 1)].$$

2.12 Other axioms. It is natural to consider some further possible axioms. First of all, note that in the separation axiom V(v) for S, it was assumed the parameters of the definition are also in S. It is possible to strengthen this, at least up to $\equiv$, and most simply for discrete sets. We shall write $A \in S$ to mean $(A, =) \in S$.

V(vi) (Discrete separation). $\forall A, B\{B \in S \Rightarrow \exists B_1(B_1 \in S \wedge B_1 \equiv B \cap A)\}$.

It will be shown in 3.6 how to get a model of $T_1(S)$ together with V(vi). The following will also be obtained in the same model:

V(vii) (Choice). $(B,\equiv_B) \in S \Rightarrow \exists C[C \subseteq B \wedge C \in S \wedge \forall x \in B \exists ! y \in C (x \equiv_B y)]$.

By choice, discrete sets serve to represent all sets.

2.13 Stratified comprehension. Another strengthening to be considered is the principle, for any stratified $\phi(x, \bar{a}, \bar{A})$:

$$\exists C[\{x | \phi(x)\} \simeq C \wedge \forall x(x \in C \Leftrightarrow \phi(x))].$$

Among other things this would allow us to introduce $\cap X[\mathrm{Clos}_\phi(X)]$ as an abstract, namely $\{x | \forall X(\mathrm{Clos}_\phi(X) \Rightarrow x \in X)\}$. (However this would not give the full strength of the elementary induction axiom, since it only yields proof by induction for stratified properties $\psi(x)$). It is also possible to model $T_1(S)$ with stratified comprehension.

Further special stronger axioms will be considered in connection with ordinals in § 5 below.

3. Models of the theories.

3.1 Outline and preliminaries. There are quite a variety of models to be considered. We describe here the general pattern of construction. By an applicative structure we mean any model $\mathfrak{A} = (A, \simeq, k, s, d, p, p_1, p_2, 0)$ of the applicative axioms (I) of $T_0(S)$. 2/ Ordinary recursion theory and its generalizations provide a wealth of examples of such structures; some familiar ones are recalled in 3.2. For our purposes, a pairing structure is any structure $\mathfrak{A}_0 = (A, P, P_1, P_2, 0)$ where $P : A^2 \xrightarrow{1-1} A$, $P_1(0) = 0$, $P_2(0) = 0$, $P_i(P(x_1, x_2)) = x_i$, and $P(x_1, x_2) \neq 0$ all x_1, x_2. Any pairing structure generates an applicative model $\bar{\mathfrak{A}}_0$, as will be described in 3.3. More generally we can incorporate any pre-assigned collection of functions $\mathfrak{F}$; the result is denoted $\bar{\mathfrak{A}}_0(\mathfrak{F})$. In particular, given any model $\mathfrak{M} = (M, \in_M)$ of set theory taken as a pairing structure in the standard way, we shall obtain applicative models $\bar{\mathfrak{M}}(\mathfrak{F})$ which range from ordinary recursion theory on $\mathfrak{M}$ to the incorporation of all set-functions (3.4).

2/ Following Friedman [Fr 1], I previously called these enumerative structures. Their source is in the Wagner-Strong axioms for abstract enumerative recursion theory.

Any pairing structure provides us with finite coding ability. First of all, the natural numbers are represented via the successor operation $x \mapsto x' = P(x,0)$. We may regard A as providing the alphabet for a symbolic system. Any word from A is represented by a code in A, then any finite sequence of such words is represented by another code, etc. We shall refer to coding procedures without giving specific details.

Given any applicative structure $\mathfrak{A}$ we shall show in 3.5 how to construct a model $\mathfrak{A}^*$ of $T_0(S)$, by interpreting the class variables to range over a certain collection of codes in A. Actually, interpretation of S and the membership relationship on S are explained first and these are then used to explain the interpretation of $C\ell$ and ϵ in general. The basic method of model building goes back to [F1]pp.104-107 for T_0.

Of special interest to us will be the case where we start with an applicative structure $\mathfrak{A}$ over a model $\mathfrak{M} = (M, \epsilon_M)$ of set theory. By feeding in a code for each set of M in S, we can arrange that $\mathfrak{A}^*$ is a model of $T_0(S)$ in which S is a system of representatives of all ordinary sets(3.6). Taking $\mathfrak{A} = \overline{\mathfrak{M}}(\mathfrak{F})$ for various $\mathfrak{F}$ from 3.4, we can thus compare ordinary recursion theory, hyperarithmetic theory and full set-function theory as operative in a full set-theoretical situation.

There is only one additional point to be made for the theory $T_1(S)$. We shall call $\mathfrak{A}$ plus e_N (in A) an $\exists^N$ - applicative model if it satisfies as well the $\exists^N$ - axiom I(vii). Here the relation $x \in N$ is to be given its standard interpretation i.e.: x belongs to the smallest subset of A which contains 0 and is closed under the "successor" operation $u \mapsto (u,0)$. (Note that N itself will appear as a code for this set in A). Then $\mathfrak{A}^*$ is automatically a model for $T_1(S)$ if it is a model of $T_0(S)$ and $\mathfrak{A}$ is $\exists^N$ - applicative.

3.2 Familiar applicative models.

(1) (Ordinary recursion theory). We write $\mathrm{Rec}(\omega)$ for the applicative model $\mathfrak{A} = (\omega, \simeq, k, s, d, p, p_1, p_2, 0)$ where $xy \simeq z \Leftrightarrow \{x\}(y) \simeq z$ and the constants are suitably chosen. It is convenient for the following to assume that all of ω is generated from 0 by the operation $x, y \mapsto pxy$ and that $p_1 0 = p_2 0 = 0$. The subset N is in effective 1-1 correspondence with ω and the functions $\lambda y.fy$ for which $f : N \to N$ are the images of the recursive functions under this correspondence.

(2) ($\exists^N$ - recursion theory). Here again $\mathfrak{A}$ has domain ω. Take a Π^1_1 predicate $\phi(x,y,z)$ which for $x = 0,1,2,\ldots$ enumerates all Π^1_1 partial functions (z as a function of y), and put $xy \simeq z \Leftrightarrow \phi(x,y,z)$. ($\phi$ may be obtained by uniformization; cf. [Ro]§ 16.5). We may choose a number e_N to satisfy

$$e_N f \simeq u \Leftrightarrow u = 0 \wedge \exists y \in N\,(fy \simeq 0) \vee u = 1 \wedge \forall y \in N\, \exists z (z \neq 0 \wedge fy \simeq z),$$

since the defining condition is arithmetical. (We assume the same effective pairing and projection functions as in (1).) Thus $\mathfrak{A}$ with e_N is an $\exists^N$ - applicative structure, denoted $\exists^N$ - $\mathrm{Rec}(\omega)$. The total functions here are exactly the hyperarithmetic functions, as are the total functions on N. (The relation $\simeq$ is not quite the same as obtained from Kleene recursion in ${}^2E$ [K] with $xy \simeq z \Leftrightarrow \{x\}({}^2E, y) \simeq z$, since ${}^2E(\lambda y.\{f\}({}^2E,y))$ is defined only when $\lambda y\{f\}({}^2E,y)$ is total. The total functions generated are the same.)

(3) (Admissible recursion theory) [Ba]. Let A be an admissible set in which we have Σ_1 uniformization, e.g. when $A = L_\alpha$ with α admissible, or more generally when A has a Σ_1 global well-ordering. Using a Σ_1 enumeration of the Σ_1 partial functions we then obtain an applicative model $\mathfrak{A} = (A, \simeq, \ldots)$. We write Σ_1 - $\mathrm{Rec}(A)$ for $\mathfrak{A}$, and Σ_1 - $\mathrm{Rec}(\alpha)$ when $A = L_\alpha$. When $\alpha > \omega$, Σ_1 - $\mathrm{Rec}(\alpha)$ is $\exists^N$ - applicative. For $\alpha = \omega_1^c$ (the least "non-constructive" ordinal) the Σ_1 partial functions from N to N in Σ_1 - $\mathrm{Rec}(\omega_1^c)$ coincide with the $\exists^N$ - partial recursive functions of (2).

3.3 Generating applicative models. Let $\mathfrak{A}_0 = (A, P, P_1, P_2, 0)$ be a pairing structure, i.e. $P: A^2 \xrightarrow{1\text{-}1} A$, $P_i(P(x_1,x_2)) = x_i$, $P(x_1,x_2) \neq 0$ and $P_1(0) = P_2(0) = 0$. Let $\mathfrak{F}$ be any family of unary partial functions on A with $\mathrm{card}(\mathfrak{F}) \leq \mathrm{card}(A)$. Choose codes $k, s, d, p, p_1, p_2, f_F (F \in \mathfrak{F})$, $k_x, s_x, s_{xy}, d_a, d_{ab}$, d_{abx}, p_x which are distinct from 0 and from each other for all a, b, x, y in A. Then we take $\simeq$ to be the least relation satisfying: $kx \simeq k_x$, $k_x y \simeq y$, $sx \simeq s_x$, $s_x y \simeq s_{xy}$, $s_{xy} z \simeq u$ whenever $xz \simeq w$, $yz \simeq v$ and $wv \simeq u$, $da \simeq d_a$, $d_a b \simeq d_{ab}$, $d_{ab} x \simeq d_{abx}$, $d_{abx} y \simeq a$ if $x = y$, $d_{abx} y \simeq b$ if $x \neq y$, $px \simeq p_x$, $p_x y \simeq P(x,y)$, $p_1 x \simeq P_1(x)$, $p_2 x \simeq P_2(x)$, and $f_F x \simeq F(x)$ for each F in $\mathfrak{F}$. The resulting structure is an applicative model, denoted $\overline{\mathfrak{A}_0}(\mathfrak{F})$, such that each F in $\mathfrak{F}$ is represented by an element f_F. Similarly we can define $\overline{\mathfrak{A}_0}(\mathrm{E}^N, \mathfrak{F})$ which is the E^N-applicative structure generated from $\mathfrak{F}$. When $\mathfrak{F}$ is empty we obtain applicative models $\overline{\mathfrak{A}_0}$ and $\overline{\mathfrak{A}_0}(\mathrm{E}^N)$, respectively.

Given $\mathfrak{A}_0$, let $A_0 \subseteq A$ and $\mathrm{Gen}_P(A_0)$ = the closure of A_0 under pairing. A_0 is said to be an atomic base for $\mathfrak{A}_0$ if $P: A^2 \to (A - A_0)$ and $A = \mathrm{Gen}_P(A_0)$ and $P_i(x) = x$ for $x \in A_0$. We get a nice mapping from $\overline{\mathfrak{A}_0}$ to $\overline{\mathfrak{B}_0}$ when both $\mathfrak{A}_0$, $\mathfrak{B}_0$ have atomic bases A_0, B_0, resp. and we have $H: A_0 \to B_0$ with $H(0) = 0$. H extends canonically to $H: A \to B$ with $H(P(x_1,x_2)) = P(H(x_1), H(x_2))$; it is seen that $H(P_i(x)) = P_i(H(x))$ for $i = 1,2$. For simplicity fix k, s, d, p, p_1, $p_2 \in \mathrm{Gen}_P(\{0\})$ ($\subseteq \mathrm{Gen}_P(A_0)$ since $0 \in A_0$), and fix $kx \simeq (k,x)$, $(k,x)y \simeq x$, $sx \simeq (s,x)$, $(s,x)y \simeq ((s,x),y)$, $((s,x),y)z \simeq xz(yz)$, etc. in the same way both in $\overline{\mathfrak{A}_0}$ and $\overline{\mathfrak{B}_0}$. Thus $H(k) = k$, $H(s) = s$, etc.; it is then proved by induction that

$$(*) \qquad xy \simeq z \Rightarrow (Hx)(Hy) \simeq Hz.$$

Hence if f determines a total function $F: N \to N$ in $\overline{\mathfrak{A}_0}$ then Hf determines the same function in $\overline{\mathfrak{B}_0}$. We apply this in particular to $\overline{\mathfrak{B}_0} = (\mathrm{Gen}_P(\{0\}), \simeq, \ldots)$, which is effectively isomorphic to $\mathrm{Rec}(\omega)$. It follows that the total functions from N to N in $\overline{\mathfrak{A}_0}$ represent just the ordinary recursive functions. In this sense $\overline{\mathfrak{A}_0}$ is a conservative lifting of ordinary recursion theory to $A = \mathrm{Gen}_P(A_0)$.

(It is really a form of Moschovakis' prime computability theory [M] on the pure domain A_0 since $Gen_P(A_0) \cong A_0^*$.)

All of the preceding is directly extended to $\overline{\mathfrak{A}_0}(\exists^N)$ and $\overline{\mathfrak{B}_0}(\exists^N)$ when $\mathfrak{A}_0, \mathfrak{B}_0$ have atomic bases A_0, B_0, resp. It is proved that (*) continues to hold by showing that e_N behaves in $\overline{\mathfrak{B}_0}(\exists^N)$ on Hf just as e_N behaves on f in $\overline{\mathfrak{A}_0}(\exists^N)$. In particular, by taking $B_0 = \{0,1\}$ we obtain that $\overline{\mathfrak{A}_0}(\exists^N)$ _is a conservative lifting of_ $\exists^N$_-recursion theory to_ $A = Gen_P(A_0)$. Hence the total functions $\lambda f.fy$ from N to N in $\exists^N$-Rec($\mathfrak{A}_0$) are just the hyperarithmetic functions.

3.4 _Applicative models on set-theoretical structures_. We now simply specialize the preceding to the pairing structure $\mathfrak{A}_0 = (M, P, P_1, P_2, 0)$ obtained from a model $\mathfrak{M} = (M, \epsilon_M)$ of set theory by taking the standard set-theoretical pairing and projection functions. Using well-foundedness it follows that $\mathfrak{A}$ has atomic base consisting of the elements of M which are not pairs. We write Rec($\mathfrak{M}$) and $\exists^N$-Rec($\mathfrak{M}$) for Rec($\mathfrak{A}_0$), $\exists^N$-Rec($\mathfrak{A}_0$), resp. These structures thus constitute conservative extensions of ordinary, resp. $\exists^N$-recursion theory to M.

In addition to the preceding we wish also to consider the applicative model $\overline{\mathfrak{A}_0}(\mathfrak{F})$ where $\mathfrak{F}$ is the collection of all partial functions from M to M whose graph is a set in M. We denote this by Set-Fun($\mathfrak{M}$). When $\mathfrak{M} = (M, \epsilon_M)$ is a standard model, say $M = V_\alpha$ for limit α, and $F = \lambda x.fx$ is a partial function in $\overline{\mathfrak{A}_0}(\mathfrak{F})$, i.e. $f \in M$ and $fx \simeq F(x)$ for $x \in dom(F)$, then the restriction of F to any set in M is in $\mathfrak{F}$. Note for $\alpha > \omega$ that Set-Fun($\mathfrak{M}$) is also an $\exists^N$-applicative model.

For illustrative purposes in the following we shall concentrate on the applicative models Rec(ω), $\exists^N$-Rec(ω) and (for standard $\mathfrak{M} = (M, \epsilon_M)$) Rec($\mathfrak{M}$), $\exists^N$-Rec($\mathfrak{M}$) and Set-Fun($\mathfrak{M}$).

3.5 _Generating models of_ $T_0(S)$ _and_ $T_1(S)$.

Let $\mathfrak{A}$ be any applicative model and let binary E_0 be given. This determines sets $\{x : xE_0 a\}$ for each $a \in A$. We shall build a model of $T_0(S)$

in which each such set is represented by some member of S. First we attend to codes. Let $i_n z = (1,n,z)$, $jz = (2,z)$, and $c_a = (3, a)$ for each a ; these are thus total operations. For any elementary $\phi(x,X^+; y_1,\ldots,y_n ; Y_1,\ldots,Y_m)$ and $\bar{y} = (y_1,\ldots,y_n)$, $\bar{a} = (a_1,\ldots,a_m)$ the object $\{x|\phi_c(x, - ; \bar{y} ; \bar{a})\}$ at $Y_j \to a_j$ (in other words, $\bigcap X[\forall x(\phi(x,X;\bar{y},\bar{a}) \Rightarrow x \in X)]$) is taken by definition to be the code $i_{\ulcorner\phi\urcorner}(\bar{y},\bar{a})$. In particular this is the code for $\{x|\phi(x;\bar{y};\bar{a})\}$ when X does not occur in ϕ. We shall define S inductively and along with this inductive definition the membership relation $x \in a$ for each a in S. When $\phi(x;\bar{y};\bar{Y})$ is elementary the variables Y_i only occur to the right of $\in$ in ϕ and no class quantifiers are used (nor does 'S' appear in ϕ). Thus if membership in a_j has been determined for each $j = 1,\ldots,m$, it is automatically determined for $\{x|\phi(x;\bar{y};\bar{a})\}$ by $z \in \{x|\phi(x;\bar{y};\bar{a})\} \Leftrightarrow \phi(z;\bar{y},\bar{a})$. More generally, membership $z \in \bigcap X[\forall x(\phi(x,X;\bar{y},\bar{a})) \Rightarrow x \in X]$ is determined to hold just in case z belongs to every subset X of A (in the real world) which satisfies the closure condition shown. In particular membership in N is determined in the standard way. Next, suppose membership in a is determined as well as membership in fx for each $x \in a$; then membership in $j(a,f)$ (or $\Sigma_{x\in a} fx$) is determined so as to satisfy the join axiom IV. From this we define $\Pi_{x\in a} fx$ and membership in it as in 2.8(1). Let $(a,=)$ be $(a,\{x|x = (p_1x, p_2x) \wedge p_1x = p_2x\})$. Define St as the smallest subset of A which (i) contains $(c_a, =)$ for each $a \in S$, (ii) contains $(N, =)$, (iii) contains $(a \times b, \equiv_{a\times b})$ and $(b^a, \equiv_{b^a})$ whenever it contains $(a, \equiv_a)$, $(b, \equiv_b)$ (iv) contains $\Sigma_{x\in a} fx$ and $\Pi_{x\in a} fx$ whenever it contains $(a, =)$ and fx for each $x \in a$, (v) contains $(\{x|x \in b \wedge \phi_c(x,-;\bar{y}; (a_i, I_i)_{i\le m}\}, \equiv_b)$ whenever it contains $(a_1,I_1),\ldots,(a_m,I_m)$ and $(b, \equiv_b)$, and (vi) contains (b,I') whenever it contains $(b, \equiv_b)$ and $I' = \{x|\phi_c(x, -; \bar{y}, (a_i,I_i)_{i\le m}\}$ and $b^2 \cap I'$ is an equivalence relation on b. Simultaneously with this inductive generation we determine $x \in a$ and $x \equiv_a y$ for each $(a, \equiv_a)$ in St by the procedure described above combined with the explanations in 2.9(3) - (6); to begin with put $x \in c_a \Leftrightarrow (x,a) \in E_o$.

Next we give St a code S in A, and put $a \in S$ if it belongs to St. To complete the construction of the model, we simply take $C\ell$ to be the smallest subset of A such that (i) $St \subseteq C\ell$, (ii) $\{x|\phi_c(x, - ; \bar{y} ; \bar{a})\}$ is in $C\ell$

whenever $\phi(x,X^+;\bar{y};\bar{Y})$ is elementary and each a_j is in $C\ell$, and (iii) $\Sigma_{x\epsilon a} fx$ is in $C\ell$ whenever a and fx are in $C\ell$, for each $x \epsilon a$. Again this is accompanied by the definition of the membership relation $x \epsilon a$ for each a in $C\ell$, by the determination procedure described above.

It is now readily checked that $(\mathfrak{a}, C\ell, \epsilon, S)$ is a model of $T_0(S)$. The only point to be observed in the inductive generation axiom III is that for $C = \{x|\phi_C(x,-)\}$ and any $\psi(x)$ we have $\mathrm{Clos}_\phi(\hat{\psi}) \Rightarrow \forall x \epsilon C.\psi(x)$. This is because $z \epsilon C$ iff z belongs to every subset of A in the real world which is closed under ϕ, and in particular to $\{x : \psi(x)\}$ when $\mathrm{Clos}_\phi(\hat{\psi})$, even if there is no member of $C\ell$ which represents that set. (To distinguish real set formation from the code $\{x|\phi(x)\}$ we write $\{x : \psi(x)\}$ in the first case.)

We write $\mathfrak{a}^*/E_0$ for the structure $(\mathfrak{a}, C\ell, \epsilon, S)$ just constructed (or simply $\mathfrak{a}^*$ if E_0 is empty). When $\mathfrak{a}$ is an E^N-applicative structure then $\mathfrak{a}^*/E_0$ is a model of $T_1(S)$. In particular

(1) $$\mathrm{Rec}(\omega)^* \models T_0(S)$$

and

(2) $$\mathrm{E}^N\text{-}\mathrm{Rec}(\omega)^* \models T_1(S).$$

Remark. By a modification of this construction using the technique given in [F1] p.134, we can obtain a model $\mathfrak{a}^+/E_0$ of $T_0(S)$ + (Stratified comprehension). The idea is to start with the full 2nd order structure over $\mathfrak{a}$, introduce Skolem functions for the formulas in this structure and then close under codes for these functions when generating $C\ell$.

3.6 Models of $T_0(S)$ and $T_1(S)$ over set-theoretical structures.

For simplicity, take $\mathfrak{m} = (M, \epsilon_M)$ with $M = V_\alpha$ where α is inaccessible. Thus full (2nd order) replacement holds in $\mathfrak{m}$, i.e. if a is a set in M and $F : a \to M$ is a subfunction of M then $F[a] = \{F(x) : x \epsilon_M a\}$ belongs to M. Let

$\mathfrak{a} = (M, \simeq, k, s, d, p, p_1, p_2, 0)$ be an applicative structure over M using the standard set-theoretical pairing and projective functions, and let E_0 be the membership relation ϵ_M on M. Thus $\mathfrak{a}^*/\epsilon_M$ is a model of $T_0(S)$ such that:

(1) <u>for each</u> a <u>in</u> M, <u>we have</u> $(c_a, =) \in S$ <u>and</u> $x \in c_a \Leftrightarrow x \in_M a$.

We shall now associate with each $a = (A, \equiv_A) \in S$ a function H_a and a set $\hat{a}$ such that

(2)
(i) $\hat{a} \in M$ and

(ii) $H_a : (A/\equiv_A) \to \hat{a}$ is one-one and onto.

The definition of $\hat{a}$ and H_a is by induction on the generation of S in 3.5. We shall only follow the former, the latter accompanying it in a natural manner. For convenience we also write $a\hat{}$ for $\hat{a}$. When $(A, =) \in S$ we write $\hat{A}$ for $\hat{a}$. The definition is:

(3)
(i) $(c_a, =)\hat{} = a$

(ii) $(N, =)\hat{} = N$ (the smallest set in M containing 0 and closed under $x \to (x, 0)$).

(iii) $(a \times b)\hat{} = \hat{a} \times \hat{b}$ and

$(b^a)\hat{} = \{F : F$ is a function in M from $\hat{a}$ to $\hat{b}$ and for some f, $\forall x \in A\,(F(H([x])) = H([fx]))\}$.

(iv) $(\Sigma_{x \in A} fx)\hat{} = \Sigma_{x \in \hat{A}} (fx)\hat{}$ and

$(\Pi_{x \in A} fx)\hat{} = \{G \in \Pi_{x \in A} (fx)\hat{} :$ for some g, $\forall x \in A(G(H_A(x)) = H_{fx}([gx]))\}$

(v) $(\{x | x \in B \wedge \phi_c(x, -; \bar{y}, \bar{a}\}, \equiv_B)\hat{}$ is the smallest subset X of $\hat{b}$ such that $\forall x[x \in_M \hat{b} \wedge \phi(x, X; \bar{y}, \bar{a}\hat{}) \Rightarrow x \in_M X]$.

(vi) $(B, I')\hat{}$, for $I' = \{x | \phi_c(x, -; \bar{y}, \bar{a})\}$ with $b = (B, I)$, $I \subseteq I'$ and $B^2 \cap I'$ an equivalence relation, is the image in M of $b\hat{}$ under the equivalence relation $x_1 \equiv x_2 \Leftrightarrow \forall x\,[\phi((x_1, x_2), X; \bar{y}, \bar{a}\hat{}) \Rightarrow (x_1, x_2) \in_M X]$.

Full replacement for M is used in (iv) and full separation is used in (v) and (vi) to show that the resulting $\hat{a}$ belongs to M.

(1) and (2) may be summarized by saying that <u>the sets of M are exactly the subsets of M represented in $\mathfrak{A}^*/\epsilon_M$ by the members of S</u>. For illustrative purposes, we shall concentrate in the following on the three structures $\mathfrak{A}^*/\epsilon_M$ obtained by starting with the three applicative models $\mathfrak{A} = \text{Rec}(\mathfrak{M})$, $\exists^N\text{-Rec}(\mathfrak{M})$ and Set-Fun($\mathfrak{M}$) of 3.4, which will be more simply designated as follows:

(4) (i) $\mathfrak{M}^*_{\text{Rec}} = (\text{Rec}(M))^*/\epsilon_M$ (which satisfies $T_0(S)$)

(ii) $\mathfrak{M}^*_{\exists^N\text{-Rec}} = (\exists^N\text{-Rec}(M))^*/\epsilon_M$ (which satisfies $T_1(S)$)

(iii) $\mathfrak{M}^*_{\text{Set-Fun}} = (\text{Set-Fun}(M))^*/\epsilon_M$ (which satisfies $T_1(S)$).

In the first of these, $(N \to N)$ consists of codes of the recursive functions, in the second it consists of codes of the hyperarithmetic functions and in the third of codes of all set-theoretical functions from N to N. In all of these, S consists of codes of all the sets of M.

We now show that $\mathfrak{A}^*/\epsilon_M$ is also a model of the further axioms of Discrete Separation and of Choice formulated in 2.12. For the first of these V(vi), note that discrete sets can be generated only from the c_a's and N by $\times, \Sigma$ and inductive separation, since $\to$, Π and (proper) coarsening never lead to discrete sets. We can then prove by induction that

(5) if $a = (A, =) \in S$ then $x \in_M \hat{A} \Leftrightarrow x \in A$.

To establish V(vi) from this, given discrete B and any class A, form the subset $a = \{x\colon x \in B \cap A\}$ of M. Then $x \in_M a \Leftrightarrow x \in_M \hat{B} \wedge x \in A$, so $a \in M$ by full separation. Then $c_a \in S$ and $c_a \equiv B \cap A$ as required. To prove the choice axiom V(vii), consider any $(B, \equiv_B) \in S$. Since $B \subseteq M$ and $B/{\equiv_B}$ is equivalent to a set $\hat{b}$ in M, there is a choice set a in M for B, i.e. $\forall x[x \in_M a \Rightarrow x \in B]$

and $\forall x \in B \exists! y \in_M a(x \equiv_B y)$. Then c_a is a discrete choice set for B in the model $\mathfrak{a}^*/\in_M$.

This completes our model-theoretic work. We now turn to an outline of several recursion-theoretic applications of the theories $T_0(S)$, $T_1(S)$ via the models of 3.5 and 3.6.

4. Bishop's constructive measure theory in $T_0(S)$.

4.1 Introduction. It was claimed in [F1] §5.1 that all of Bishop's constructive analysis [B1] could be formalized in T_0, where Bishop's basic notion of operation f applied to an element x is read fx and where one takes for the notion of set $(A, =_A)$ ($=_A$ being an equality relation on A) pairs (A,E) with A,E classes for which $E \subseteq A^2$ is an equivalence relation on A. In other words, in the terminology of 2.9 above we are dealing with members of $C\ell$-Eq. Bishop's notion of function

$$f: (A, =_A) \to (B, =_B)$$

is formally expressed by

$$(f : A \to B) \wedge \forall x \in A \; \forall y \in A[x =_A y \Rightarrow fx =_B fy] ,$$

i.e. by $f \in B^A$ in the sense of 2.9(3) above. It is a direct matter to proceed from this basis to transcribe the work of [Bi 1] into T_0. This will be modified to an interpretation of [Bi 1] into $T_0(S)$ in 4.2; the reason for passing to $T_0(S)$ will be given in a moment. Some elaboration of general approach and points involved has been given in [F3]. [3] We wish here to concentrate on aspects of the constructive theory of measure and so only relevant preliminary notions will be mentioned in 4.2. The treatment in [Bi1] was superseded by that in Bishop and Cheng [Bi,C], which is both more natural and more powerful. It was also claimed in [F1] that the latter could be formalized in T_0. Literally speaking this is not correct, since as will be seen below the abstract notion of integration space

[3] Unpublished notes, a published version of which is eventually planned.

in [Bi,C] requires, prima facie, a power set operation $X \mapsto \mathcal{P}(X)$. That has also been an obstacle to other formal representations of Bishop's work such as given by Myhill [My] and Friedman [Fr 2] in extensional systems, in consequence of which they argued for modifying the mathematics to fit the systems. (In any case, there is no problem for T_0 or these other systems to deal with concrete constructive measure and integration theories such as Lebesgue measure on Euclidean spaces $\mathbb{R}^n$, because only the notions of being measurable and integrable are then needed.) It will be shown here how to formalize the abstract theory of [Bi,C] in $T_0(S)$, using a form of the operation $X \mapsto \mathcal{P}_S(X)$. The possible significance of this for constructive and recursive mathematics will be discussed in 4.5 - 4.6.

4.2 Basic concepts. We shall work informally within $T_0(S)$, calling members of $C\ell$ the classes and members of S the sets. Following [Bi 1] we shall write $=_A$ instead of $\equiv_A$ and we shall talk about sets A rather than $(A, =_A)$ (as is frequent in mathematics, one designates a structure by its domain). Thus, instead of using lower case letters for sets as in 2.9, we here use capital letters and write $A \times B$, B^A for the operations defined in 2.9(3) and $\Sigma_{x \in A} B_x$ $\Pi_{x \in A} B_x$ for the operations defined in 2.9(4) when A is discrete. Also $A \subseteq B$ is, as defined in 2.9(1), given by a function $i \in B^A$ such that $ia_1 =_B ia_2 \Rightarrow a_1 =_A a_2$; i is called an inclusion map in this case and A a subset of B. Our classes do not need to have equality relations attached to them, though every class A of mathematical interest does in fact have an $=_A$ prescribed for it. We now describe how various further notions from [Bi 1] are to be treated in $T_0(S)$.

First, from [Bi 1] Ch.2, the integers Z are defined by separation from $N \times N$ with $=_Z$ a coarsening of $=_{N \times N}$, and then the rationals Q are defined by separation from $Z \times Z$ with $=_Q$ the usual coarsening of $=_{Z \times Z}$. The arithmetical operations are extended to Z and Q. Z^+ can be identified with the discrete set $\{x \mid x \in N \wedge x \neq 0\}$. Given any set X, the sequences $x = \{x_n\}_{n \in Z^+}$ or $x = \{x_n\}$ from X are simply the members of X^{Z^+} (writing

x_n for xn). The class of all these sequences thus forms a set. The set $\mathbb{R}$ of real numbers is defined to be the set of regular sequences of rationals $x = \{x_n\}$, i.e. for which $\forall m, n \in Z^+ (|x_m - x_n| \leq \frac{1}{m} + \frac{1}{n})$. $\mathbb{R}$ is thus defined by separation from Q^{Z^+}. Equality of reals $\{x_n\} =_{\mathbb{R}} \{y_n\}$ is defined by $\forall n \in Z^+(|x_n - y_n| < \frac{2}{n})$ (which is a coarsening of equality of sequences of rationals). $\mathbb{R}^+$ is the subset of $\mathbb{R}$ which consists of the pairs (x,n) where $x \in \mathbb{R}$, $n \in Z^+$ and $x_n > \frac{1}{n}$; the inclusion map is $i(x,n) = x$. In other words, these are reals with an explicit positive lower bound $\frac{1}{n}$. Bishop continually stresses the requirement of such explicit witnessing or side information, but for notational simplicity mostly does not show it in practice. This is potentially ambiguous, e.g. when we speak about a real number x being in $\mathbb{R}^+$ without specifying n for the lower bound. However, the context determines what additional information is to be understood as supplied - e.g. when talking about reals in $\mathbb{R}^+$. We shall follow [Bi 1] in this practice of casual designation.

The relation $y < x$ (or $x > y$) is defined to hold if $(x-y)$ is in $\mathbb{R}^+$, and $x \neq y$ if $y < x$ or $x < y$; clearly, both of these relations require witnessing information, e.g. $y < x$ (by n) if $(x-y, n) \in \mathbb{R}^+$. For each pair of real numbers a,b with $a < b$, the open interval (a,b) is defined as a subset of $\mathbb{R}$; its members are those x with $a < x$ and $x < b$ (together with the appropriate witnessing information). Closed or partially closed intervals are treated similarly. The set of all sequences of real numbers $x = \{x_n\}$ is $\mathbb{R}^{Z^+}$. The relation $\lim_{n \to \infty} x_n = y$ is defined to hold with the side-information $\{n_k\}$ when $\forall n \geq n_k(|x_n - y| \leq \frac{1}{k})$; then $\sum_{n=1}^{\infty} x_n = y$ is defined as usual. The remainder of [Bi 1] Ch.2 is devoted to a constructive development of the calculus, which we do not need to follow.

We next look at some set-theoretical notions from Ch.3 of [Bi 1]. Some of these have already been dealt with in 2.9 and at the beginning of this section. A family of subsets of a set B with index set A is given by an operation f which associates with each $x \in A$ a subset B_x of B, in such a way that equal subsets are associated with equal elements of A. To be more precise we are to be supplied with an i which gives for each x an inclusion map i_x (or ix) of B_x in B and a j which gives for each $x,y \in A$ with $x =_A y$ a map $j_{x,y}$ (or $j(x,y)$) which is an inclusion map of Bx in By, and where all these maps commute appropriately. When A is discrete only the maps i_x are needed; we shall only have to deal with families over discrete sets. What Bishop denotes by $\bigcup_{x \in A} B_x$ is here written $\Sigma_{x \in A} B_x$, which is a subset of B by the inclusion map $h(x,y) = i_x y$. On the other hand, $\bigcap_{x \in A} B_x$ is defined to consist of the members g of $\Pi_{x \in A} B_x$ such that $\forall x,y \in A\,(i_x(gx) = i_y(gy))$.

Given a set X and an apartness relation $\neq$ on X (i.e. one which satisfies conditions like those of $\neq$ on $\mathbb{R}$) we call a pair of subsets (A^1, A^2) of X <u>complemented</u> if $(x \in A^1 \wedge y \in A^2 \Rightarrow x \neq y)$. (For example if $a < b < c$ then the pair consisting of the open intervals (a,b), (b,c) is complemented.) For each complemented set (A^1, A^2) we can associate a <u>characteristic function</u> χ_A to $A = A^1 \cup A^2$, i.e. to $A = \Sigma_{n \in \{1,2\}} A^n$. This is simply given by $\chi_A(x,1) = 1$ and $\chi_A(x,2) = 0$. The <u>complement</u> of a complemented set (A^1, A^2) is (A^2, A^1). The <u>union</u> of a countable family $A_n = (A^1_n, A^2_n)$ of complemented sets is defined to be $\vee_n A_n = (\cup_n A^1_n, \cap_n A^2_n)$ where $\cup$ and $\cap$ are as defined in the preceding paragraph. Similarly the <u>intersection</u> is taken to be $\wedge_n A_n = (\cap_n A^2_n, \cup_n A^2_n)$. The class of <u>Borel sets</u> generated from a given class B_o of complemented sets is the smallest class B which includes B_o and which is closed under countable unions and intersections $\vee_n$ and $\wedge_n$.

4.3 <u>Abstract constructive integration theory</u>. We now turn to [Bi,C] p.1. It is assumed that X is a set with an apartness relation $\neq$. The following definition is given loc. cit. : " let $F(X)$ be the set of all ordered pairs $(f,D(f))$ such that $D(f)$ is a subset of X and f is a function from $D(f)$ to the set of real numbers $\mathbb{R}$, with the property that $x \neq y$ whenever $f(x) \neq f(y)$." Here is where the operation $X \mapsto \mathcal{P}(X)$ makes a <u>prime-facie</u> appearance. In place of it we shall use in $T_o(S)$ what will be defined as the <u>class of all subsets of</u> X, where we are taking the notion of subset in its wider mapping sense, i.e. as a pair (a,i) with $a = (A, =_A) \in S$ and $a \subseteq_i X$. To define this class, we make use of the membership relation on S given by

(1) $$E_S^1 = \sum_{a \in S} p_1 a \ .$$

Thus for $a = (A, =_a)$ or $(A, =_A)$ we have $(a,x) \in E_S^1 \Leftrightarrow a \in S \wedge x \in A$. We shall also write $x \in a$ for $x \in p_1 a$ in this case. The corresponding sum of equality relations gives us

(2) $$E_S^2 = \sum_{a \in S} p_2 a \ ,$$

so that for $a = (A, =_a)$ we have $(a,(x,y)) \in E_S^2 \Leftrightarrow a \in S \wedge x =_a y$. It follows that the property (of a and i) $a \subseteq_i X$ is expressible by the elementary formula

(3) $$\forall x[(a,x) \in E_S^1 \Rightarrow ix \in X] \wedge \forall x,y[(a,x) \in E_S^1 \wedge (a,y) \in E_S^1 \Rightarrow \{(a,(x,y)) \in E_S^2 \Leftrightarrow ix =_X iy\}].$$

Hence the class

(4) $$\mathcal{P}_S(X) = \{(a,i) \mid a \in S \wedge a \subseteq_i X\}$$

exists by the Elementary Comprehension axiom schema III. However, it need not exist as a set since it is not obtained by separation and since it involves the parameter S.

The definition of $F(X)$ in $T_o(S)$ is given similarly:

(5) $$F(X) = \{(f,(a,i)) \mid a \in S \wedge a \subseteq_i X \wedge \forall x[(a,x) \in E_S^1 \Rightarrow fx \in \mathbb{R}] \wedge \forall x,y((a,x) \in E_S^1 \wedge (a,y) \in E_S^1 \Rightarrow [(a,(x,y)) \in E_S^2 \Rightarrow fx =_{\mathbb{R}} fy] \wedge [fx \neq_{\mathbb{R}} fy \Rightarrow ix \neq iy])\} \ .$$

Again $F(X)$ exists as a class in $T_o(S)$, which we designate more simply by

(5)' $F(X) = \{(f,(a,i)) | a \in S \wedge a \subseteq_i X \wedge f: a \to \mathbb{R} \wedge$

$$\forall x,y[x \in a \wedge y \in a \wedge fx \neq_{\mathbb{R}} fy \Rightarrow ix \neq iy]\}.$$

As in [Bi,C], we write $D(f)$ for (a,i) when $(f,(a,i)) \in F(X)$ and we write $x \in D(f)$ for $\exists y(y \in a \wedge iy = x)$. We also write f alone for $(f, D(f))$.

Following this on p.2 in [Bi,C] we meet the basic definition (1.1) of <u>integration space</u>. This is read in $T_0(S)$ as follows: a triple (X,L,I) is an integration space if X is a non-empty set with an apartness relation and L is a subclass of $F(X)$ and $I: L \to \mathbb{R}$ has the properties (1)-(4) of [Bi,C]1.1. Here I is a partial function in the sense of $T_0(S)$ (we are using a capital letter so as to follow the notation of [Bi,C]). The idea of 1.1 is that L is an initial stock of integrable functions each defined "almost everywhere" and that $I(f)$ is the integral of f. (More precisely this is $I(f,(a,i))$ where $(f,(a,i)) \in L$.) An <u>integrable function</u> is then defined (1.6) to be a pair $(f,\{f_n\}_{n=1}^{\infty})$ for which $f \in F(X)$, $\{f_n\}$ is a sequence in L, $\Sigma_n I(|f_n|)$ exists and $fx = \Sigma_n f_n x$ holds whenever $\Sigma_n |f_n x|$ converges. Let L_1 be the class of all such pairs; $I_1(f,\{f_n\})$ is defined on L_1 by $I_1 f = \Sigma_n I_1(f_n)$. The first main result is that (X, L_1, I_1) is again an integration space; furthermore it is shown to have good completeness properties. Constructive <u>Lebesgue integration theory</u> falls out as the special case $(\mathbb{R}, L_1, I_1)$ for suitable initial $(\mathbb{R},L,I)$.

4.4 <u>Measure theory in arbitrary integration spaces</u> (X,L,I) ([Bi,C] § 2). A complemented set $A = (A^1,A^2)$ is said to be <u>integrable</u> (or <u>measurable</u>) if its characteristic function χ_A is integrable; in this case the <u>measure</u> of A is defined to be $\mu(A) = I(\chi_A)$. Calculations with μ may be carried out as usual. Two of the results of [Bi,C] worth noting are 2.6 and 2.10:

(1) <u>if</u> $A = (A^1,A^2)$ <u>is an integrable set with</u> $\mu(A) > 0$ <u>then</u> A^1 <u>contains at least one element, and</u>

(2) _if_ $\{A_k\}_{k=1}^{\infty}$ _is a sequence of integrable sets and_ $\alpha = \lim_{n\to\infty} \mu(\vee_{k=1}^{n} A_k)$ _exists then_ $\vee_k A_k$ _is integrable with measure_ α.

4.5 $T_0(S)$ _and constructive mathematics_. Following these lines, all of [Bi,C] can be formalized in $T_0(S)$. Indeed, only intuitionistic logic need be used in the process. Thus if $T_0(S)^{(i)}$, i.e. intuitionistic $T_0(S)$ is regarded as constructively justified, it provides us with a constructive formalization of the whole of [Bi 1] and [Bi,C]. While I argued in [Fl] that $T_0^{(i)}$ is constructively justified, I am a little hesitant about extending this claim to $T_0(S)^{(i)}$, though I think a case can also be made for that.[5/] However, I do believe it has the character of a theory which would both be recognized on direct grounds as constructively valid and be adequate to the body of mathematics in [Bi 1], [Bi,C] and the further publications continuing Bishop's mathematical program.

4.6 $T_0(S)$ _and recursive measure theory_. Using models of $T_0(S)$ such as $(\mathrm{Rec}(\omega))^*$ (3.5) and $\mathfrak{m}^*_{\mathrm{Rec}}$ (3.6) in which the functions in $(N \to N)$ are just the recursive functions, every notion and result of $T_0(S)$ has a recursion-theoretic interpretation. In particular, the members of $\mathbb{R}$ are just the recursively regular sequences of rational numbers, which are one form of the _recursive real numbers_. Borel sets over $\mathbb{R}$ are what would otherwise be called _recursively coded Borel sets_. Every such set (regarded as a complemented set) is definable in $T_0(S)$. Hence we can apply the conclusion of [Bi,C] 2.6 noted in 4.4 above to obtain:

(1) _if a recursive Borel set_ $A = (A^1, A^2)$ _is integrable with_ $\mu(A) > 0$ _then_ A^1 _contains some recursive real number_.

The potential utility of (1) is limited by the hypothesis of integrability, which is strong. It is not true (as might first be expected) that every recursive Borel set is integrable. Indeed, let x_n be a monotone increasing recursive

[5/] It is only weak evidence that $T_0(S)$ is consistent with Church's thesis $\forall f \in N \to N\, \exists e \in N \forall x \in N(fx \simeq \{e\}x)$, by the model $(\mathrm{Rec}(\omega))^*$.

sequence of recursive reals $0 \leq x_n \leq 1$ such that $\lim_{n \to \infty} x_n$ is not recursive. Take disjoint open intervals A_n with $\mu(A_n) = x_{n+1} - x_n$. Then $\vee_n A_n$ is recursively open but not measurable. To verify integrability of a recursive Borel set A in general, one must verify recursiveness of μ applied at each step of the build-up of A. (No doubt, with this understanding, one could give a relatively simple direct statement and proof of (1) which does not need to pass through the formalization in $T_0(S)$ outlined above.) It is of interest to compare (1) with recursion-theoretic basis results, e.g. [S], where one gets information about existence for definable A of definable members when $\mu(A) > 0$. In these cases the conclusions are much weaker than (1) (we cannot get recursive members) but so also are the hypotheses, since μ is read there in terms of standard measure theory.

4.7 Remarks on $T_0(S)$ and recursive mathematics. Some expectations about the relations of T_0 to recursive mathematics were formulated in [Fl] 5.2 ; these are continued and in certain respects better borne out by $T_0(S)$. As explained in the introduction, the idea is simply that a theorem $\phi^{(set)}$ of classical set-theoretic mathematics which has a positive recursive analogue $\phi^{(rec)}$ can be assimilated to a common general form ϕ which is provable in $T_0(S)$. The classical result can be then read off by the interpretation of $T_0(S)$ in $\mathfrak{M}^*_{Set\text{-}Fun}$ while the recursive one is given by the specialization to $(Rec(\omega))^*$ or $\mathfrak{M}^*_{Rec}$. On the other hand, if $\phi^{(rec)}$ turns out negatively, use of the latter models shows the independence of ϕ from $T_0(S)$. Note that there is no reason to expect intuitionistic logic to play any special role here, and none to suppose - contrary to [Fl] - that in the positive case ϕ is already provable in $T_0(S)^{(i)}$. This point is relevant to attempts to give a constructive redevelopment of classical algebra, which has turned out to be surprisingly difficult. To begin with, as mentioned in [Bi 2], it is not constructively true that the ring Z is Noetherian. Given an ideal A in Z we cannot in general decide whether $A \equiv \{0\}$ or not, and even if $A \not\equiv \{0\}$ is known, we cannot find a finite basis for A. On the other hand, if we allow use of classical logic in $T_0(S)$, then we can prove that Z is a principal ideal domain. This follows from the statement in $T_0(S)$ that every non-empty subset A of N contains a least element $\min(A)$. Of course this is weaker than the would-be constructive statement, since it is not asserted that the map $A \mapsto \min(A)$ is provided by a function of the system. While these distinctions are clear in principle, it is a matter of detailed study to see how much mathematics of recursion-theoretic interest can be formulated in $T_0(S)$ which is not already derivable in $T_0(S)^{(i)}$.

5. Accessible ordinals and number classes in $T_0(S)$ and $T_1(S)$.

5.1 Introduction. Various recursion-theoretic analogues of the set theoretical notion of accessible ordinal have been developed, most extensively by Richter [R]. This involves defining classes $\mathfrak{O}_n$ of numbers by a complicated inductive definition which regulates the choice of n from "previous classes" as well as the generation of each class separately. We provide here an abstract development in $T_0(S)$ which is much like that in set-theory: we first define the class $\mathfrak{O}_S$ of accessible ordinals and then a map $x \mapsto \Omega_x$ for $x \in \mathfrak{O}_S$ where Ω_x is the ordinal number class associated with x (5.2). Classical and recursive ordinal number theories come out as special cases in the models of $T_0(S)$ (5.3). For further applications of ordinals we show that a selection operator for Ω_1 can be found in certain of these models which satisfy $T_1(S)$. This and other principles and applications are discussed at the conclusion in 5.4.

5.2 Definitions of the concepts. Throughout this section we work informally in $T_0(S)$. We shall first set up the general definition of $\mathfrak{O}_A$, where A is any class of classes with equality. $\mathfrak{O}_A$ is supposed to be the closure under 0, $'$ and $\sup_{x \in a} fx$ for a in A and unbounded $f: a \to \mathfrak{O}_A$. This will then be specialized to $A = S$.

(1) Suprema. These are given by codes using the definition

$$\sup_a f = (f, a, 1),$$

for any f, a. The '1' distinguishes the result from that of successor $x' = (x, 0)$. By projections we can uniquely recover both f, a from $\sup_a f$. We also write

$$\sup_{x \in a} fx = \sup_a f \ .$$

(2) <u>The class</u> $\mathfrak{G}_A$ <u>of ordinals with suprema from</u> A. Suppose A is a class of classes with equality $a = (p_1 a, \equiv_a)$. We continue to write $x \in a$ for $x \in p_1 a$. The classes $\mathfrak{G}_A$ and L_A will be defined by a simultaneous inductive definition. We write $x \leq_{\mathfrak{G}_A} y$ or simply $x \leq y$ for $(x,y) \in L_A$. Then we put $x < y \Leftrightarrow x \leq y \wedge y \leq x$ and $x < y \Leftrightarrow x' \leq y$. The clauses of the inductive definition are as follows:

(i) $0 \in \mathfrak{G}_A$

(ii) $x \in \mathfrak{G}_A \Rightarrow x' \in \mathfrak{G}_A$

(iii) $a \in A \wedge \forall x \in a(fx \in \mathfrak{G}_A) \wedge \forall x,y \in a[x \equiv_a y \Rightarrow fx \equiv fy]$

$\wedge \forall x \in a \exists y \in a\,(fx < fy) \Rightarrow \sup_a f \in \mathfrak{G}_A$.

(iv) $x \in \mathfrak{G}_A \Rightarrow 0 \leq x$

(v) $x,y \in \mathfrak{G}_A \wedge x \leq y \Rightarrow x' \leq y'$

(vi) $y \in \mathfrak{G}_A \wedge \sup_a f \in \mathfrak{G}_A \wedge \forall x \in a(fx \leq y) \Rightarrow \sup_a f \leq y'$

(vii) $y \in \mathfrak{G}_A \wedge \sup_a f \in \mathfrak{G}_A \wedge \exists x \in a(y \leq fx) \Rightarrow y' \leq \sup_a f$

(viii) $\sup_a f \in \mathfrak{G}_A \wedge \sup_b g \in \mathfrak{G}_A \wedge \forall x \in a \exists y \in b(fx \leq gy) \Rightarrow \sup_a f \leq \sup_b g$.

We may write $\mathfrak{G}_A = \{x \mid (x,0) \in W_A\}$ and $L_A = \{z \mid (z,1) \in W_A\}$ where $W_A = \{x \mid \phi_c(x, - ; A, E_A^1, E_A^2)\}$ for suitable elementary $\phi_c(x, X^+ ; A, B, C)$. Here $E_A^1 = \sum_{a \in A} p_1 a$ and $E_A^2 = \sum_{a \in A} p_2 a$ so for $a \in A$ we can replace the unstratified conditions $(x \in a)$ and $(x \equiv_a y)$ by the stratified conditions $(a,x) \in E_A^1$ and $(a,(x,y)) \in E_A^2$, resp.

(3) Induction on $\mathfrak{G}_A$. If we fix $\leq$ as $\leq_{\mathfrak{G}_A}$ in (i) - (iii), then $\mathfrak{G}_A$ is equivalent to the least class which satisfies the closure conditions (i)-(iii). Thus we can carry out proof by induction to show $\forall x \in \mathfrak{G}_A . \psi(x)$ for any property ψ

which satisfies the closure conditions (i) - (iii) in place of $\mathfrak{O}_A$. In particular we can prove the converses of (i)-(iii), i.e.

(i) $z \in \mathfrak{O}_A \Rightarrow z = 0$ or $z = x'$ for $x \in \mathfrak{O}_A$ or $z = \sup_a f$ where $a \in A$, $f \in (\mathfrak{O}_A)^a$ and $\forall x \in a \exists y \in a(fx < fy)$.

Of course we can also carry out proof by induction on $\leq_{\mathfrak{O}_A}$. Using (i) just established, the conclusions of (2) (iv)-(viii) give all possible comparisons of elements of the form $0, x', \sup_a f$; in particular, we obtain $x \leq 0$ only if $x = 0$. Some properties that can be proved of $\leq$, $<$ and $\equiv$ are:

(ii) $x \in \mathfrak{O}_A \Rightarrow x \leq x$

(iii) $x,y,z \in \mathfrak{O}_A \wedge x \leq y \wedge y \leq z \Rightarrow x \leq z$.

(iv) $\equiv$ is an equivalence relation on $\mathfrak{O}_A$

(v) $x,y \in \mathfrak{O}_A \Rightarrow x < x' \wedge (y < x' \Rightarrow y \leq x)$

(vi) $y \in \mathfrak{O}_A \wedge \sup_a f \in \mathfrak{O}_A \Rightarrow [\sup_a f \leq y \Leftrightarrow \forall x \in a(fx \leq y)]$
$\wedge [y < \sup_a f \Leftrightarrow \exists x \in a(y \leq fx)]$.

(4) Recursion on $\mathfrak{O}_A$. We can carry out definition by transfinite recursion on $\mathfrak{O}_A$ by applying the recursion theorem 2.6(3). Given any g_o, g_1, g_2 we find h such that

(i) $ho = g_o$

(ii) $hx' \simeq g_1(x,h)$

(iii) $h(\sup_a f) \simeq g_2(a,f,h)$

for any x,a,f. (If g_1, g_2 are total then h can be chosen total.) This can also be done uniformly in a parameter:

(i)' $h(z,0) \simeq g_o z$

(ii)' $h(z,x') \simeq g_1(z,x,h)$

(iii)' $h(z, \sup_a f) \simeq g_2(z,a,f,h)$.

Usually for applications the functions g_1, g_2 take the form $g_1(x,h) = g_1^*(x,hx)$ and $g_2(a,f,h) = g_2^*(a,f,\lambda x.h(fx))$, and similarly in (ii)', (iii)'.

For example we can define ordinal addition by these means, giving:

(iv) $z + 0 = z$

$z + x' \simeq (z + x)'$

$z + \sup_a f \simeq \sup_{x \in a} (z + fx) = \sup_a \lambda x(z + fx)$.

Then for any (class of classes) A we can prove by induction on $\mathfrak{O}_A$:

(v) $$z, x \in \mathfrak{O}_A \Rightarrow (z + x) \in \mathfrak{O}_A \ .$$

Similarly further familiar ordinal functions may be introduced and treated as usual.

Remark. The process of recursion is independent of A in (i)-(iii) or (i)'-(iii)'. Of course it is only for suitable g_1, g_2 that we will be sure that $\mathfrak{O}_A$ is closed under h , as in (v).

(5) The accessible ordinals $\mathfrak{O}_S$ and the regular number classes $\Omega_x^{(r)}$.

$\mathfrak{O}_S$ is simply the special case of $\mathfrak{O}_A$ for $A = S$.

The regular number classes $\Omega_x^{(r)}$ with equality relations $\equiv_x^{(r)}$ are defined by recursion as follows:

$$\text{(i)} \quad \begin{cases} \Omega_0^{(r)} = N \text{ and } \equiv_0^{(r)} \text{ is the relation } = ; \\ \text{for } x \in \mathfrak{O}_S \text{ with } x \neq 0,\ \Omega_x^{(r)} = \mathfrak{O}_{B_x} \text{ and } (\equiv_x^{(r)}) \text{ is } (\equiv_{\mathfrak{O}_{B_x}}) \\ \text{where } B_x = \{(\Omega_z^{(r)}, \equiv_z^{(r)}) \mid z < x\}. \end{cases}$$

Here $<$ is the relation $<_{\mathfrak{O}_S}$. This recursion is justified by the principles in (4). We are to obtain a function h such that for each $x \in \mathfrak{O}_S$,

(ii) $$hx \simeq (\Omega_x^{(r)} , \equiv_x^{(r)}) \ .$$

Recall from (2) that for any class B of classes with equality $a = (p_1 a, \equiv_a)$, $\mathfrak{O}_B = \{x \mid (x,0) \in W_B\}$ and $L_B = \{z \mid (z,1) \in W_B\}$ where $W_B = \{x \mid \not{\phi}_c (x, -; B, E_B^1, E_B^2)\}$ for suitable elementary ϕ, and where $E^i_B = \Sigma_{a\epsilon B}\, p_1 a$ for $i = 1,2$. Further, $x \leq_{\Theta_B} y \Leftrightarrow (x,y) \epsilon L_A$ and $x \equiv_{\Theta_B} y \Leftrightarrow (x,y) \epsilon L_B \wedge (y,x) \epsilon L_B$. Thus we want

$$\text{(iii)} \quad \begin{cases} ho \simeq (N, =) \text{ and, for } x \neq 0, \\ hx \simeq (\{x \mid (x,0) \epsilon W_{Ax}\}, \{(x,y) \mid ((x,y,1) \epsilon W_{Ax} \wedge ((y,x),1) \epsilon W_{Ax}\}) \\ \text{where } Ax \simeq \{u \mid \exists z(z < x \wedge u \simeq hz)\}. \end{cases}$$

Now for any b, whether or not it is a class of classes with equality, both

$$\text{(iv)} \qquad e^1 b = j(b,p_1) = \Sigma_{x\epsilon b}\, p_1 a \quad \text{and} \quad e^2 b = j(b,p_2) = \Sigma_{a\epsilon b} p_2 a$$

are always defined by axiom II(ii). Further

$$\text{(v)} \qquad wb = i_{\ulcorner\phi\urcorner}(b, e^1 b, e^2 b)$$

is defined by the same axiom. Thus for any class B of classes with equality, certainly

$$\text{(vi)} \qquad wB \simeq W_B .$$

Further, for suitable elementary θ_1 and θ_2 we have

$$\text{(vii)} \qquad i_{\ulcorner\theta_1\urcorner}(wb) = \{x \mid (x,0) \epsilon\ wb\} \quad \text{and}$$

$$i_{\ulcorner\theta_2\urcorner}(wb) = \{(x,y) \mid ((x,y),1) \epsilon\ wb \wedge ((y,x),1) \epsilon\ wb\}.$$

Finally, for suitable elementary ψ we have

$$\text{(viii)} \qquad i_{\ulcorner\psi\urcorner}(x,h) = \{u \mid \exists z(z <_{\Theta_S} x \wedge u \simeq hz)\}.$$

Combining these we put

$$\text{(ix)} \qquad g(x,h) = (i_{\ulcorner\theta_1\urcorner}(w\, i_{\ulcorner\psi\urcorner}(x,h)),\ i_{\ulcorner\theta_2\urcorner}(w\, i_{\ulcorner\psi\urcorner}(x,h))),$$

so that (iii) can be rewritten as

$$\text{(iii)}' \quad \begin{cases} ho \simeq (N, =) & \text{and} \\ hx \simeq g(x,h) & \text{for } x \neq 0. \end{cases}$$

Such h is found directly by the recursion theorem. Then it is proved by induction on $\mathfrak{O}_S$ that for each $x \in \mathfrak{O}_S$, hx is a class with equality and $B_x = \{hz | z < z\}$ is a class of such classes, interrelated as required by (i).

A case from the definition (i) of special interest below is $\Omega_1^{(r)}$ which is $\mathfrak{O}_{\{(N,=)\}}$, i.e. is the closure under successor and countable suprema, starting with 0.

Remark. It is not claimed as might be expected that the $(\Omega_x^{(r)}, \equiv_x^{(r)}) \in S$ for $x \in \mathfrak{O}_S$. The reasons are twofold. First the parameter S is used in the definition (i) with the $<$ relation. Secondly, the classes $\mathfrak{O}_{B_x}$ are not obtained from given sets by inductive separation, but rather by full inductive comprehension. This leads us to the following:

Question. Is there a reasonable extension of $T_0(S)$ (or of $T_1(S)$) which has the same kinds of models as that theory and in which we have $(\Omega_x^{(r)}, \equiv_x^{(r)}) \in S$ for each $x \in \mathfrak{O}_S$?

(6) The accessible number classes Ω_x. Obviously if $y \leq x$ then $B_y \subseteq B_x$ in (5)(i) (here, inclusion under the identity injection). We thus have $\Omega_y^{(r)} \subseteq \Omega_x^{(r)}$ and the relation $(\equiv_y^{(r)})$ is contained in $(\equiv_x^{(r)})$. This permits us to define the number classes Ω_x as follows.

$$\text{(i)} \quad \begin{cases} \Omega_x = \Omega_x^{(r)} & \text{for } x \in N \\ \Omega_{x'} = \Omega_x^{(r)} & \text{for } x \notin N \text{ and } x \in \mathfrak{O}_S \\ \Omega_{\sup_a f} = \bigcup_{x \in a} \Omega_{fx}^{(r)} & \text{for } \sup_a f \in \mathfrak{O}_S . \end{cases}$$

Thus

$$\text{(ii)} \qquad \Omega_{\sup_a f} \equiv \bigcup_{y < \sup_a f} \Omega_y^{(r)} \equiv \bigcup_{y < \sup_a f} \Omega_y .$$

The relations $\equiv_x$ on Ω_x are defined correspondingly.

5.3 Interpretation in the models of $T_o(S)$, $T_1(S)$.

(1) Assignment of set-theoretic ordinals. Given any model $\mathcal{K}$ of $T_o(S)$ and any A in $\mathcal{K}$ which is a class of classes with equality relations we can inductively assign to each $x \in \mathfrak{O}_A$ a set-theoretic ordinal $|x|_{\mathfrak{O}_A}$, or simply $|x|$, by the expected conditions:

(i) $$|0| = 0, \quad |x'| = |x| + 1, \quad |\sup_a f| = \sup_{x \in a} |fx| .$$

Then it is proved by induction on W_A that

(ii) $$x \leq_{\mathfrak{O}_A} y \Leftrightarrow |x| \leq |y| , \text{ hence } x <_{\mathfrak{O}_A} y \Leftrightarrow |x| < |y| \text{ and } x \equiv_{\mathfrak{O}_A} y \Leftrightarrow |x| = |y|,$$

for $x, y \in \mathfrak{O}_A$. It follows that

(iii) $$a = (p_1 a, \equiv_a) \in A \wedge x,y \in a \wedge x \equiv_a y \wedge \sup f \in \mathfrak{O}_A \Rightarrow |fx| = |fy| .$$

Next, given any $B \subseteq \mathfrak{O}_A$ closed under successor, define

(iv) $$|B| = \sup_{x \in B} |x| .$$

We then obtain from 5.2 (5), (6)

(v) $$\begin{cases} |\Omega_o^{(r)}| = \omega \\ |\Omega_x^{(r)}| = |\mathfrak{O}_{\{(\Omega_z^{(r)},\ \equiv_z^{(r)}) | z < x\}}| \quad \text{for } x \in \mathfrak{O}_S,\ x \neq 0. \end{cases}$$

(vi) $$\begin{cases} |\Omega_o| = \omega \\ |\Omega_{x'}| = |\mathfrak{O}_{\{(\Omega_z,\ \equiv_z) | z \leq x\}}|, \quad \text{for } x \in \mathfrak{O}_S \\ |\Omega_{\sup_a f}| = \sup_{x \in a} |\Omega_{fx}| , \quad \text{for } \sup_a f \in \mathfrak{O}_S . \end{cases}$$

(2) Interpretation in the full set-theoretical models. Let $\mathcal{K} = \mathfrak{M}^*_{\text{Set-Fun}}$, where $\mathfrak{M} = (V_{\alpha_o}, \in)$ for some inaccessible α_o. It may be seen in this case that

(i) $|\Theta_S|$ = the least inaccessible ordinal

and

(ii) $|\Omega_x| = \omega_{|x|}$ for each $x \in \Theta_S$,

while

(iii) $|\Omega_x^{(r)}|$ enumerates the accessible regular ordinals.

In the proof of $|\Omega_x| \leq \omega_{|x|}$ we use that every $a \in S$ represents a set $a/\equiv_a$ in M (3.6) and in the proof that $\omega_{|x|} \leq |\Omega_x|$ we use that every set-function in $\mathfrak{M}$ is represented by a function in $\mathcal{K}$ (3.4).

(3) Interpretation in the models. $\mathrm{Rec}^*(\omega)$, $\mathfrak{M}^*_{\mathrm{Rec}}$. In $\mathrm{Rec}^*(\omega)$ all the functions met are partial recursive, so Ω_1 is just another form of the first Church-Kleene number class O_1 , and so

(i) $|\Omega_1| = \omega_1^c = \tau_1$,

where τ_α lists the admissible (or recursively regular) ordinals [Ba]. Further it may be seen, e.g. by Richter [R] that

(ii) $|\Omega_n| = |\Omega_n^{(r)}| = \tau_n$ for $n \in N$.

This suggests our first conjecture here:

(C_1) $|\Omega_x^{(r)}| = \tau_{|x|}$ for each $x \in \Theta_S$.

The second conjecture is that

(C_2) $|\Theta_S|$ = the least recursively inaccessible ordinal.

The situation in $\mathfrak{M}^*_{\mathrm{Rec}}$ should come out the same, using the homomorphic mapping of the applicative structure $\mathrm{Rec}(\mathfrak{M})$ onto $\mathrm{Rec}(\omega)$ from 3.3 and 3.4. This is clear for Ω_1 , but the details remain to be worked out for the higher number classes. We may read $(C_1),(C_2)$ equally well as comjectures applying to $\mathfrak{M}^*_{\mathrm{Rec}}$.

(4) Interpretation in the models $\mathrm{E}^N\text{-Rec}^*(\omega)$, $\mathfrak{m}^*_{\mathrm{E}^N\text{-Rec}}$. In these cases all the functions are partial Π^1_1 and the total ones on N are hyperarithmetic. The version of O_1 obtained using hyperarithmetic functions gives no new ordinals so again we have

(i) $|\Omega_1| = \omega^c_1 = \tau_1$ and $|\Omega_n| = |\Omega^{(r)}_n| = \tau_n$ for $n \in N$.

One would expect the conjectures C_1 and C_2 to hold for the models $\mathrm{E}^N\text{-Rec}^*(\mathfrak{m})$, $\mathfrak{m}^*_{\mathrm{E}^N\text{-Rec}}$, if they hold in (3).

Remark. If the conjectures of (3) (or (4)) are correct, the development of 5.2 provides a conceptually superior way of introducing the recursively accessible ordinals and initial numbers: first, because it follows the set-theoretic pattern and second, because it provides a simultaneous generalization of both the classical and recursion-theoretic cases.

5.4 Further principles and applications.

(1) The ancestor relation in Ω_1. Define $\prec$ to be the least relation such that

(i) $x \prec x'$

(ii) $x \in N \wedge y \prec fx \Rightarrow y \prec \sup_N f$

(iii) $x \prec y \wedge y \prec z \Rightarrow x \prec z$.

Write $y \preccurlyeq x$ for $y \prec x \vee y = x$. It may be seen that we have a function e ("enumeration") so that

(iv) $x \in \Omega_1 \Rightarrow ex : N \underset{\text{onto}}{\rightarrow} \{y \mid y \preccurlyeq x\}$.

Further for $x \in \Omega_1$, $\{y \mid y \preccurlyeq x\}$ is linearly ordered by the $\prec$ relation, and finally

(v) $x, y \in \Omega_1 \wedge y \leq x \Rightarrow \exists y_1 (y_1 \preccurlyeq x \wedge y \equiv y_1)$.

The relation $\prec$ corresponds to the r.e. relation fequently used on O_1.

(2) <u>Consequences in</u> $T_1(S)$. In this theory we have $\underset{-}{e}_N$ which provides for quantification over N as an operation; using (1)(iv), this transfers uniformly to quantification over any set $\{y \mid y \preccurlyeq x\}$. Then further if $f: \Omega_1 \to V$ and $x_1 \equiv x_2 \Rightarrow fx_1 = fx_2$ we can decide (by an operation) whether $\exists y \leq x(fy = 0)$ since this is equivalent to $\exists y \preccurlyeq x\ (fy = 0)$. Hence bounded quantification on Ω_1 is decidable in this sense in $T_1(S)$.

(3) <u>The selection axiom for</u> Ω_1 is the following statement, which makes use of a new constant $\underset{-}{c}$ ("choice"):

$$(\mathrm{Sel}_{\Omega_1}) \qquad x \in \Omega_1 \wedge fx \simeq 0 \Rightarrow (\underset{-}{c}f) \in \Omega_1 \wedge f(\underset{-}{c}f) \simeq 0.$$

We claim that for suitable choice of $\underset{-}{c}$,

(i) Sel_{Ω_1} <u>is true in</u> $\exists^N\text{-Rec}^*(\omega)$ <u>and in</u> $\mathfrak{m}^*_{\exists^N\text{-Rec}}$

The reason is that in these models Ω_1 is a Π^1_1 set; then Sel_{Ω_1} is a consequence of Π^1_1-uniformization.

(4) <u>Relations with the theory</u> $T_1^{(\Omega)}$. The theory $T_1^{(\Omega)}$ set up in [F2] is an extension of T_1 in which the class Ω of ordinals is linearly ordered by $<$. It has a model in $\exists^N\text{-Rec}^*(\omega)$ or $\mathfrak{m}^*_{\exists^N\text{-Rec}}$ obtained by taking a Π^1_1 path thru O as a system of unique representatives. The theory $T_1(S) + (\mathrm{Sel}_{\Omega_1})$ can be used for all the same purposes as $T_1^{(\Omega)}$ (cf. (5) next). (We do not have an evident translation of the latter theory into the former since there are no obvious abstract means of defining a path through Ω_1 ordered by $\prec$.) As with the theory $T_1^{(\Omega)}$, a form of the continuum hypothesis CH is satisfied in the $\exists^N$-models, i.e. there is a map $g: \Omega_1 \xrightarrow[\text{onto}]{} N^N$ which takes equivalent ordinals onto equivalent functions. This is by a Π^1_1 enumeration of the hyperarithmetic functions.

(5) <u>Actual and possible applications to model theory</u>. It was shown in [F2] how certain portions of set-theoretical model theory dealing with models which are

countable or Ω_1 - enumerated can be carried out in $T_1^{(\Omega)} \pm (CH)$, thereby generalizing both the classical case and certain hyperarithmetic versions due to Cutland. The same can of course be done here using the theory $T_1(S) + (Sel_{\Omega_1}) \pm (CH)$. The advantage offered by the latter theory now is a systematic way of dealing with the higher number classes. Thus one would hope that this could be used as a means to draw further hyperarithmetic consequences. Promising areas of investigation would be Morley-Shelah theory, the logics $L(Q_\alpha)$, and the stationary logic of [B,K,M]. (I have learned from Barwise that his student E. Wimmers has recently given an abstract treatment of the logic $L(Q_1)$ in $T_1^{(\Omega)}$ which generalizes both the classical case and admissible versions found by Bruce and Keisler [Br, K].)

(6) <u>Possible applications to long hierarchies of ordinal functions.</u> Bachmann had introduced hierarchies of normal "critical" functions on higher number classes which were used eventually to define "large" countable ordinals. In unpublished work by myself and Aczel, new and somewhat simplified hierarchies were proposed as substitutes. Match-up with the Bachmann ordinals in various cases was accomplished by Bridge [Br] and Buchholz [Bu]. Further, it was shown in these special cases that the countable ordinals generated are recursive, by detailed work with explicit order relations on the terms. One would like to obtain a theoretical reason for this outcome. It is possible that the development of the number classes Ω_x in $T_1(S)$ initiated in 5.2 provides a means to do this. The idea would be to show that the hierarchies in question can be established abstractly on the basis of these principles as a continuation of 5.2(5). Then under the interpretation of $T_1(S)$ in $\exists^N\text{-Rec}^*(\omega)$ or $\mathfrak{m}^*_{\exists^N\text{-Rec}}$, every specific countable ordinal generated is $< \omega_1^c$, hence recursive. Schütte [Sch] has isolated the set-theoretic properties of ordinals and functions which are sufficient for a development of the "long" hierarchies. This work may be a useful starting point for the proposal just made.

Bibliography

[Ba] J. Barwise, Admissible sets and structures (Springer, Berlin, 1975).

[B,K,M] J. Barwise, M. Kaufmann and M. Makkai, Stationary logic, to appear in Annals of Math. Logic.

[Bi 1] E. Bishop, Foundations of constructive analysis (McGraw-Hill, New York, 1967).

[Bi 2] E. Bishop, Mathematics as a numerical language, in Intuitionism and proof theory (North-Holland, Amsterdam, 1970)53-71.

[Bi,C] E. Bishop and H. Cheng, Constructive measure theory, Memoirs A.M.S. No. 116(1972).

[Br] J. Bridge, A simplification of the Bachmann process for generating large countable ordinals, J. Symbolic Logic 40(1975) 171-185.

[Br,K] K. Bruce and H.J. Keisler, $L_A(\dashv)$, to appear.

[Bu] W. Buchholz, Normal functionen und konstruktive Systemen von Ordinalzahlen, in Proof Theory Symposion Kiel 1974, Lecture Notes in Maths. 500(1975), 4-25.

[C] N. Cutland, Π_1^1-models and Π_1^1-categoricity, in Conference in Mathematical Logic-London '70 (Lecture Notes in Maths.255, 1972) 42-63.

[F 1] S. Feferman, A language and axioms for explicit mathematics, in Algebra and Logic (Lecture Notes in Maths. 450, 1975) 87-139.

[F 2] S. Feferman, Generalizing set-theoretical model theory and an analogue theory on admissible sets, to appear in Proc. 4th Scandinavian Logic Symposium (Jyväskyla, Finland, 1976).

[F 3] S. Feferman, Notes on the formalization of Bishop's constructive mathematics (unpublished).

[Fr 1] H. Friedman, Axiomatic recursive function theory, in Logic Colloquium '69 (North-Holland, Amsterdam 1971)113-137.

[Fr 2] H. Friedman, Set theoretic foundations for constructive analysis, Annals of Maths. 105(1977)1 - 28.

[K] S.C. Kleene, Recursive functionals and quantifiers of finite types I, Trans. A.M.S. 91(1959)1-52.

[M] Y.N. Moschovakis, Abstract first order computability I, Trans. A.M.S. 138 (1969), 427-464.

[My] J. Myhill, Constructive set theory, J. Symbolic Logic 40(1975)374-382.

[R] W. Richter, Constructively accessible ordinal numbers, J. Symbolic Logic 33(1968)43-55.

[Ro] H. Rogers, Jr. Theory of recursive functions and effective computability (McGraw-Hill, New York 1967).

[S] G. Sacks, Measure-theoretic uniformity, Trans. A.M.S. 30(1969)381-420.

[Sch] K. Schütte, Proof theory (Springer-Verlag, Berlin, 1977).

J.E. Fenstad, R.O. Gandy, G.E. Sacks (Eds.)
GENERALIZED RECURSION THEORY II

ON THE FOUNDATION OF GENERAL RECURSION THEORY: COMPUTATIONS VERSUS INDUCTIVE DEFINABILITY.

Jens Erik Fenstad

General recursion theory can be approached in a number of different ways. One starting point is to analyze the relation

$$\{a\}(\sigma) \simeq z$$

which expresses that the "computing device" named or coded by a and acting on the input sequence $\sigma = (x_1, \ldots, x_n)$ gives z as output.

The history of this notion goes back to the very foundation of the theory of general recursion in the mid 1930's. It can be traced from the theory of Turing and other types of idealized machine computability via Kleene's indexing and normal form theorems to present day generalizations.

It was Kleene who in 1959 took this relation as basic in developing his theory of recursion in higher types [7]; subsequently it was adopted by Moschovakis [9] in his study of prime and search computability over more general domains.

Indexing was also behind various other abstract approaches. We mention the axiomatics of Strong [15], Wagner [16], and H. Friedman [5], and the computation theories of Y. Moschovakis [10].

But "computations" is not the only possible way of doing general recursion theory. It is the purpose of this note to set out briefly some other approaches based on inductive definability and fixed point operators and to compare them to theories based on computations.

First we have to recall some notions in connection with computation theories. We shall be brief and refer the reader to our previous expositions [2] and [3] , and the forthcoming book [4] if further details are desired.

COMPUTATIONS. Let there be given a notion of computation $\{a\}(\sigma) \simeq z$ over some domain A. Let us from this abstract the set of all computation tuples $\Theta = \{(a,\sigma,z); \{a\}(\sigma) \simeq z\}$. A function f on A is computable by the given notion if there is an index or code a such that

$$f(\sigma) \simeq z \quad \text{iff} \quad (a,\sigma,z) \in \Theta .$$

The axiomatic approach reverses this procedure. Let there be given a set Θ of tuples (a,σ,z) over a domain A. A function on A is called Θ-computable if there is a code a such that the equivalence above holds.

Not any set Θ is a reasonable computation theory. We must put in some basic functions and require closure of Θ under some reasonable properties; the details are given in any of the references [2],[3],[4]. Usually we shall require that there is a Θ-computable coding scheme on the domain A.

Let $\underset{\sim}{f} = f_1,\ldots,f_m$ be any list of partial functions on the domain A. In a standard way $\underset{\sim}{f}$ generates a "least" computation theory on A, which we will denote by $PR[\underset{\sim}{f}]$ (= the prime computation theory generated by $\underset{\sim}{f}$), in which the functions $\underset{\sim}{f}$ are computable.

$PR[\underset{\sim}{f}]$ is the fixed point for a certain inductive operator. Thus the following basic representation theorem effects, at least for some purposes, a reduction of the computation theoretic approach to the theory of inductive definability.

Simple Representation. Let Θ be a (pre-)computation theory on a domain A with a Θ-computable coding scheme. There exists a Θ-computable partial function f such that Θ is equivalent to $PR[f]$.

At this point the confirmed "inductivist" may take leave of us. Let the others proceed.

PLATEK'S THESIS. The aim of this approach is to study definability/computability over an arbitrary domain Ob using the fixed point operator. But fixed points at one level may be obtained from fixed points of higher levels. This leads to the hierarchy HC of hereditarily consistent functionals over Ob as the natural domain of the general theory.

Remark. Platek's thesis was never published, we follow the discussion in J. Moldestad Computations in Higher Types [8].

The hierarchy HC is defined inductively as follows. Any type symbol $\tau \neq 0$ can be written in the form $\tau_1 \to (\tau_2 \to \ldots(\tau_k \to 0)\ldots)$. The level of τ is defined by $\ell(\tau) = \max\{\ell(\tau_i) + 1\}$. Monotonicity for partial functions is defined as usual. Then $HC(\tau)$ is defined to be the set of all partial monotone functions defined on a subset of $HC(\tau_1) \times \ldots \times HC(\tau_k)$ and with values in Ob.

The fixed point operator at type τ is an element $FP \in HC((\tau \to \tau) \to \tau)$, such that when FP is applied to an element $f \in HC(\tau \to \tau)$ it produces the least fixed point $FP(f)$ of f, which will be an element of $HC(\tau)$.

Platek's indexfree approach to recursion theory can now be introduced. Let $\mathcal{B} \subseteq HC$. Then the recursion theory generated by $\mathcal{B}$, which we will denote by $\mathcal{R}_\omega(\mathcal{B})$, is the least set extending $\mathcal{B}$, closed under composition, and containing the function DC (definition by cases), the combinators $I(f) = f$, $K(f,g) = f$, and $S(f,g,h) = f(h)(g(h))$, and the fixed point operator FP .

To set out the relationship with computation theories we quote two results from Moldestad's study. The first is Platek's reduction theorem.

Reduction Theorem. Let $\mathcal{B}$ contain some basic functions (a coding scheme, the characteristic function of the natural numbers, the successor and predecessor functions). If $\mathcal{B} \subseteq HC^{\ell+2}$, then

$$\mathcal{R}_\omega(\mathcal{B})^{\ell+3} = \mathcal{R}_{\ell+1}(\mathcal{B})^{\ell+3}$$

If we are interested in objects of level at most $\ell+3$, then we need only apply the fixed point operator up to type $\ell+1$.

Equivalence Theorem. There exists a computation theory Θ (derived from Kleene's schemata S1 - S9 in [7]) such that

$$\mathcal{R}_\omega(\{h_1,\dots,h_k\}) = \Theta[h_1,\dots,h_k]$$

for all finite lists of HC objects $h_1,\dots,h_k$.

We shall in a moment return to the reduction theorem. But first we spell out the content of the equivalence theorem. It seems that we can draw the following conclusions. Let $\mathcal{B} \subseteq HC$.

1 $\mathcal{R}_\omega(\mathcal{B}) = \cup\{\mathcal{R}_\omega(\mathcal{B}_0);\ \mathcal{B}_0$ finite subset of $\mathcal{B}\}$.

2 For finite $\mathcal{B}_0$, $\mathcal{R}_\omega(\mathcal{B}_0) = \Theta[\mathcal{B}_0]$

3 $\mathcal{R}_\omega(\mathcal{B}) \subseteq \Theta[\mathcal{B}]$, but in general $\subsetneq$.

Only the third assertion requires a comment. The notation $\Theta[\mathcal{B}]$

is somewhat ambiguous. We must assume that $\mathcal{B}$ is given as a list, i.e. with a specific enumeration. This means that in any precise version of $\Theta[\mathcal{B}]$ we have the enumeration function of the list $\mathcal{B}$. But this enumeration function is not necessarily in $\mathcal{R}_{\omega}(\mathcal{B})$.

Back to the reduction theorem. This result shows that the framework of computation theories is adequate if the aim is to study computability/definability over some given domain. We need not climb up through the hierarchy HC. A computation theory Θ can be considered as a set of functions $\Theta \subseteq HC^1$. Then, by the reduction theorem

$$\mathcal{R}_{\omega}(\Theta)^1 = \mathcal{R}_1(\Theta)^1 = \Theta,$$

the last equality being true since Θ satisfies the first recursion theorem.

RECENT DEVELOPMENTS IN INDUCTIVE DEFINABILITY. The index free approach of Platek is conceptually of great importance in the development of generalized recursion. The theory has, however, some weak points. Recently improved and largely equivalent versions have been published independently of Y. Moschovakis and S. Feferman.

The relevant papers of Moschovakis is the joint contribution with Kechris, Recursion in higher types [6], and the paper On the basic notions in the theory of induction [11]. Feferman's version is presented in Inductive schemata and recursively continuous functionals [1].

Feferman summarizes his critisism of Platek in the following points.

a The structure of natural numbers is included as part, - there could be more general situations, e.g. applications in algebra.

b Inductive definability of relations is not accounted for.

c Platek's pull-back of Kleene's theory of recursion in higher types from HC is complicated and ad hoc.

We shall return to a and b below. In connection with c we just note our complete agreement.

As a basis for a comparison with computation theories we shall discuss a result from Recursion in higher types.

A partial monotone functional $\Phi(\underset{\sim}{x}, f, \underset{\sim}{g})$ defines in the usual way a fixed point $\Phi^{\infty}(\underset{\sim}{x}, \underset{\sim}{g})$. Let $\mathcal{F}$ be a class of functionals. If Φ belongs to $\mathcal{F}$, we call Φ^{∞} an $\mathcal{F}$-fixed point. The class Ind($\mathcal{F}$) will consist of all functionals $\Psi(\underset{\sim}{x}, \underset{\sim}{g})$ for which there exists an $\mathcal{F}$-fixed point $\Phi^{\infty}(\underset{\sim}{h}, \underset{\sim}{x}, \underset{\sim}{g})$ and constants $\underset{\sim}{n}$ from ω such that

$$\Psi(\underset{\sim}{x}, \underset{\sim}{g}) = \Phi^{\infty}(\underset{\sim}{n}, \underset{\sim}{x}, \underset{\sim}{g})$$

The induction completeness theorem tells us that Ind($\mathcal{F}$) is closed under definability.

Not every collection $\mathcal{F}$ gives a reasonable recursion theory. A class $\mathcal{F}$ is called suitable if it contains the following initial objects: characteristic function of ω, the identity on ω, the successor function on ω, the characteristic function of equality on ω, and the evaluation functional. In addition $\mathcal{F}$ is required to be closed under addition of variables, composition, definition by cases, substitution of projections, and functional substitution.

Ind($\mathcal{F}$) is said to have the enumeration property (is ω-parametrized) if for each $n \geq 1$ there is some $\Phi(e, x_1, \ldots, x_n) \in$ Ind($\mathcal{F}$) such that a function f belongs to Ind($\mathcal{F}$) iff there exists some $e \in \omega$ such that

$$f(\underset{\sim}{x}) = \Phi(e, \underset{\sim}{x}) .$$

Enumeration theorem [6]. Let $\mathcal{F}$ be a suitable class of functionals on a domain A including ω. If $\mathcal{F}$ is finitely generated and admits a coding scheme, then $\mathrm{Ind}(\mathcal{F})$ has the enumeration property.

Remark. One notices that the machinery provided by the requirement of suitability of $\mathcal{F}$ corresponds to a large extent to what we have put into our general notion of a computation theory. There is one difference, on the computation theoretic approach we have adopted the enumeration property and proved the first recursion theorem, on the inductive approach the first recursion theorem is an axiom and enumeration a theorem. One may argue what is "philosophically" the most basic or natural, mathematically they serve the same purposes. Provided, we should add, there is enough coding machinery available.

The enumeration theorem for $\mathrm{Ind}(\mathcal{F})$ leads to the same situation as pointed out in connection with Platek's $\mathcal{R}_\omega(\mathcal{B})$. The enumeration theorem leads to a computation theory Θ, such that $\mathcal{F}_0$ is a finite basis for the finitely generated class $\mathcal{F}$ (and $\mathcal{F}$ is suitable and admits a coding scheme), then

$$\mathrm{Ind}(\mathcal{F}) = \Theta[\mathcal{F}_0]$$

Both Moschovakis [11] and Feferman [1] include inductive definability in relations. Moschovakis' approach is based on the notion of induction algebra, which is a structure

$$\mathbb{F} = \langle \{X^\alpha\}_{\alpha\in I}, \{\leq^\alpha\}_{\alpha\in I}, \{v^\alpha\}_{\alpha\in I}, \mathcal{F}\rangle$$

where each $\leq^\alpha$ is a partial ordering on X^α in which every chain has a least upper bound, and v^α is supremum operator on X^α, i.e. $x \leq x \vee y$ and if $x \leq y$, then $x \vee y = y$, for all $x,y \in X^\alpha$. $\mathcal{F}$ is a class of operations, i.e. maps of the form $f: X^{\alpha_1}\times\ldots\times X^{\alpha_n} \to X^\alpha$.

The two main examples are the induction algebras of relations and induction algebras of partial functions. The case of partial functions was discussed above. In the case of relations we start with a domain A, let X^0 equals the set of truth values $\{T,F\}$, and for $n \geq 1$, set X^n = all n-ary relations on A. In this case $\leq$ is set inclusion and $\vee$ is set union.

Let Φ be a class of second order relations on A. To each $\varphi(R,s)$ in Φ we associate an operation f by

$$f(\underset{\sim}{R}) = \{\underset{\sim}{s};\ \varphi(\underset{\sim}{R},\underset{\sim}{s})\}$$

(Conversely, an operation f determines a relation by the equivalence $\varphi(\underset{\sim}{R},\underset{\sim}{s})$ iff $\underset{\sim}{s} \in f(\underset{\sim}{R})$.)

In this way - provided suitable conditions are imposed on Φ - we get an induction algebra of relations on A.

Furthermore, provided the classes we start with are "rich enough" in structure, there will be an enumeration theorem for the finitely generated algebras, and, hence, a computation theoretic eqiuvalent. But there could, in principle, be more general situations.

COMPUTATIONS IN ALGEBRA. In computing over an algebraic system $\mathfrak{A} = \langle A,f_1,\ldots,f_k\rangle$ one should clearly consider all the operations f_i, and hence all "polynomials" in the operations computable. But more could be computable. We shall illustrate by discussing an example due to Dag Normann.

Example (D. Normann). Let $A = \{\langle n,m\rangle;\ n < m,\ m > 1\}$, and define an operation $f(\langle n,m\rangle)$ for $\langle n,m\rangle \in A$ by

$$f(\langle n,m\rangle) = \begin{cases} \langle n+1,m\rangle & \text{if } n+1 < m. \\ \langle 0,m\rangle & \text{if } n+1 = m. \end{cases}$$

The algebraic structure $\mathfrak{A} = \langle A,f\rangle$ consists of all finite cycles with f as a "local" successor operation.

Let $g(x)$ be the predecessor of x under f. g is not a polynomial, but can be computed by the following algorithm: Compute in succession $f(x), f^{(2)}(x), \ldots, f^{(n)}(x)$ up to the least n such that $f^{(n)}(x) = x$. Then $g(x) = f^{(n-1)}(x)$.

This computation uses the natural numbers as an auxiliary. It is, however, not difficult to capture g in the framework of induction algebras. Noting that f is a "successor" function and g the corresponding "predecessor", we introduce a functional $\Phi(h')$ by

$$\Phi(h')(x,y) = \begin{cases} y & \text{if } f(x) = y \\ h'(x, f(y)) & \text{if } f(x) \neq y \end{cases}$$

If h is the least fixed-point for Φ, then $g(x) = h(x,x)$.

So far, so good. But consider the following function. Let $s(x)$, for $x \in A$, be the length of the cycle to which x belongs. Let $h^*: \mathbb{N} \to \{0,1\}$ be recursive but not primitive recursive. Define

$$g^*(x) = \begin{cases} x & \text{if } h^*(s(x)) = 0 \\ f(x) & \text{if } h^*(s(x)) = 1 \end{cases}$$

Clearly g^* should be computable over $\mathfrak{A} = \langle A,f \rangle$.

But does g^* belong to the induction algebra generated by f over $\mathfrak{A}$? It does <u>not</u> for the following reason. Any finitely generated substructure of $\mathfrak{A}$ is finite (a finite collection of cycles). Normann has proved in his note [13] that if $\mathfrak{B}$ is a substructure of $\mathfrak{A}$ and g is an inductively defined function on $\mathfrak{A}$, then the restriction of g to $\mathfrak{B}$ can be inductively defined inside $\mathfrak{B}$. In our case the finitely generated substructures are finite, and for each type level τ the type structure up to τ is finite and can be primitively recursively described. Hence g^* cannot be inductive, since otherwise h^* would be primitive recursive.

It is a common experience that computations in algebra operate both on elements of the structure and on natural numbers. In the first case above an explicit reference to natural numbers could be avoided, the "predecessor" g is inductive. The second case, however, seems to point to a limitation of the inductive approach. The g* there constructed is obviously computable, but it is not inductive. And g* is a rather "natural" computable function: Divide the cycles of $\mathfrak{A}$ in two classes; if x belongs to a cycle of the first class, do nothing, if x belongs to a cycle of the second class, take the successor.

With this example in mind it seems to us that the critisism of Feferman refered to in point a above, that the structure of natural numbers is included as part, is not well taken, at least when the emphasize is on algebraic computations and not on inductive definability over an algebraic structure. Both elements of the structure and natural numbers enter into algebraic computations in general, and neither can be eliminated.

This fact can be easily accommodated in the framework of computation theories. We may e.g. construct an appropriate computation theory over the expanded domain $A^* = A \cup \mathbb{N}$. Or, as suggested in Normann [13], we may use set recursion (E-recursion, see [12]) on HF($\mathfrak{A}$), the hereditarily finite sets over the structure $\mathfrak{A}$.

Of course, we can also use an induction algebra over $A^* = A \cup \mathbb{N}$. But then, in a rather explicit way, we have adjoined the structure of the natural numbers.

There is, perhaps, one more comment to make. Codes, indices are usually claimed to be *ad hoc* and, hence conceptually unsatisfactory. And a comparison with an intrinsic versus coordinate based treatment in geometry is often made. But is the analogy really to the point? In our discussion of Platek's $\mathcal{R}_\omega(\mathcal{B})$, - and a similar result holds for $\mathrm{Ind}(\mathcal{F})$, we concluded that

1 $\mathcal{R}_\omega(\mathcal{B}) = \cup\, \mathcal{R}_\omega(\mathcal{B}_0)$

where $\mathcal{B}_0$ is a finite subset of $\mathcal{B}$, and for finite $\mathcal{B}_0$

2 $\mathcal{R}_\omega(\mathcal{B}_0) = \Theta[\mathcal{B}_0]$.

In 2 the codes are introduced as a systematic, even canonical way of refering to the objects in the finite list $\mathcal{B}_0$ and to the operations generating $\mathcal{R}_\omega(\mathcal{B}_0)$ out of $\mathcal{B}_0$. 1 and 2 together say that the global theory $\mathcal{R}_\omega(\mathcal{B})$ admits natural local coordinates $\Theta[h_1,\ldots,h_k]$ suitable for more "delicate", i.e. computation theoretic investigations of the theory, - degree structure, computation in higher types etc.

This seems to be a reasonable analogy with e.g. geometry. But is the analogy complete? Are there any properties in the large of generalized recursion theory? Or is everything local, i.e. computation theoretic?

Remark. This has been a discussion of the elementary part of the theory. For more advanced topics we refer to [2],[3], and [4]. We would also like to point out that the notion of a computation theory on two types as developed by Moldestad [8] is easily seen to be equivalent to the recent notion of a Spector 2-class, and hence is an equally adequate setting for discussing 2nd order inductive definability.

REFERENCES.

1. Feferman, S., Inductive schemata and recursively continuous functionals, Gandy, Hyland (eds.), Logic Colloquium 76, North-Holland, 1977.

2. Fenstad, J.E., Computation theories: An axiomatic approach to recursion on general structures, Müller, Oberschelp, Potthoff (eds.), Logic Conference Kiel 1974, Lecture Notes vol 499, Springer Verlag, 1975.

3. Fenstad, J.E., Between recursion theory and set theory, Gandy, Hyland (eds.), Logic Colloquium 76, North-Holland, 1977.

4. Fenstad, J.E., Recursion theory: An axiomatic approach, Springer Verlag (to appear).

5. Friedman, H., Axiomatic recursive function theory, Gandy, Yates (eds.) Logic Colloquium 69, North-Holland, 1971.

6. Kechris, A. and Moschovakis, Y., Recursion in higher types, Barwise (eds.), Handbook of Mathematical Logic, North-Holland, 1977.

7. Kleene, S.C., Recursive functionals and quantifiers of finite type I, Trans. Am. Math. Soc. 91 (1959).

8. Moldestad, J., Computations in higher types, Lecture Notes vol 574, Springer Verlag 1977.

9. Moschovakis, Y., Abstract first order computability I, II, Trans. Amer. Math. Soc. 138 (1969).

10. Moschovakis, Y., Axioms for computation theories - first draft, Gandy, Yates (eds.), Logic Colloquium 69, North-Holland, 1971.

11. Moschovakis, Y., On the basic notions in the theory of induction, Butts, Hintikka (eds.), Logic, Foundations of Mathematics, and Computability Theory, Reidel, 1978.

12. Normann, D., Set recursion, this volume.

13. Normann, D., Recursion on algebraic structures, (unpublished note).

14. Platek, R.A., Foundation of recursion theory, Stanford thesis, 1966.

15. Strong. H.R., Algebraically generalized recursive function theory, IBM J. Res. Devel. 19 (1968).

16. Wagner, E.G., Uniform reflexive structures: on the nature of Gödelizations and relative computability, Trans. Amer. Math. Soc, 144 (1969).

J.E. Fenstad, R.O. Gandy, G.E. Sacks (Eds.)
GENERALIZED RECURSION THEORY II

An Introduction to β-Recursion Theory

Sy D. Friedman
Department of Mathematics
University of Chicago
Chicago, Illinois 60637

Section 1. Background for the Theory.

α-Recursion Theory, or Recursion Theory on (Σ_1)-admissible ordinals, is certainly a very successful generalization of Recursion Theory on ω. It has demonstrated that constructions exploiting the Σ_2 or Σ_3 admissibility of ω can be replaced by finer constructions which rely only on Σ_1 admissibility.

The minimal degree construction however has resisted generalization to arbitrary admissible ordinals, though important progress was made by Richard Shore ([14]) who treated the Σ_2-admissible case. It is instructive to examine his argument, as the attempt to adapt it to all admissible ordinals led to the development of β- Recursion Theory.

Shore's construction can be outlined as follows: If α is Σ_2 admissible, then the structure $< L_\alpha, \epsilon, C >$ is admissible where C is a complete α-recursively enumerable set. A minimal α-degree can then be constructed by applying the α-finite injury method to this structure, much in the way Sacks and Simpson ([13]) first used it in their solution to Post's Problem.

What if α is only Σ_1 admissible ? Then the structure $< L_\alpha, \epsilon, C >$ is no longer admissible. Nonetheless, is it still possible to apply the α-finite injury method to this structure, yielding a minimal α-degree ?

This leads to the study of Recursion Theory on possibly inadmissible structures $< L_\beta[A], \epsilon, A >$, or β-Recursion Theory. There are two goals for this theory:

1) To produce new constructions of recursively enumerable sets which are not dependent on any admissibility assumption.

2) To clarify the concepts and techniques used in Admissible Recursion Theory. In both cases we hope to provide new data in the search for the axioms needed to do Recursion Theory. In this paper we report on the progress that has been made in these directions.

Section 2. Basic Notions.

The correct general setting for β-Recursion Theory is Jensen's S-Hierarchy for L. For limit ordinals β, S_β has all of the important properties shared by limit levels of Gödel's L-Hierarchy. We proceed to define the S-Hierarchy and list these properties, referring the reader to [1] for a more thorough treatment. For ordinals β such that ω^ω divides β, we have $S_\beta = L_\beta$, so in this case one may work with the usual L-Hierarchy.

A function $f: V^n \to V$ is rudimentary if it can be generated from the following schemata:

$$\left.\begin{array}{ll} \text{i)} & f(x_1, \ldots, x_n) = x_i \\ \text{ii)} & f(\overline{x}) = \{x_i, x_j\} \\ \text{iii)} & f(\overline{x}) = x_i - x_j \end{array}\right\} 1 \leq i, j \leq n$$

$$\text{iv)}\quad f(\overline{x}) = h(g_1(\overline{x}), \ldots, g_k(\overline{x}))$$

$$\text{v)}\quad f(y, \overline{x}) = \bigcup_{z \in y} g(z, \overline{x})$$

$R \subseteq V^n$ is rudimentary if for some rudimentary function f, $\overline{x} \in R \iff f(\overline{x}) = \emptyset$. If X is transitive, the rudimentary closure of X = the closure of X under the rudimentary functions. Also define $\text{rud}(X)$ = rudimentary closure of $X \cup \{X\}$.

Lemma. There is a rudimentary function $\underline{S}$ such that for transitive X, $\underline{S}(X)$ is transitive, $X \cup \{X\} \subseteq \underline{S}(X)$ and $\text{rud}(X) = \bigcup_{n \in \omega} \underline{S}^n(X)$.

The S-Hierarchy is now defined by:

$$S_0 = \emptyset ,$$

$$S_{\alpha+1} = \underline{S}(S_\alpha),$$

$$S_\lambda = \bigcup_{\alpha < \lambda} S_\alpha .$$

Properties of the S-Hierarchy

1) β limit $\to S_\beta$ is closed under rudimentary functions, $On(S_\beta) = \beta$.

2) $S_{\beta+\omega} \cap \mathcal{P}(S_\beta) = \{X \subseteq S_\beta \mid X$ is first-order definable over $\langle S_\beta, \epsilon \rangle\}$.

3) $L_\beta = S_\beta \leftrightarrow \omega^\omega$ divides β, $\bigcup_\beta S_\beta = L$.

4) Suppose β is a limit ordinal and $Tr_n = \{\langle i, x_1, \ldots, x_n \rangle \mid$ the ith Σ_n formula φ is n-ary and $\langle S_\beta, \epsilon \rangle \models \varphi(x_1, \ldots, x_n)\}$. Then Tr_n is Σ_n over $\langle S_\beta, \epsilon \rangle$, uniformly in β.

5) There is a well-ordering $<$ of L such that for limit β, $< \mid S_\beta \times S_\beta$ is Σ_1 over $\langle S_\beta, \epsilon \rangle$, uniformly in β.

The above properties are sufficient to safely adapt the definitions of α-Recursion Theory to an arbitrary $\langle S_\beta, \epsilon \rangle$, β limit.

For $A \subseteq S_\beta$, we define:

A is β-recursively enumerable (β-r.e.) $\leftrightarrow$ A is Σ_1-Definable over $\langle S_\beta, \epsilon \rangle$.

A is β-recursive (β-rec) $\leftrightarrow A, \overline{A} = S_\beta - A$ are both β-r.e.

A is β-finite $\leftrightarrow A \in S_\beta$.

By virtue of property 4) above, there is a universal β-r.e. set $W(e,x)$ such that any β-r.e. set is of the form $W_e = \{x \mid W(e,x)\}$ for some e.

The reducibilities of β-Recursion Theory are also derived from the admissible case. For any $A \subseteq S_\beta$, let $A^* = \{\langle x,y \rangle \mid x \subseteq A,\ y \subseteq \overline{A}$, x, y β-finite$\}$ and $A^*_f = \{\langle x,y \rangle \mid x \subseteq A,\ y \subseteq \overline{A},\ x, y$ finite$\}$. Then:

A is finitely β-reducible to B $\leftrightarrow$ (for some β-r.e. R,

$(A \leq_{f\beta} B)$ $\qquad x \in A^*_f \leftrightarrow \exists\, y \in B^*_f [\langle x, y \rangle \in R])$

A is weakly β-reducible to B $\leftrightarrow$ $A_f^* \leq_{f\beta} B^*$

$(A \leq_{w\beta} B)$

A is β-reducible to B $\leftrightarrow$ $A^* \leq_{f\beta} B^*$

$(A \leq_{\beta} B)$

Both finitely β-reducible and β-reducible imply weakly β-reducible (but not conversely in general) and these two notions are in general incomparable. β-reducibility is the fundamental reducibility for β-Recursion Theory. β-reducibility is reflexive and transitive and thus we may define the β-degree of $A = \{B \mid A \leq_{\beta} B,\ B \leq_{\beta} A\}$; the β-degree are partially-ordered by $\leq_{\beta}$. There is a smallest β-degree 0 and a largest β-degree of a β-r.e. set, denoted 0'.

Tameness and Regularity

β-r.e. sets are constructed in stages. As computations from a set A are determined by pairs $<x, y>$, where x and y are β-finite and satisfy $x \subseteq A$, $y \subseteq \overline{A}$, it is convenient to know that such pairs are satisfied by some stage of the construction; i.e., if A^{σ} denotes the amount of A enumerated by stage σ, we would like to know that:

(*)x β-finite, $x \subseteq A \longrightarrow$ For some σ, $x \subseteq A^{\sigma}$.

In this case, the enumeration $\{A^{\sigma}\}_{\sigma<\beta}$ is said to be tame. More precisely, if $\exists y\, \varphi(x, y)$ (φ a Δ_0 formula) defines A over $<S_{\beta}, \epsilon>$, then it gives rise to the enumeration $\{A^{\sigma}\}$ where $x \in A^{\sigma} \leftrightarrow \exists y \in S_{\sigma}\, \varphi(x, y) \wedge x \in S_{\sigma}$. A is tamely-r.e. if A has such an enumeration with the property (*) above. This is equivalent to the assertion that $\{\sigma\text{-finite } x \mid x \subseteq A\}$ is β-r.e. From this it follows that $\deg A = 0 \longleftrightarrow A, \overline{A}$ are both tamely-r.e.

Theorem 1 ([3]). Assume that β is inadmissible. Let W be a universal β-r.e. set. Then there is a β-recursive set A such that $0 <_{\beta} A <_{\beta} W$ and every β-recursive or tamely-r.e. set is β-reducible to A.

It follows that in the inadmissible case there are β-recursive sets of nonzero β-degree. In particular, t.r.e. ≠ r.e. and $\leq_{w\beta} \neq \leq_{\beta}$. Theorem 1 provides a weak solution to Post's Problem in β-Recursion Theory (the question of the existence of β-r.e. degrees between 0 and 0'). The solution is weak because it does not provide incomparable β-r.e. degrees; moreover, in this case we have $W \leq_{w\beta} A \leq_{w\beta} 0$. Denoting deg(A) by $0^{1/2}$, we are led to the following picture of the β-r.e. degrees:

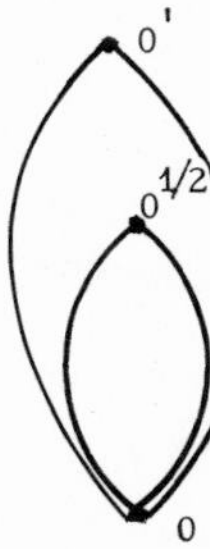

In general, however, nonzero tamely-r.e. degrees will not exist (though $0^{1/2}$ provides an example of a nonzero β-recursive degree). Simple questions regarding these degrees remain unsettled; in particular, it is not known if the t.r.e. degrees or the β-recursive degrees form an initial segment of the β-r.e. degrees. It follows from [10], though, that the t.r.e. degrees are always contained in the β-recursive degrees when β is inadmissible.

$A \subseteq S_\beta$ is <u>regular</u> if $A \cap x$ is β-finite whenever x is β-finite. It is a theorem of Sacks ([12]) in the case that β is admissible that every β-r.e. degree has a regular β-r.e. representative. Regular β-r.e. sets are more rare for inadmissible β; in fact, for some β's, every regular β-r.e. set (even every regular set) has degree 0 (see [3]). However, a slight extension of t.r.e.-ness is enough to guarantee the existence of regular, β-r.e. representatives:

Theorem 2 ([3],[10]). If $\{x \mid x \text{ is } \beta\text{-finite}, x \cap A \neq \emptyset\}$ is t.r.e., then A has the same β-degree as a regular β-r.e. set.

However, for some β's there are t.r.e. sets which do not lie in the same β-degree as some regular set. Thus, there appears to be no simple characterization of the regular β-r.e. degrees.

The reader is referred to [10] for proofs of the above facts as well as further information concerning the β-r.e., t.r.e., and regular β-r.e. degrees.

In the further development of the theory, the limit ordinals fall into two classes determined by their degree of admissibility. This split into cases was first revealed in Jensen's proof of Σ_2-Uniformization for S_β, and is determined by the values of certain key parameters which we now proceed to define.

A relation on S_β is Σ_n if it can be defined over $\langle S_\beta, \epsilon \rangle$ by a formula consisting of n alternating unbounded quantifiers (beginning with an existential) followed by a limited formula. A function is Σ_n if its graph is.

The first type of parameter that we define measures the extent to which β is not a cardinal. The Σ_n-projectum of β, ρ_n^β, is the least ordinal γ such that there is a Σ_n injection of β into γ. Jensen shows ([9]) that this is the same as the least γ such that some Σ_n subset of γ is not β-finite. As there is always a Σ_1 bijection between β and S_β (see [1]), we can in fact inject S_β into ρ_n^β via a Σ_n function.

Our second set of parameters describes the extent to which β is singular. The Σ_n-cofinality of β, $\Sigma_n \mathrm{cf}\, \beta$, is the least γ such that some Σ_n function with domain γ has range unbounded in β. If β is Σ_{n-1}-admissible, then this is the same as the least γ such that some Σ_n function with domain γ is not β-finite (though this equivalence is not true for all β).

In case $n = 1$, ρ_1^β and $\Sigma_1 \mathrm{cf}\,\beta$ are alternatively written β^* and $\mathrm{Rcf}\,\beta$, respectively. (Rcf abbreviates "Recursive Cofinality".) We shall be mostly concerned with β^*, $\mathrm{Rcf}\,\beta$, ρ_2^β and $\Sigma_2 \mathrm{cf}\,\beta$. Note that if β is admissible and A is a regular β-r.e. set of degree $0'$, then ρ_2^β and $\Sigma_2 \mathrm{cf}\,\beta$ are just the Σ_1-projectum and Σ_1 cofinality of the relativized structure $< L_\beta, \epsilon, A >$.

In case $\mathrm{Rcf}\,\beta \geq \beta^*$ we say that β is <u>weakly admissible</u>. In this case, many of the arguments from admissibility theory apply. The reason for this is that many priority arguments use β^* to index a listing of requirements and the above assumption allows one to perform Σ_1 inductions of length β^*. Σ_2-Uniformization is also easy in this case. If β is admissible and $\Sigma_2 \mathrm{cf}\,\beta \geq \rho_2^\beta$, then we say that β is <u>weakly Σ_2-admissible.</u> In this case, one can carry out the construction of minimal β-degrees, minimal pairs of β-r.e. degrees and major subsets of β-r.e. sets.

If $\mathrm{Rcf}\,\beta < \beta^*$ we say that β is <u>strongly inadmissible.</u> In this case, the arguments of admissibility theory do not apply and new techniques are needed. This is the difficult case of Σ_2-Uniformization. If β is admissible and $\Sigma_2 \mathrm{cf}\,\beta < \rho_2^\beta$, then we say that β is <u>strongly Σ_2-inadmissible.</u> The constructions of minimal β-degrees, minimal pairs of β-r.e. degrees and major subsets of β-r.e. sets are all very difficult for such β and have only been accomplished in very special cases. However, the techniques of β-Recursion Theory are now beginning to apply themselves to this case (see Section 5).

Section 3. <u>Weak Admissibility</u>

As mentioned before, the methods of α-Recursion Theory apply in this case. In particular, the method of Shore blocking (see [17]) was used in [3] to prove:

Theorem 3. If β is weakly admissible, then there are regular t.r.e. sets A, B such that $A \not\leq_{w\beta} B$, $B \not\leq_{w\beta} A$.

W. Maass (in [10]) has found a technique for transferring many results from α-Recursion Theory to arbitrary weakly admissible ordinals. He associates to each weakly admissible β an admissible structure $\mathfrak{A}$ such that $\mathfrak{A}$-r.e. degrees embed into the β-r.e. degrees. In this way, known results about the admissible structure $\mathfrak{A}$ have consequences about the β-r.e. degrees. We now describe his construction in more detail.

Let $\kappa = \Sigma_1 \operatorname{cf} \beta$. As $\kappa \geq \beta^*$, there certainly is a Σ_1 injection of β into κ. In fact, more is true: there is a Σ_1 bijection of β onto κ (see [3], p. 15). Let $f: \beta \to \kappa$ be such a bijection. Let

$$<e, x, \sigma> \in \tilde{T} \longleftrightarrow x \in W_e^\sigma$$

and $T = f[\tilde{T}]$. Then $T \subseteq \kappa$, T is β-recursive and $\mathfrak{A} = <L_\kappa, \in, T>$ is admissible. Moreover, if $A \subseteq \kappa$, then A is β-r.e. if and only if A is $\mathfrak{A}$-r.e. (Σ_1 over $\mathfrak{A}$). Define $\leq_{\mathfrak{A}}$ analogously to $\leq_\beta$. Then these two reducibilities do not necessarily agree on subsets of κ. $A \subseteq \kappa$ is β-immune if

$$x \ \beta\text{-finite}, \ x \subseteq A \longrightarrow x \in L_\kappa ,$$

$$x \ \beta\text{-finite}, \ x \subseteq \kappa - A \longrightarrow x \in L_\kappa .$$

Then $\leq_{\mathfrak{A}}$ and $\leq_\beta$ do agree on β-immune sets. Maass shows that every $\mathfrak{A}$-r.e. degree has a β-immune $\mathfrak{A}$-r.e. representative. This gives an embedding E of the $\mathfrak{A}$-r.e. degrees 1-1 into the β-r.e. degrees.

Theorem 4 (Maass) (β weakly admissible). The range of E = the t.r.e. degrees = the recursive degrees. E (complete $\mathfrak{A}$-r.e. set) = $0^{1/2}$.

An application of admissibility theory to $\mathfrak{A}$ ([15] and [16]) yields:

Corollary. Any nonzero t.r.e degree is the join of two lesser t.r.e. degrees. If one t.r.e. degree is below another, then there is a t.r.e. degree in between.

Section 4. Strong Inadmissibility

This is the most challenging case for β-Recursion Theory, for the lack of admissibility is now so strong that many of the ideas from the admissible case become useless. The alternative is to employ deeper techniques from the Fine Structure of L as developed initially by Gödel [8] and more extensively by Jensen [9]. All of these techniques emanate from two basic lemmas due to Godel:

Lemma. For each limit ordinal β, there is a partial function $h: \omega \times S_\beta \to S_\beta$ which is Σ_1 over S_β such that for any Σ_1 formula $\varphi(x, p)$,

$$\langle S_\beta, \epsilon\rangle \models \exists x\varphi(x, p) \longrightarrow \exists i \in \omega\varphi(h(i, p), p) .$$

Proof. Recall the canonical Σ_1 well-ordering $<$ of S_β. Then if the ith Σ_1 formula is $\exists y\psi(x, p, y)$, define $h'(i, p) \simeq$ least (in the sense of $<$) pair $\langle x, y\rangle$ such that $\psi(x, p, y)$. Then $h(i, p)$ = first component of $h'(i, p)$. ⊣

The h above is called the canonical Σ_1 skolem function for S_β.

Transitive Collapse Lemma. If $X \prec_1 S_\beta$ (i.e., $X \subseteq S_\beta$ and any Σ_1 formula with parameters from X and a solution in S_β has a solution in X) then $\langle X, \epsilon\rangle$ is isomorphic to a unique $\langle S_\gamma, \epsilon\rangle$.

Using these two lemmas, we can now illustrate in a simple example how Fine Structure technique can be used to generalize to arbitrary β a result whose "recursion-theoretic" proof only succeeds for admissible β.

Proposition (Jensen). Suppose $A \subseteq \gamma < \beta^*$ and A is β-r.e. Then A is β-finite.

Proof Number 1, β admissible. If A is not β-finite, then it has a 1-1 β-recursive listing $f: \beta \to A$. But then $\beta^* \leq \sup A \leq \gamma < \beta^*$, contradiction.

Proof Number 2, β arbitrary. Let $p \in S_\beta$ be a parameter defining A as a set Σ_1 over S_β. Form X = Range h on $\omega \times (\gamma \cup \{p\})$, where h is from the Lemma. Then $X \prec_1 S_\beta$, so apply the Transitive Collapse Lemma to get $j: X \simeq S_\delta$. Let $g = j \circ h$.

Now g is Σ_1 over S_δ (simply transfer the Σ_1 definition for h over X to S_δ). Then so is g^{-1}. But if f uniformizes g^{-1}, f Σ_1 over S_δ, we see that f injects S_δ into $\omega \times (\gamma \cup \{p\})$; hence into γ. Since $\gamma < \beta^*$, we have proved that $\delta < \beta$.

But A is Σ_1 definable over S_δ, so $A \in S_\beta$. ⊣

Further ideas of Jensen, in particular an effectivized version of his ◇ principle, were used in [4] to establish:

Theorem 5. If β^* is regular with respect to β-recursive functions, then there are β-r.e. sets $A, B \subseteq \beta^*$ such that $A \not\leq_{w\beta} B$, $B \not\leq_{w\beta} A$.

This is the best solution to Post's Problem so far known in the strongly inadmissible case. This covers the case where $S_\beta \models$ "β^* is a successor cardinal".

Open Problem. Does the conclusion of Theorem 5 hold for arbitrary strongly inadmissible β?

Forcing can be used to achieve a stronger and more model-theoretic incomparability than that in Theorem 5. The following result will appear in [5]:

Theorem 6. Assume β^* is regular with respect to β-recursive function and $S_\beta \models$ "β^* is the largest cardinal." Then there are β-r.e. sets $A, B \subseteq \beta^*$

such that

$$A \text{ is not } \Delta_1 \text{ over } \langle S_\beta[B], \epsilon \rangle ,$$

$$B \text{ is not } \Delta_1 \text{ over } \langle S_\beta[A], \epsilon \rangle .$$

($S_\beta[A]$ is the βth level of the S[A]-hierarchy. This hierarchy is defined exactly as the S-hierarchy except the function $f(x) = A \cap x$ is added to the schemes for the rudimentary functions.)

We conclude this section by sketching the proof of a theorem which illustrates the use of Skolem Hulls and $\diamondsuit$ in β-Recursion Theory.

Theorem 7. There are β-r.e. sets A, B such that $A \not\leq_{f\beta} B$, $B \not\leq_{f\beta} A$.

The proof of this theorem is not uniform in the sense that it divides into cases depending on the nature of β. Thus, the sets A, B will be defined relative to the choice of a parameter $p \in S_\beta$.

Open Problem. Can Theorem 7 be made uniform in that the sets A, B have parameter-free Σ_1 definitions independent of β?

We believe that the answer is "yes." In fact, we

Conjecture. There are integers m, n such that for all limit ordinals β,

$$W_m^\beta \text{ is not } \Delta_1 \text{ over } \langle S_\beta[W_n^\beta], \epsilon \rangle ,$$

$$W_n^\beta \text{ is not } \Delta_1 \text{ over } \langle S_\beta[W_m^\beta], \epsilon \rangle ,$$

where W_n^β = the nth parameter-free β-r.e. set.

Before giving our proof sketch of Theorem 7, we make some preliminary definitions and remarks. In view of Theorem 3, it suffices to prove Theorem 7 in the strongly inadmissible case. Choose β-recursive functions $f_0: S_\beta \xrightarrow{1-1} \beta^*$ and $g_0: \mathrm{Rcf}\,\beta \to \beta$ such that g_0 is order-preserving, Range g_0 unbounded in β. Let $p_0' \in S_\beta$ be such that both f_0 and g_0 are Σ_1 over S_β in the parameter p_0'. Let $p_0 = \langle p_0', \beta^* \rangle$.

Let $h(i, p)$ be the canonical Σ_1 skolen function for S_β and for $\gamma < \beta^*$ define $H(\gamma) = \{h(i, <\gamma', p_0>) \mid i \in \omega, \gamma' < \gamma\}$. Thus $H(\gamma)$ is the "Σ_1 Skolem Hull of $\gamma \cup \{p_0\}$". Note that $\bigcup_{\gamma < \beta^*} H(\gamma) = S_\beta$. In our construction, $H(\gamma)$ consists of those reduction procedures e of priority γ. (Thus the construction is redundant in that each reduction procedure is assigned a final segment of different priorities.) Of special importance are those $\gamma < \beta^*$ such that $\gamma \notin H(\gamma)$.

<u>Claim.</u> Let $\kappa < \beta^*$ be a β-cardinal (i.e., $S_\beta \models$ "κ is a cardinal"). Let $\kappa^+ =$ the next β-cardinal. Then $\{\gamma < \kappa^+ \mid \gamma \notin H(\gamma)\}$ is closed, unbounded in κ^+.

<u>Proof.</u> See [4], Page 24. $\dashv$

Let $\mathcal{S} = \{\gamma < \beta^* \mid \gamma \notin H(\gamma)$ and γ is not a β-cardinal$\}$. The form of $\diamondsuit$ that we need (which more resembles $\diamondsuit'$ in fact) reads as follows:

$\diamondsuit_{\beta^*}$: There is a sequence $<D_\gamma \mid \gamma < \beta^*>$, Σ_1-definable without parameter over L_{β^*}, such that

1) $D_\gamma \subseteq$ Power Set of γ
2) $D_\gamma \in L_{\gamma^+}$ (where $\gamma^+ =$ least β-cardinal $> \gamma$)
3) If $A \subseteq \beta^*$ is β-r.e. with parameter p_0, then

$$\gamma \in \mathcal{S} \longrightarrow A \cap \gamma \in D_\gamma .$$

<u>Proof.</u> See [4], Page 25. $D_\gamma = \{x \subseteq \gamma \mid x \in L_{\hat{\gamma}}\}$ where $\hat{\gamma} =$ least δ such that γ is not a δ-cardinal. $\dashv$

We are now ready to outline the construction. We wish to satisfy the requirements:

$$R_e^A : \overline{B} \neq \{x \mid \exists \text{finite } y \subseteq \overline{A} \ (<x, y> \in W_e)\}$$
$$R_e^B : \overline{A} \neq \{x \mid \exists \text{finite } y \subseteq \overline{B} \ (<x, y> \in W_e)\}.$$

It is easy to see that satisfying these requirements for each $e \in S_\beta$ guarantees $A \not\leq_{f\beta} B$, $B \not\leq_{f\beta} A$. Here, $W_e = e^{th}$ β-r.e. set. Our method for attempting to satisfy R_e^A is to put some x into B at stage σ if some finite $y \subseteq \overline{A^\sigma}$ satisfies $<x, y> \in W_e^\sigma$. ($A^\sigma =$ part of A enumerated by stage σ, similarly for W_e^σ.)

If we can succeed in guaranteeing $y \subseteq \overline{A^{\sigma'}}$ for all $\sigma' > \sigma$, R_e^A will be satisfied.

These attempts at the above requirements conflict with each other. The solution is to order the requirements in a list, the requirements lower on the list having higher priority. In this construction, each requirement R_e^A (or R_e^B) is assigned all of the priorities $\gamma < \beta^*$ such that $e \in H(\gamma)$.

The construction proceeds in $R\,cf\,\beta$ steps. Recall the function $g_0: R\,cf\,\beta \to \beta$. At each stage σ we will use an approximation to the set $\mathcal{S}$ and the function H. Let $H^\sigma(\gamma) = \{h^\sigma(i, <\gamma', p_0>) \mid i \in \omega, \gamma' < \gamma\}$ where h^σ is the canonical Σ_1 skolem function for $S_{g_0(\sigma)}$. (If β is of the form $\beta' + \omega$, let $H^\sigma(\gamma) = H(\gamma) \cap g_0(\sigma)$.) Then define $\mathcal{S}_\sigma = \{\gamma < \beta^* \mid \gamma \in H^\sigma(\gamma)\}$.

<u>Stage σ.</u> Do the following for each $\gamma \in \mathcal{S}_\sigma$: Form all pairs $<R_e^A, z>$, $<R_e^B, z>$ where $z \in D_\gamma$, $e \in H^\sigma(\gamma)$. Order such pairs in a list and choose the least β-finite bijection j between $H^\sigma(\gamma)$ and γ. Given that earlier pairs have been considered, attack $<R_e^A, z>$ as follows: See if there is an $x \notin H^\sigma(\gamma)$, $x > \gamma$, x not being restrained from entering B by γ, and a finite $y \subseteq \overline{A^\sigma}$ such that $<x, y> \in W_e^\sigma$ and $y \cap j^{-1}[z] = \emptyset$. Then for the least such pair $<x, y>$, put x into B and have γ restrain the members of y from entering A. The pairs $<R_e^B, z>$ are handled similarly. This ends the construction.

The idea, then, is that the members of D_γ (via the bijection $j: H^\sigma(\gamma) \to \gamma$) provide "guesses" at $A \cap H^\sigma(\gamma)$, $B \cap H^\sigma(\gamma)$. Of course, since the parameter p_0 defines the entire construction, $\diamondsuit_{\beta^*}$ implies that $j[A \cap H^\sigma(\gamma)]$, $j[B \cap H^\sigma(\gamma)] \in D_\gamma$ if $\gamma \in \mathcal{S}$. So for $\gamma \in \mathcal{S}$, one of the "guesses" is correct. Then these guesses are each used to search for an x and y which attempt to satisfy R_e^A (or R_e^B).

Now each pair $<R_e^A, z>$, $e \in H(\gamma)$, $z \in D_\gamma$ is attacked at most once at each stage of the construction; thus, any x put into A or B by γ and any y restrained from intersecting A or B by γ must belong to $H(\gamma+1)$.

Lemma. Suppose $y \in H^\sigma(\gamma+1) - H^\sigma(\gamma)$ and $\gamma \in \mathcal{S}$. Then $y \notin H(\gamma)$.

Proof. Otherwise, let $y \in H^{\sigma'}(\gamma')$, $\gamma' < \gamma$, σ' least. Assume that γ' and γ have the same β-cardinality κ. Let δ_0 be the least $\delta < \kappa^+$ so that for some $\tau < \sigma'$, $y \in H^\tau(\delta) - H^\tau(\gamma')$. Then $\delta \in H^{\sigma'}(\gamma'+1)$ and $\delta \geq \gamma$. But as $\kappa \leq \gamma'$, $\delta \subseteq H^{\sigma'}(\gamma'+1)$ and so $\gamma \in H^{\sigma'}(\gamma'+1)$. This contradicts $\gamma \in \mathcal{S}$. $\dashv$

Now any y restrained by γ at stage σ must belong to $H^\sigma(\gamma+1) - H^\sigma(\gamma)$. Thus, by the Lemma, if $\gamma \in \mathcal{S}$, we have $y \notin H(\gamma)$. But then as each $\gamma' < \gamma$ only puts members of $H(\gamma)$ into A or B, no member of y can ever be put into A or B. If in addition the attempt associated with y used a correct guess z for $A \cap H^\sigma(\gamma)$ (or $B \cap H^\sigma(\gamma)$), then this attempt will succeed and the corresponding requirement R_e^A (or R_e^B) will be satisfied.

Lastly, note that no $\gamma \in \mathcal{S}$ can ever be put into A or B, by construction Thus we may argue for $B \not\leq_{f\beta} A$ as follows ($A \leq_{f\beta} B$ is similar): If $\overline{B} = \{x \mid \exists \text{ finite } y \subseteq \overline{A}, <x, y> \in W_e\}$, then choose $\gamma \in \mathcal{S}$ and σ so that if γ' = least member of $\mathcal{S}$ greater than γ,

1) $\exists$ finite $y \subseteq \overline{A^\sigma}$, $<\gamma', y> \in W_e^\sigma$,

2) $y \cap (A \cap H^\sigma(\gamma)) = \emptyset$,

3) $e \in H^\sigma(\gamma)$.

Such a γ and σ exist since $\mathcal{S} \subseteq \overline{B}$. Then there must be an attempt made at this stage or an earlier stage for the pair $<R_e^A, j[A \cap H^\sigma(\gamma)]>$. By earlier Remarks, this attempt will succeed. End of proof sketch.

Section 5. Minimal α-Degrees Revisited

We return now to the original problem which motivated our study. Have we learned anything new concerning minimal α-degrees through the study of β-Recursion Theory? The following result gives an affirmative answer:

Theorem 8. If $\alpha^* = \alpha$ and ρ_2^α is a successor α-cardinal, then there is a minimal α-degree which is α-r.e. in 0'.

The proof, which applies the techniques of β-Recursion Theory to the structure $\langle L_\alpha, \epsilon, C\rangle$ (C a complete α-r.e. set), will appear in [6].

References

1. Devlin, Keith J. Aspects of Constructibility, Springer Lecture Notes in Math. #354, 1973.

2. Friedberg, R. Two r.e. sets of incomparable degrees of unsolvability, Proceedings N.A.S. 43, 1957.

3. Friedman, Sy D. β-Recursion Theory, to appear.

4. Friedman, Sy D. Post's Problem Without Admissibility, to appear.

5. Friedman, Sy D. Forcing and the Fine Structure of L, in preparation.

6. Friedman, Sy D. On Minimal α-Degrees, in preparation.

7. Friedman, Sy D. and Sacks, Gerald E. Inadmissible Recursion Theory, Bulletin AMS Vol. 83 No. 2 1977.

8. Gödel, K. Consistency Proof for the Generalized Continuum Hypothesis, Proceedings N.A.S. 25, 1939.

9. Jensen, Ronald B. The Fine Structure of the Constructible Hierarchy, Annals of Math. Logic 4, 1972.

10. Maass, Wolfgang. Inadmissibility, Tamely RE Sets and the Admissible Collapse, to appear.

11. Maass, Wolfgang. On Minimal Pairs and Minimal Degrees in Higher Recursion Theory, to appear.

12. Sacks, Gerald E. Post's Problem, Admissible Ordinals and Regularity, Transactions AMS 124, 1966.

13. Sacks, Gerald E. and Simpson, S. G. The α-Finite Injury Method Annals of Math. Logic 4, 1972.

14. Shore, Richard A. Minimal α-Degrees, Annals of Math. Logic 4, 1972.

15. Shore, Richard A. Splitting an α-RE Set, Transactions AMS 204, 1975.

16. Shore, Richard A. The R. E. α-Degrees are Dense, Annals of Math. Logic 9, 1976.

17. Simpson, Steve G. Degree Theory on Admissible Ordinals, in <u>Generalized Recursion Theory,</u> edited by Fenstad, Hinman 1974.

E. Fenstad, R.O. Gandy, G.E. Sacks (Eds.)
NERALIZED RECURSION THEORY II
North-Holland Publishing Company (1978)

Negative Solutions to Post's Problem, I

Sy D. Friedman
Department of Mathematics
University of Chicago
Chicago, Illinois 60637

0. Introduction

For background in β-Recursion Theory, see [2] and our earlier paper in his volume. In [2], [3] the following version of Post's Problem is solved for a arge class of ordinals β:

(*) There are β-r.e. sets A, B s.t. $A \not\leq_{w\beta} B$, $B \not\leq_{w\beta} A$.

It was conjectured in [2] that (*) holds for arbitrary limit ordinals β. It is the purpose of this note to exhibit a failure of (*) for some primitive-recursively closed β. The results of [2], [3] imply that such a β must be strongly inadmissible and for such a β, $\beta^* = \Sigma_1$ projectum of β must be singular with respect to β-recursive functions.

Thus the priority method can be applied to many but not all limit ordinals. We are not at present able to determine exactly for which ordinals (*) holds, but make the

Conjecture (*) holds if and only if either β is weakly admissible or β^* is regular with respect to β-recursive functions.

Thus, we feel that the positive results of [2], [3] are best possible. A conceptual explanation for our Conjecture is as follows: Define $K \subseteq \beta$ to be β^*-finite if K is β-finite of β-cardinality less than β^*. Then our Conjecture says that (*) holds if and only if β cannot be written as the β-recursive union of β^*-finitely many β^*-finite sets.

A key ingredient in our proof is a use of stationary sets and Fodor's Theorem much in the way Silver used them in his work ([7]) on the Generalized Continuum Hypothesis at singular cardinals of uncountable cofinality. We have found Prikry's proof ([5]) of Silver's Theorem extremely useful.

§1. Statement of Theorem and Preliminaries

Fix $\beta = \omega^{\text{th}}$ primitive-recursively closed ordinal greater than $\aleph^L_{\omega_1}$. Let $f: \omega \to \beta$ be defined by $f(n) = n^{\text{th}}$ primitive-recursively closed ordinal greater than $\aleph^L_{\omega_1}$. Then f is β-recursive, so β is strongly inadmissible an $\Sigma_1 \operatorname{cf} \beta = \omega$. It now follows that $\beta^* = \aleph^L_{\omega_1}$ and thus β^* is singular with resp respect to the β-finite function $d: \omega^L_1 \to \beta^*$ given by $d(\alpha) = \aleph^L_\alpha$. Fix $C = \{<e,x> \mid \{e\}(x)\downarrow\}$, a complete β-r.e. set.

Theorem. If A is β-r.e. then either $A =_\beta \emptyset$ or $C \leq_{w\beta} A$.

As B β-r.e. implies that $B \leq_{f\beta} C$, the Theorem shows that (*) fails for β. Moreover, any β-recursive set is β-reducible to C using only finite neighborhood conditions on C and thus β-reducible to any set A s.t. $C \leq_{w\beta} A$. So if $\underset{\sim}{d}$ is a β-r.e. degree then $\underset{\sim}{0} < \underset{\sim}{d} \longrightarrow \underset{\sim}{0}^{1/2} \leq \underset{\sim}{d}$. In a future paper we shall exhibit a primitive-recursively closed ordinal where $\underset{\sim}{0}, \underset{\sim}{0}^{1/2}$ and $\underset{\sim}{0}'$ are the only β-r.e. degrees.

We end this section by reducing our Theorem to a lemma. This lemma has as its forerunner a theorem of Simpson ([8], page 71) who established it when $\beta = (\aleph^L_\omega)^+$, the first admissible greater than $\aleph^L_\omega$:

Main Lemma. If $A \subseteq \beta^*$ is β-r.e. then either A is β-finite or $C \leq_{w\beta} A$.

Proof of Theorem from Lemma. Let $A \subseteq \beta$ be β-r.e. If $A \cap f(n)$ is not β-finite for some n, then an application of the Lemma shows that $C \leq_{w\beta} A \cap f(n) \leq_{\beta} A$ so we are done. Otherwise, let $K: L_{\beta} \xrightarrow{1-1} \beta^{*}$ be β-recursive and define $\ell(n) = K(A \cap f(n))$. Then $\ell : \omega \to \aleph^{L}_{\omega_1}$ and as ℓ is constructible, ℓ is β-finite. But then $K^{-1}\ell$ is a β-recursive function listing $A \cap f(0), A \cap f(1), \ldots$. From this it is easily seen that $A =_{\beta} \emptyset$. ⊣

§ 2. Proof of Main Lemma.

Let $A \subseteq \beta^{*}$ be β-r.e. There is an injection $L_{\beta} \xrightarrow{1-1} \beta^{*}$ which is Σ_1 over L_{β} with parameter $\aleph^{L}_{\omega_1}$. This implies that there is a complete β-r.e. set $C^{*} \subseteq \beta^{*}$ which is Σ_1 over L_{β} with parameter $\aleph^{L}_{\omega_1}$ and that A is Σ_1 definable over L_{β} with parameter of the form $p = \langle \aleph^{L}_{\omega_1}, p_0 \rangle$, $p_0 \in L_{\aleph^{L}_{\omega_1}}$. We will show that either A is β-finite or $C^{*} \leq_{w\beta} A$.

Let $h(i, x)$ be a Σ^{p}_{1} Skolem Function for L_{β}; i.e., h is a partial function from $\omega \times L_{\beta}$ into L_{β}, h is Σ_1 over L_{β} with parameter p and if $\varphi(x, y)$ is a Σ_1 formula with parameter p, then for some i and all $x \in L_{\beta}$,

$$L_{\beta} \models \exists y \varphi(x, y) \longrightarrow h(i, x) \text{ is defined and } L_{\beta} \models \varphi(x, h(i, x)) .$$

Now fix $\lambda_0 < \omega^{L}_{1}$ such that $p_0 \in L_{\lambda_0}$. If $\lambda_0 \leq \lambda < \omega^{L}_{1}$ then $h[\omega \times \aleph^{L}_{\lambda}]$ is a Σ^{p}_{1}-elementary substructure of L_{β} and so $C^{*} \cap h[\omega \times \aleph^{L}_{\lambda}]$ is Σ^{p}_{1} Definable over $h[\omega \times \aleph^{L}_{\lambda}]$. The function h has a natural approximation $h_n = (h)^{L_{f(n)}}$ and then h_n is a Σ^{p}_{1} Skolem Function for $L_{f(n)}$.

Now for each $\lambda \geq \lambda_0$, $n < \omega$, $h_n[\omega \times \aleph^{L}_{\lambda}] \cap \aleph^{L}_{\lambda+1}$ is an ordinal. Call it S^{n}_{λ}. Then $\aleph^{L}_{\lambda} < S^{1}_{\lambda} < S^{2}_{\lambda} < \ldots$ and if $S_{\lambda} = \bigcup_n S^{n}_{\lambda}$ then $S_{\lambda} = h[\omega \times \aleph^{L}_{\lambda}] \cap \aleph^{L}_{\lambda+1}$.

Lemma 1. Let $X \subseteq \omega^{L}_{1}$ be unbounded, $Y = \{S_{\lambda} \mid \lambda \in X\}$. Then $C^{*} \leq_{f\beta} Y$.

Proof. Let $L_{\hat{S}_\lambda}$ be the transitive collapse of $h[\omega \times \aleph_\lambda^L]$ for $\lambda \geq \lambda_0$. Let $p_\lambda = \langle (\aleph_{\omega_1})^{L_{\hat{S}_\lambda}}, p_0 \rangle$. If $C^* = \{x \in L_\beta \mid L_\beta \models \varphi(x, p)\}$ where φ is Σ_1, then $C^* \cap \aleph_\lambda^L = \{x < \aleph_\lambda^L \mid L_{\hat{S}_\lambda} \models \varphi(x, p_\lambda)\}$. So it suffices to show that $\{\hat{S}_\lambda \mid \lambda \in X, \lambda \geq \lambda_0\} \leq_{f\beta} Y$. But $S_\lambda = (\aleph_{\lambda+1})^{L_{\hat{S}_\lambda}}$ so $L_{\hat{S}_\lambda} \models S_\lambda$ is regular while $(S_\lambda)^* = \aleph_\lambda^L$ so $L_{\hat{S}_\lambda+1} \models S_\lambda$ is singular. Thus $\hat{S}_\lambda$ can be found β-recursively from S_λ. ⊣

The idea of the proof is to compare the "growth rate" of A with that of the sequence $\{S_\lambda \mid \lambda \geq \lambda_0\}$. The "growth rate" of A is measured by $f_A : \omega_1 \to \aleph_{\omega_1}^L$ defined by:

$$f_A(\lambda) = \mu\gamma[A \cap \aleph_\lambda^L \text{ is definable over } L_\gamma].$$

Lemma 2. $f_A(\lambda) \leq \hat{S}_\lambda$ for all $\lambda \geq \lambda_0$.

Proof. It is enough to show that $A \cap \aleph_\lambda^L$ is definable over $L_{\hat{S}_\lambda}$. But as in the proof of Lemma 1, if $A = \{x \mid L_\beta \models \varphi(x, p)\}$, $\varphi\ \Sigma_1$, then:

$$A \cap \aleph_\lambda^L = \{x < \aleph_\lambda^L \mid L_{\hat{S}_\lambda} \models \varphi(x, p_\lambda)\}, \ \lambda \geq \lambda_0. \quad ⊣$$

Now there are two cases. Either $f_A(\lambda) \geq S_\lambda$ for unboundedly many $\lambda < \omega_1^L$, or $f_A(\lambda) < S_\lambda$ for sufficiently large $\lambda < \omega_1^L$. In the first case we show that $C^* \leq_{w\beta} A$. In the second case, A is β-finite.

Lemma 3. If $f_A(\lambda) \geq S_\lambda$ for unboundedly many $\lambda < \omega_1^L$ then $C^* \leq_{w\beta} A$

Proof. Let $X = \{\lambda \mid f_A(\lambda) \geq S_\lambda\} \subseteq \omega_1^L$. For $\lambda \in X$, $L_{f_A(\lambda)} \models S_\lambda = \aleph_{\lambda+1}$. Since X is β-finite, this shows that $Y = \{S_\lambda \mid \lambda \in X\} \leq_{w\beta} A$ and so we are done by Lemma 1. ⊣

Lemma 4. If $f_A(\lambda) < S_\lambda$ for sufficiently large $\lambda < \omega_1^L$, then A is β-finite.

Proof. Suppose $f_A(\lambda) < S_\lambda$ for $\lambda \geq \lambda_1 \geq \lambda_0$. Define:

$$g(\lambda) = \mu n [f_A(\lambda) < S_\lambda^n], \quad \text{for } \lambda \geq \lambda_1.$$

Then for some fixed n_0,

$$X = \{\lambda \mid f_A(\lambda) < S_\lambda^{n_0}\}$$

is stationary in ω_1^L (with respect to constructible closed, unbounded sets).

At this point, the proof proceeds much as in [5].

For $\lambda \in X$, let $j_\lambda : L_{S_\lambda^{n_0}} \xrightarrow{1-1} \aleph_\lambda^L$ be the $<_L$-least injection. We assume that $\lambda \in X \to \lambda$ a limit ordinal. Then

$$\lambda \in X \to j_\lambda(A \cap \aleph_\lambda^L) < \aleph_\lambda^L \to j_\lambda(A \cap \aleph_\lambda^L) < \aleph_{\lambda'}^L \quad \text{some } \lambda' < \lambda.$$

We therefore have a regressive function j on the stationary set X defined by:

$$j(\lambda) = \mu\lambda'[j_\lambda(A \cap \aleph_\lambda^L) < \aleph_{\lambda'}^L].$$

By Fodor's Theorem, there is an unbounded $X_0 \subseteq X$ and $\lambda_2 < \omega_1^L$ such that

$$\lambda \in X_0 \to j(\lambda) < \lambda_2 \to j_\lambda(A \cap \aleph_\lambda^L) < \aleph_{\lambda_2}^L.$$

Now the function $f : X_0 \to \aleph_{\lambda_2}^L$ defined by:

$$f(\lambda) = j_\lambda(A \cap \aleph_\lambda^L), \quad \lambda \in X_0$$

is β-finite. Moreover, the sequence $\{j_\lambda\}_{\lambda \in X}$ is definable over $L_{f(n_0)}$ and hence β-finite. But we have:

$$x \in A \iff \exists \lambda \in X_0 [x \in j_\lambda^{-1}(f(\lambda))]$$

so A itself is β-finite. $\dashv$

§3. Extensions

The techniques used here can be used to obtain further information abou the α- and β-degrees. The following results will appear in future papers:

1) Assume $V = L$ and let $\alpha = \aleph_{\omega_1}$. Then the α-degrees $\geq 0'$ are well-ordered by $\leq_\alpha$ with successor given by α-jump. The α-jump operation on α-degrees is definable in terms of $\leq_\alpha$.

2) Assume $V = L$ and let $\beta = \omega_1^{\underline{st}}$ p. r. closed ordinal greater than $\aleph_{\omega_1}$. Then the β-degrees are well-ordered with successor given by the $1/2$-β-jump. The only β-r. e. degrees are $\underset{\sim}{0}, \underset{\sim}{0}^{1/2}, \underset{\sim}{0}'$.

3) Let α = first stable ordinal greater than $\aleph_{\omega_1}$ and $\beta = \aleph_{\omega_1}{}^{\underline{st}}$ p. r. closed ordinal greater than α. Then there are incomparable β-r. e. degrees if and only if there are incomparable α-degrees α-r. e. in and above $0'$ if and only if $0^{\#}$ exists.

4) If $0^{\dagger}$ does not exist, then the Turing Degrees and the $\beth_{\omega_1}$-degrees are not elementarily equivalent as partial-orderings. If κ is a Strongl Compact Cardinal, then the Turing Degrees and the $\beth_{\omega_1}(\kappa)$-degrees are not elementarily equivalent as partial-orderings.

References

1. Friedman, Sy D., An Introduction to β-Recursion Theory, this volume.

2. ______________, β-Recursion Theory, to appear.

3. ______________, Post's Problem without Admissibility, to appear.

4. ______________, Negative Solutions to Post's Problem, II, in preparation.

5. Prikry, Karel, On a Theorem of Silver, handwritten notes

6. Sacks, Gerald E. and Simpson, Stephen G., The α-Finite Injury Method, Annals of Mathematical Logic 4, 1972

7. Silver, Jack, On the Singular Cardinals Problem, Proceedings of the International Congress of Mathematicians in Vancouver, 1974.

8. Simpson, Stephen G., Thesis, M.I.T., 1971

J.E. Fenstad, R.O. Gandy, G.E. Sacks (Eds.)
GENERALIZED RECURSION THEORY II

The intrinsic recursion theory on the countable or continuous functionals

J.M.E. Hyland

§0. Introduction. The collection $\mathcal{C} = \{C_\sigma \mid \sigma \text{ is a type symbol}\}$ of countable (or continuous) functionals was first considered in the pioneering papers of Kleene [9] and Kreisel [10]. Already in these papers, an intrinsic notion of recursive countable functional was defined. Since that time smoother approaches to the subject have been made by Ershov [2], [3], Feferman [4] and Hyland [7], [8] and a natural "intrinsic recursion theory" has begun to be studied. The purpose of this paper is to describe the present state of knowledge and the major unsolved problems in this area.

What I call the intrinsic recursion theory on the countable functionals is an attractive area to study for the following reasons.

(i) It arises naturally out of the various definitions (Ershov [2], [3], Hyland [7], [8].) of $\mathcal{C}$; and $\mathcal{C}$ is itself the natural collection of continuous objects of higher type defined over the natural numbers (Hyland [8]).

(ii) It has pleasing applications to questions of constructivity (Kreisel [10].

(iii) As is sensible for recursion on objects given by a countable amount of information, every countable functional is recursive in a function (from natural numbers to natural numbers).

(iv) Elementary arguments about weaker notions of recursion turned out to be arguments about the intrinsic recursion theory (see §2).

(v) It provides an instructive challenge to generalized recursion theory as that subject has been developed over the last ten years (see especially §4); note that this challenge is already implicit in Kreisel [10].

Of course $\mathcal{C}$ is also a suitable domain for Kleene's generalized recursion theory via the schemes S1-S9, and a survey of that area, <u>computability</u> on $\mathcal{C}$, is in Gandy and Hyland [5]. Indeed, despite (iv) above, it has proved easier to obtain results about computability. In particular, since the writing of [5], work on 1-sections (initiated by Wainer) and on 2-envelopes (initiated by Bergstra) has been carried to a very satisfactory conclusion by Norman [11]. This work directly contradicts what seemed plausible at the time I wrote my thesis [7]. I discuss the relation between it and what we know about the intrinsic recursion theory in §3.

Work on computability is indirectly related to the main outstanding question concerning the countable functionals: whether the intrinsic recursion theory can be characterized by one of the approaches of generalized recursion theory (most plausibly by inductive definitions see Feferman [4]). This question enables one to look critically at generalized recursion theory. A negative answer, indicating the limitations of the "generalizations" of ordinary recursion theory, would have considerable philosophical interest. However as Feferman [4] observed, it is hard to imagine how to obtain a proof of this kind of impossibility. In §4, I give a vague outline of what appears to be the only natural possibility of an inductive definability approach to the intrinsic recursion theory on the countable functionals. My view is that if this fails, there can be no sensible positive answer to the fundamental question.

§1. <u>Basic definitions.</u> Various definitions of <u>partial recursive</u> functional on $\mathcal{C}$ have been proposed. That given here is from Hyland [7].

For countable functionals $G_1,\ldots,G_n$ and F, $\{e\}_c(G_1,\ldots,G_n) = F$ iff for all associates $\alpha_1,\ldots,\alpha_n$ for $G_1,\ldots,G_n$ $\lambda x.\{e\}(\alpha_1,\ldots,\alpha_n,x)$ is an associate for F.

The partial functionals $\{e\}_c$, are the partial recursive countable functionals. That this definition is equivalent to others in the literature (Ershov [3], Feferman [4]) follows readily from work in Hyland [8]. (In this definition I have omitted codings for the types of the arguments and prospective value of the functional, but never mind). A clear discussion of the relation of this notion to the notion of partial computable functional (i.e. partial S1-S9 recursive) can be found in Feferman [4].

The partial recursive countable functionals give rise to a notion of countably recursive in: F is countably recursive in G iff there is a partial recursive countable functional mapping G to F (cf. Gandy and Hyland [5] §3.8). This gives rise to the countable degrees as follows. The degree of a countable functional F,

$$\deg(F) = \{G \mid G \text{ is countably recursive in } F\}.$$

These form the countable degrees $\mathcal{D}_1$, ordered by inclusion. A degree is of type n iff it is of form deg(F) with F of type n; $\mathcal{D}_n$ denotes the collection of degrees of type n. Clearly,

$$\mathcal{D} = \bigcup \{\mathcal{D}_n \mid n \geq 1\}.$$

A functional of type n is said to be irreducible (with respect to countable recursion) iff it does not have the same degree as a functional of type less than n.

We define the countable k-section of F to be

$$\text{ct-k-sc}(F) = \{G \mid G \text{ of type k and countably recursive in } F\}.$$

Partial recursive countable functionals with numerical values define a notion of semi-recursive (s.r) or recursively enumerable set. The subsets of C_k s.r. in F are those of the form

$$\{G \mid G \text{ of type k and } \{e\}_c\{F,G\} = 0\},$$

for some index e. The countable (k+1)-envelope of F is defined to be

$$\text{ct-(k+1)-env}(F) = \{A \subseteq C_k \mid A \text{ s.r. in } F\}.$$

§2. The countable degrees. In this section we survey what little is known about the countable degrees. Many open problems remain and we discuss a few of these.

The proofs of some simple results can be easily derived from the literature.

Theorem 2.1. There is a type 1 degree minimal amongst all the countable degrees.

Proof:- By an easy adaptation of Spector [12].

Corollary 2.2. (answering a question in Hinman [6]) $\mathcal{D}$ is not dense in $\mathcal{D}_1$; whence the type 2 Kleene degrees of countable functionals are not dense in $\mathcal{D}_1$ (ordinary degrees).

(2.1) has some content beyond Spector's result in view of the next result.

Theorem 2.3. There are irreducible type 2 objects.

Proof:- It is this result which is really proved in Hinman [6].

Remark (2.3) can be strengthened in various ways:

1) (Folklore?) One can obtain an irreducible F whose Kleene one section is the recursive functions.

2) With difficulty one can obtain an irreducible F with ct-1-sc(F) = the recursive functions and (Harrington) with F not the continuous extension of an effective operation.

3) Since Hinman's proof involves a simple spoiling argument, one can modify elementary degree theory constructions to get irreducible type 2 objects in place of type 1 objects.

A k-section which is not the k-section of any type k object is called topless. Bergstra [1] showed the existence of topless sections for Kleene computations (S1-S9); a different approach to this result is provided by Norman [11]. Bergstra's method adapts with difficulty to countable recursion; we only get a result at the very first level.

Theorem 2.4. There exists F of type 2 such that ct-1-sc(F) is topless.

A basic problem in the theory of the countable degrees would be to determine the relation between the familiar structure $\mathcal{D}_1$ of the ordinary degrees, and the $\mathcal{D}_n$'s and $\mathcal{D}$ which include $\mathcal{D}_1$. The difficulties are indicated by the following long-standing question (which also shows the limitations of 3) of the above remark).

OPEN PROBLEM Are there minimal degrees which are irreducible of type 2?

Though by (2.3), we know that $\mathcal{D}_2$ properly includes $\mathcal{D}_1$, nothing is known of the corresponding result for higher levels of the type structure.

OPEN PROBLEM For $n \geq 2$, does $\mathcal{D}_{n+1}$ properly include $\mathcal{D}_n$?

Remarks 1) A formulation of this question for Kleene's computations on $\mathcal{C}$ was answered (by Bergstra [1]) in the affirmative. For computability, the relation between the degree structures at different types is complicated by the existence of elements (e.g. the fan functional - see Gandy and Hyland [5]) at type 3, which are not 1-obtainable (i.e. computable from a function).

2) It seems most likely that the answer is "yes". For the C_n's become more complicated as n increases in the sense of the

Proposition If $m > n$, then there is no continuous onto map from C_n to C_m.

This is essentially Cantor's Theorem (together with type-changing manipulations). As observed by Norman [11], the proposition can also be obtained from (3.2).

3) Results on countable sections, or extensions of (2.4) might be contributions to the solution of this problem. But neither possibility seems very easy.

§3. Associates and envelopes. The aim of thissection is to determine what are the countable 2-envelopes of countable functionals. The result will be in marked contrast to what is known about Kleene 2-

envelopes (Norman [11]). It is clear from the definitions that we will need to know something about associates.

Write Ass(F) for the set of associates (the details of the definition will be of no significance) for a countable functional F, and set

$$\mathrm{Ass}_n = \bigcup \{\mathrm{Ass}(F) \mid F \in C_n\}.$$

It seems to be a matter of folklore that for $n \geq 2$, Ass_n is a complete Π^1_{n-1} set of functions, but as I do not know a published proof, I sketch the simple basis for this result here.

<u>Lemma 3.1</u>. The sets of the form

$$\{\vec{x} \mid (\forall \alpha)(\alpha \in \mathrm{Ass}_n \longrightarrow P(\vec{x},\alpha))\},$$

where P is Σ^0_1, $\vec{x}$ denotes a sequence of variables of types 0 and 1 and $n \geq 1$, are closed under universal qualification (effectively in an index for P).

<u>Proof</u>:- This follows immediately from the fact that there exist recursive (indeed elementary) maps from Ass_n <u>onto</u> $\mathbb{N}^{\mathbb{N}} \times \mathrm{Ass}_n$. (Observe that for any σ, C_σ is isomorphic to $C_\sigma \times C_\sigma$. So for $n \geq 1$ C_n is isomorphic to $C_n \times C_n$ which easily maps onto $C_1 \times C_n$. The map so constructed will be onto at the level of associates).

<u>Proposition 3.2</u>. For $n \geq 2$, Ass_n is a complete Π^1_{n-1} set.

Proof:- The case $n = 2$, the basis for an induction, is easy. Suppose result true for n. Then an arbitrary Σ^1_n set is of the form

$$\{\vec{x} \mid (\exists \alpha)\ \Phi(\vec{x},\alpha) \in \mathrm{Ass}_n\},$$

for some recursive functional Φ, i.e. of the form

$$\{\vec{x} \mid (\exists \beta)(\beta \in \mathrm{Ass}_n \ \&\ (\exists \alpha)(\beta = \Phi(\vec{x},\alpha)))\}.$$

Thus an arbitraty Π^1_n set A is of the form

$$\{\vec{x} \mid (\forall \alpha)(\forall \beta)(\beta \in \mathrm{Ass}_n \longrightarrow \beta \neq \Phi(\vec{x},\alpha))\},$$

so using (3.1) of the form

$$\{\vec{x} \mid (\forall \beta)(\beta \in \mathrm{Ass}_n \longrightarrow (\exists y)T|e, \overline{\vec{x}(y)}, \overline{\beta(y)})\}$$

for a suitable T-predicate.

Now define Ψ by

$$\Psi(\vec{x})(u) = \begin{cases} 0 & \text{if not } (\exists v \subseteq u)T(e, \overline{\vec{x}(lh(v)}, v) \\ 1 & \text{otherwise.} \end{cases}$$

Then $\vec{x} \in A$ iff $\Psi(\vec{x}) \in \mathrm{Ass}_{n+1}$. Since Ass_{n+1} is clearly Π^1_n, we have completed the induction step.

There are some immediate corollaries of the above result and proof. Let ${}^n0$ be the everywhere zero functional of type n.

<u>Corollary 3.3</u>. For $n \geq 2$, $\mathrm{Ass}({}^n0)$ is a complete Π^1_{n-1} set.

Proof:- By the above argument, $\vec{x} \in A$ iff $\Psi(\vec{x}) \in \mathrm{Ass}({}^n0)$.

<u>Corollary 3.4</u>. For $n \geq 2$, ct-2-env$({}^n0) = \Pi^1_n$.

Proof:- ct-2-env$({}^n0)$ consists of all sets of the form

$$\{\vec{x} \mid (\forall \alpha)(\alpha \in \mathrm{Ass}({}^n0) \longrightarrow P(\vec{x},\alpha))\}, \text{ where } P \text{ is } \Sigma^0_1.$$

So clearly ct-2-env$({}^n0) \subseteq \Pi^1_n$ and is closed under substitution of recursive functionals. It remains to show that ct-2-env$({}^n0)$ contains a complete Π^1_n set. But this follows by the first part of the argument for (3.2). (The existence of a recursive onto map from $\mathrm{Ass}({}^n0)$ to $\mathbb{N}^{\mathbb{N}} \times \mathrm{Ass}({}^n0)$ is much as in the proof of (3.1) - though it lacks the structural motivation of that result).

The generalizations of (3.3) and (3.4) to arbitrary F (of type $n \geq 2$), involves ugly coding problems; my proofs rely on equivalences from Hyland [8] so I do not give them here, but simply state the results. Suppose F is of type n, $n \geq 2$ and let h_F give the value of F on some recursive dense sequence in C_{n-1}.

<u>Theorem 3.5</u>. (a) Ass(F) is a complete $\Pi^1_{n-1}(h_F)$ set

(b) ct-2-env(F) $= \Pi^1_n(h_F)$.

<u>Remarks</u> 1) (3.5)(a) is proved in full detail from a completely different point of view in Norman [11]; (3.5)(b) could also be obtained using his methods.

2) (3.5)(b) should be contrasted with the result of Norman [11], that in the sense of Kleene (S1-S9) recursion, 2-env(F) = $\Pi^1_{n-2}(h_F)$ (F of type 3 or more).

(3) The most significant feature of Norman [11] is his ability to handle 1-sections (see his Theorem 3). At the moment there is nothing corresponding for countable 1-sections (cf. Remark 3 of §2).

§4. Definitions by recursion on the inductive definition of C_2. The outstanding question concerning the countable functionals is whether one can obtain the intrinsic recursion theory by applying the usual ideas of generalized recursion theory. This problem was raised in embryonic form by Kreisel [10]. It was considered in Hyland [7], but the tentatively negative conclusion reached there was based in part on conjectures which have since been disproved. It is discussed in detail in Feferman [4], where a positive answer to the corresponding question for the partial continuous functionals is indicated. At first sight the problem seems to be one of finding the "right structure" to put on $\mathcal{C}$ (cf. the final section of Feferman [4]). However $\mathcal{C}$ doesn't seem to have any structure in the sense of model theory apart from the usual structure on C_o, the natural numbers, and the type structure (essentially, evaluation and λ-abstraction); and this much gives rise to Kleene's computations (S1-S9) on $\mathcal{C}$. One would appear to search in vain for further natural inductive schemeta, while there must exist suitable unnatural ones since the partial recursive countable functionals can clearly be enumerated. This is the impasse reached by Feferman [4]. In this section, I sketch the lines of what seems to be the only plausible way out.

My suggestion is based on two observations.

1) Once we have C_o, C_1 and C_2, the rest of $\mathcal{C}$ is determined by demanding closure under explicit definition (i.e. avoiding ${}^2E$).

2) (A point made to me by Gandy). The natural numbers are inductively defined and thereby carry a good recursion theory; but C_2 is also inductively defined (Brouwer, Konig) and should carry a good

recursion theory in virtue of this fact.
Since there is no problem with the recursion theory on C_o and C_1, it would seem that the inductive definition of C_2 is the one element of the structure of $\mathcal{C}$ missing from what is described above. If we can add an appropriate process of definition by recursion over the inductive definition of C_2, we ought to get <u>the</u> natural recursion theory on $\mathcal{C}$ from the point of view of computation schemes or inductive definitions. If this does not coincide with the intrinsic recursion theory, one could conclude that the intrinsic recursion theory on the countable functionals falls outside the scope of the main developments of generalized recursion theory.

For F in C_2 and u a sequence (number) define F_u by

$$F_u(\beta) = F(u*\beta),$$

where * denotes concatenation. Both the fan functional and Gandy's functional Γ of [5], can be defined as functionals Δ for appropriate (primitive recursive) $\mathcal{F}$, in the following simple way:

$$\Delta(\lambda\beta,k,\alpha) = \alpha(k)$$

$$\Delta(F,\alpha) = \mathcal{F}(F,\lambda u \neq \langle\ \rangle.\Delta(F_u,\alpha),\alpha).$$

But the fact that such a definition uniquely determines Δ, depends on the fact that for the corresponding $\mathcal{F}$,

$$\alpha(k) = \mathcal{F}(\lambda\beta.k,\ \lambda u \neq \langle\ \rangle.\alpha(k),\alpha).$$

In other words we must take account of the fact that

(*) is not decidable (countably recursive) whether of not an element of C_2 is a member of the basis for the inductive definition of C_2 (viz. the constant functionals).

It is not hard to give a formulation of a "computation scheme" SI (with which one could augment S1-S9) which would define functionals as above.

If $\{e\}(\lambda\beta.k,\ \lambda u \neq \langle\ \rangle.\alpha(k) = \alpha(k)$ then $\{e'\}(\lambda\beta.k,\alpha) = \alpha(k)$.

(SI)

If for all $n \neq \langle\ \rangle$, $\{e'\}(F_u,\alpha)$ is defined, then

$$\{e'\}(F,\alpha) = \{e\}(F,\lambda u \neq \langle\ \rangle.\{e'\}(F_u,\alpha)).$$

(Here e' is the new index which codes up e together with other appropriate information). It seems clear that (SI) will not close the gap indicated in §3 between countable and S1-S9 semi-recursion (2 envelopes). However I have been able to obtain no evidence against the following conjecture.

<u>CONJECTURE</u>. S1-S9 + SI suffice to generate the recursive countable functionals of type 3.

(Of course, SI may not be quite right for the job).

The augmentation of Kleene's schemes by SI is rather crude. It would be more satisfactory (both for general reasons and particularly since we are trying to use the inductive definition of C_2) to use the approach of inductive schemata as described in Feferman [4]. However from this point of view it is not at all obvious how to take account of (*) above. One seems to get involved either with non-monotone schemata, or with "partial" schemata, and I have not been able to devise a convincing formulation with either. There seems to be a genuine conceptual problem here:

What inductive schemata encapsulate the idea of definition by recursion on the inductive definition of C_2.

I hope that I have said enough in this section to show that the problem whether or not there is a natural inductive definability approach to the intrinsic recursion theory on the countable functionals is an accessible one. While I am optimistic about the specific conjecture above, I feel that the overall answer is likely to be "no".

References

[1] J.A. Bergstra, Computability and continuity in finite types, Dissertation, Utrecht (1976).

[2] Y.L. Ershov, Maximal and everywhere defined functionals, Algebra and Logic 13 (1974), 210-255 (374-397 in Russian).

[3] Y.L. Ershov, Models of partial continuous functionals, in Logic Colloquium 76, North-Holland (1977).

[4] S. Feferman, Inductive schemata and recursively continuous functionals, in Logic Colloquium 76, North-Holland (1977).

[5] R.O. Gandy and J.M.E. Hyland, Computable and recursively countable functions of higher type, in Logic Colloquium 76, North-Holland (1977).

[6] P.G. Hinman, Degrees of continuous functionals, J.S.L. 38 (1973), 393-395.

[7] J.M.E. Hyland, Recursion theory on the countable functionals, Dissertation, Oxford (1975).

[8] J.M.E. Hyland, Filter spaces and continuous functionals, submitted to Ann. Math. Logic.

[9] S.C. Kleene, Countable functionals, in Constructivity in Mathematics, North-Holland (1959).

[10] G. Kreisel, Interpretation of Analysis by means of functionals of finite type, in Constructivity in Mathematics, North-Holland (1959).

[11] D. Normann, Countable functionals and the analytic hierarchy, Oslo preprint (1977).

[12] C. Spector, On degrees of recursive unsolvability, Ann. Math. 64 (1956), 581-592.

J.E. Fenstad, R.O. Gandy, G.E. Sacks (Eds.)
GENERALIZED RECURSION THEORY II

SPECTOR SECOND ORDER CLASSES AND REFLECTION

Alexander S. Kechris[1)]
Department of Mathematics
California Institute of Technology
Pasadena, California

TABLE OF CONTENTS

INTRODUCTION

PART I

THE CONCEPT OF A SPECTOR SECOND ORDER CLASS AND SOME EXAMPLES

A. The basic notions.

§1. Preliminaries.

§2. Review of inductive second order relations.

§3. Spector 2-classes.

B. Some important examples.

§4. Positive elementary induction in a quantifier.

§5. Non-monotone induction.

§6. Recursion in type 2 objects.

§7. Recursion in normal type 2 functionals.

§8. The second order hierarchy.

§9. Recursion in type 3 objects.

§10. Recursion in normal type 3 functionals.

§11. Picture of some examples for $\mathcal{M} = \underset{\sim}{\omega}$.

PART II

REFLECTING SPECTOR 2-CLASSES

A. The basic notions.

§12. Reflecting and rigid Spector 2-classes.

[1)] Research and preparation for this paper were partially supported by NSF Grant MGS 16-17254

B. Rigid Spector 2-classes.

§13. General results and examples.

§14. The Harrington Representation Theorem.

§15. Characterization and classification problems.

C. Reflecting Spector 2-classes.

§16. General theory and examples.

§17. The smallest reflecting Spector 2-class.

PART III

REFLECTION IN RECURSION IN HIGHER TYPES

§18. Inductive analysis of the 2-envelope of a type 3 object.

§19. Reflecting ordinals.

§20. Applications.

REFERENCES

INTRODUCTION

The purpose of this paper, which grew out of a series of lectures (short course) delivered at the Oslo Conference, is to give a survey of the theory of reflection in the context of Spector second order classes. We have mainly aimed at presenting a global view of a rather extensive part of definability theory and at the same time illustrate how the main concepts and results to be presented below unify a large number of diverse areas in this theory and also clarify many of its important aspects.

The paper is divided into three major parts. Part I explains the notion of a Spector second order class, introduced by Moschovakis [Mos 6] and reviews some important examples in various branches of definability, including inductive definability, descriptive set theory and recursion in higher types. In Part II the concept of reflection for Spector second order classes, due independently to Harrington [Ha 1] and the author [Ke 2], is introduced and the basic classification of Spector second order classes into reflecting and rigid ones is presented. The structure of each of these categories is then examined separately in some detail. Finally, in Part III we concentrate on the theory of reflection in the particular context of Spector second order classes arising in recursion in higher types (i.e. envelopes of higher type objects), mostly due to Harrington [Ha 1], together with some of its applications.

PART I

THE CONCEPT OF A SPECTOR SECOND ORDER CLASS AND SOME EXAMPLES

A. The basic notions.

§1. Preliminaries

Let $\mathcal{M} = \langle M, R_1 \dots R_\ell \rangle$ be a structure. We will assume in the following that M contains a copy N of ω and that both N and the relation $\leq^N$ which is the copy of the natural ordering $\leq$ on ω are among M, $R_1 \dots R_\ell$. As usual we shall identify N, $\leq^N$ with ω, $\leq$ in the following. In general we shall use lower case letters a, b, c, ..., x, y, z as variables over M and capital letters A, B, C, ... X, Y, Z as variables over relations (of any number of arguments) on M. We shall reserve e, i, j, k, ℓ, m, n for members of ω. Finally, barred letters will denote finite sequences, as for example $\bar{x} = (x_1 \dots x_n)$ or $\bar{P} = (P_1 \dots P_m)$.

A second order relation on M is a relation of the form

$$\mathcal{P}(\bar{x}, \bar{P}) \equiv \mathcal{P}(x_1 \dots x_n, P_1 \dots P_m),$$

where P_i varies over k_i-ary relations on M. We call $\nu = (n, k_1 \dots k_m)$ the signature of $\mathcal{P}$. Thus every relation $P(x_1 \dots x_n)$ is also a second order relation of signature (n).

A 1-class or just class on M is a collection of relations on M, while a 2-class or second order class is a collection of second order relations on **M**.

§2. Review of inductive second order relations.

By way of motivating the notion of Spector second order class we shall review here some of the properties of the 2-class of inductive second order relations as established in Moschovakis [Mos 1].

2.1. Definition. A second order relation $\theta(x_1 \dots x_n, \bar{Y}, S)$ is operative (in the distinguished variable S) if S is n-ary, so that $S(x_1 \dots x_n)$ makes sense. For such θ define inductively:

$$\theta^{\xi}(\bar{x},\bar{Y}) \Leftrightarrow \theta^{<\xi}(\bar{x},\bar{Y}) \vee \theta(\bar{x},\bar{Y}, \{\bar{x}':\theta^{<\xi}(\bar{x}',\bar{Y})\}),$$

where $\theta^{<\xi}(\bar{x},\bar{Y}) \Leftrightarrow \exists\, \eta < \xi\, \theta^{\eta}(\bar{x},\bar{Y})$. We view of course here θ as a (nonmonotone in general) operator on M, with $\bar{Y}$ being carried through in the above induction as a parameter. Let also

$$\theta^{\infty}(\bar{x},\bar{Y}) \Leftrightarrow \exists\, \xi\ \theta^{\xi}(\bar{x},\bar{Y})$$

be the second order relation inductively definable by θ.

If $\mathcal{C}$ is a 2-class let

$$_2\mathrm{IND}(\mathcal{C}) = \{\mathcal{O}(\overline{x}, \overline{Y}) : \text{For some } \theta(\overline{u}, \overline{x}, \overline{Y}, S) \text{ operative in } \mathcal{C} \text{ and some constants } \overline{n} \text{ from } \omega: \mathcal{O}(\overline{x}, \overline{Y}) \Leftrightarrow \theta^\infty(\overline{n}, \overline{x}, \overline{Y})\}$$

Let also

$$_1\mathrm{IND}(\mathcal{C}) = \text{all relations in } {_2\mathrm{IND}(\mathcal{C})}.$$

Thus $_2\mathrm{IND}(\mathcal{C})$ is a 2-class, while $_1\mathrm{IND}(\mathcal{C})$ is a 1-class.

If we now take as $\mathcal{C}$ the 2-class

$\mathcal{E\ell}^+ \equiv$ all operative elementary in $\mathfrak{M}$ without parameters from M $\theta(\overline{x}, \overline{Y}, S_+)$ in which S occurs positively,

then we obtain

$$_2\mathrm{IND}(\mathfrak{M}) \equiv^{\text{def}} {_2\mathrm{IND}(\mathcal{E\ell}^+)}$$

$$\equiv \underline{\text{inductive}} \text{ second order relations on } \mathfrak{M}.$$

To avoid confusion we should emphasize here that in $\mathcal{E\ell}^+$ we have not allowed parameters from M to occur in the positive elementary operators and also, by our definition of $_2\mathrm{IND}(\mathcal{C})$, only constants from ω are allowed to be used in obtaining sections of the fixed points θ^∞. This is in contrast with Moschovakis' definitions in [Mo 1], where arbitrary parameters from M are allowed in the operators and sections. For distinction we shall explicitly call notions in which parameters are allowed bold face. We shall be mostly concerned in this paper with the light face versions of these and other notions to be introduced below.

2.2. General assumption. From now on and without explicit mentioning we shall assume that $\mathfrak{M}$ admits a hyperelementary (light face!) pairing function or equivalently that it admits a tuple coding function $\langle\ \rangle : M^{<\omega} \xrightarrow{1-1} M$ such that the corresponding relations and functions

$$\mathrm{Seq}(x) \Leftrightarrow x \in \mathrm{range}\ (\langle\ \rangle),$$

$$\ell h(x) = \text{length of the sequence coded by } x, \text{ if } \mathrm{Seq}(x); \quad = 0, \text{ otherwise,}$$

$$f(x,i) = (x)_i = i^{\text{th}} \text{ coordinate of the sequence coded by } x, \text{ if } \mathrm{Seq}(x) \text{ and } i < \ell h(x) \quad = 0, \text{ otherwise}$$

are hyperelementary (i.e. both inductive and coinductive). This coding apparatus will be also fixed in the sequel.

We can now collect the basic structural properties of ${}_2\mathrm{IND}(\mathfrak{M})$ as follows:

2.3. Theorem (Moschovakis [Mos 1]). Let $\Gamma = {}_2\mathrm{IND}(\mathfrak{M})$. Then

i) Γ contains $=$, $R_1 \ldots R_\ell$ and their negations.

ii) Γ is closed under $\wedge$, $\vee$, $\exists^M$, $\forall^M$.

iii) Γ is normed i.e. if $\mathcal{P}(\bar{x},\bar{Y})$ is in Γ there is $\sigma : \mathcal{P} \to$ Ordinals such that the corresponding relations

$$(\bar{x},\bar{Y}) \leq_\sigma^* (\bar{x}',\bar{Y}') \Leftrightarrow (\bar{x},\bar{Y}) \in \mathcal{P} \wedge \sigma(\bar{x},\bar{Y}) \leq \sigma(\bar{x}',\bar{Y}')$$

$$(\bar{x},\bar{Y}) <_\sigma^* (\bar{x}',\bar{Y}') \Leftrightarrow (\bar{x},\bar{Y}) \in \mathcal{P} \wedge \sigma(\bar{x},\bar{Y}) < \sigma(\bar{x}',\bar{Y}')$$

are in Γ, where $\sigma(\bar{x},\bar{Y}) = \infty$ if $\neg\mathcal{P}(\bar{x},\bar{Y})$. Such a σ is called a Γ-norm (on $\mathcal{P}$).

iv) Γ is ω-parametrized i.e. if $\nu = (n, k_1 \ldots k_m)$ is a signature there is $\mathcal{U} \in \Gamma$ of signature $(n+1, k_1 \ldots k_m)$ which is universal for the second order relations in Γ of signature ν; in other words if $\mathcal{U}_e = \{(\bar{x},\bar{Y}) : \mathcal{U}(e, \bar{x}, \bar{Y})\}$, then $\{\mathcal{U}_e : e \in \omega\} = \{\mathcal{P} \in \Gamma : \mathcal{P}$ has signature $\nu\}$.

To express the final property of inductive second order relations that we want to isolate we need the following notion. Let $\mathcal{P}(\bar{x}, Y_1 \ldots Y_k)$ be a second order relation, Γ a 2-class. We say that Γ is uniformly closed under Δ substitutions in $\mathcal{P}$ or simply that $\mathcal{P}$ is Γ on Δ if for any $\mathcal{Q}_i$, $\mathcal{R}_i$, $1 \leq i \leq k$, in Γ there is $\mathcal{S} \in \Gamma$ such that: For every $\bar{x}_1, \bar{Y}_1, \ldots, \bar{x}_k, \bar{Y}_k$ such that

$$\forall i \leq k\ \forall \bar{z}_i\ (\mathcal{Q}_i(\bar{x}_i, \bar{z}_i, \bar{Y}_i) \Leftrightarrow \neg\mathcal{R}_i(\bar{x}_i, \bar{z}_i, \bar{Y}_i))$$

we have

$$\mathcal{S}(\bar{x},\bar{x}_1,\bar{Y}_1 \ldots \bar{x}_k,\bar{Y}_k) \Leftrightarrow \mathcal{P}(\bar{x}, \{\bar{z}_1 : \mathcal{Q}_1(\bar{x}_1,\bar{z}_1,\bar{Y}_1)\}, \ldots, \{\bar{z}_k : \mathcal{Q}_k(\bar{x}_k,\bar{z}_k,\bar{Y}_k)\}).$$

We now have

v) Every second order relation $\mathcal{P} \in \Gamma$ is Γ on Δ.

It is not hard here to see that this last property v) is equivalent in the presence of i) - iv) to the following one:

v*) Γ is closed under Δ-bounded existential quantification

In detail this means the following:

Let $\breve{\Gamma} = \{\neg \mathcal{P} : \mathcal{P} \in \Gamma\}$, $\Delta = \Gamma \cap \breve{\Gamma}$. For each 2-class Γ and each $\bar{x}, \bar{Y}$ let $\Gamma(\bar{x},\bar{Y}) = \{\mathcal{P}(\bar{u},\bar{V}) : \text{There is } \mathcal{R} \in \Gamma \text{ such that } \mathcal{P}(\bar{u},\bar{V}) \Leftrightarrow \mathcal{R}(\bar{x}, \bar{Y}, \bar{u}, \bar{V})\}$ and $\Delta(\bar{x},\bar{Y}) = \Gamma(\bar{x},\bar{Y}) \cap \breve{\Gamma}(\bar{x},\bar{Y})$. In this notation $v^*)$ asserts that if $\mathcal{P}$ is in Γ so is

$$\mathcal{S}(\bar{x},\bar{Y}) \Leftrightarrow \exists Z \in \Delta(\bar{x},\bar{Y})\ \mathcal{P}(\bar{x}, \bar{Y}, Z).$$

§3. Spector 2-classes

The following concept was introduced by Moschovakis [Mos 6].

3.1. Definition. A 2-class Γ is called a Spector 2-class on $\mathfrak{M}$ if it satisfies properties i) - v) above.

Correspondingly (see Moschovakis [Mos 1]) a 1-class Γ is called a Spector (1-) class on $\mathfrak{M}$ if it satisfies (the obvious restrictions to relations of) i) - iv) above. The reader should be cautioned again that these are lightface Spector classes as opposed to the boldface ones that Moschovakis defines in [Mos 1], which are additionally closed under substitutions by arbitrary constants from M and are M-parametrized instead of ω-parametrized.

In general if Γ is a 2-class we let

$$_1\Gamma \equiv \text{the 1-reduct of } \Gamma$$
$$\equiv^{\text{def}} \text{all relations in } \Gamma.$$

Then it is easy to check that if Γ is a Spector 2-class so is $\Gamma(\bar{x},\bar{Y})$ for each $\bar{x},\bar{Y}$ while $_1\Gamma$ is a Spector 1-class. On the other hand if for any (1- or 2-)class Γ we put

$$\underset{\sim}{\Gamma} = \bigcup_{p \in M} \Gamma(p),$$

then for any Spector class Γ, $\underset{\sim}{\Gamma}$ is a boldface Spector class.

Clearly $_2\text{IND}(\mathfrak{M})$ is a Spector 2-class on $\mathfrak{M}$. The next result provides a minimality characterization of $_2\text{IND}(\mathfrak{M})$ among all Spector 2-classes on $\mathfrak{M}$. Later on we will see many more such minimality-type characterizations of other important Spector 2-classes. Results of this sort provide simple and transparent descriptions of interesting Spector 2-classes whose original definitions might involve complicated constructions and concepts. Moreover, they are extremely useful in many areas of definability theory since they provide an elegant and powerful method for comparing and many times identifying Spector 2-classes arising in various branches of this theory.

3.2. Theorem (Moschovakis [Mos 1]).

i) The 2-class ${}_2\mathrm{IND}(\mathfrak{M})$ is the smallest Spector 2-class on $\mathfrak{M}$.

ii) The 1-class ${}_1\mathrm{IND}(\mathfrak{M})$ is the smallest Spector 1-class on $\mathfrak{M}$.

The proof of i) [ii) being similar] is based on a closure property of every Spector 2-class which is called the First Recursion Theorem. In its full generality this will be stated in Section 9. Here one only needs a simple version of it. We need a definition first.

3.3. Definition. Let $\theta(\bar{x}, \bar{Y}, S)$ be an operative second order relation, Γ a 2-class. We call Γ uniformly closed under substitutions in θ (in the distinguished variable S) or simply we say that θ is Γ on Γ if for every $\mathcal{P}(\bar{x}', \bar{u}, \bar{Z})$ in Γ the relation

$$\mathcal{S}(\bar{x}, \bar{Y}, \bar{u}, \bar{Z}) \Leftrightarrow \theta(\bar{x}, \bar{Y}, \{\bar{x}' : \mathcal{P}(\bar{x}', \bar{u}, \bar{Z})\})$$

is also in Γ.

3.4. Theorem (Simple 1st Recursion Theorem - Moschovakis [Mos 1]). Let Γ be a Spector 2-class and let $\theta(\bar{x}, \bar{Y}, S)$ be operative and monotone on S. If θ is Γ on Γ, then $\theta^\infty \in \Gamma$.

B. Some important examples.

§4. Positive elementary induction in a quantifier.

4.1. Definition. A quantifier Q on M is a set $\emptyset \subsetneq Q \subsetneq p(M) \equiv$ power (M) which is monotone i.e. if $X \in Q$ and $X \subseteq Y$ then $Y \in Q$. As usual we write interchangeably

$$QxP(x) \Leftrightarrow Q(\{x : P(x)\}) \Leftrightarrow P \in Q.$$

Let also $\check{Q}$ be the dual quantifier of Q:

$$\check{Q}xP(x) \Leftrightarrow \neg Qx \neg P(x).$$

If $\mathcal{EL}^{Q,+} \equiv$ all operative $\theta(\bar{x}, \bar{Y}, S_+)$ which are elementary (lightface!) in the language of $\mathfrak{M}$ enlarged by the quantifiers $Q, \check{Q}$, and in which S occurs positively, then we write

$${}_2\mathrm{IND}(\mathfrak{M}, Q) \equiv^{\mathrm{def}} {}_2\mathrm{IND}(\mathcal{EL}^{Q,+}).$$

The results of the preceding section have been generalized in this context by Aczel.

4.2. Theorem (Aczel [Ac 2].

i) The 2-class ${}_2\mathrm{IND}(\mathfrak{M}, Q)$ is the smallest Spector 2-class Γ closed under both Q and $\check{Q}$ (i.e. if $\mathcal{P}(\bar{x}, z, \bar{Y})$ is in Γ so are $Q\, z\, \mathcal{P}(\bar{x}, z, \bar{Y})$ and

$\breve{Q}\, z\, \theta(\bar{x}, z, \bar{Y}))$.

ii) The 1-class ${}_1\mathrm{IND}(\mathcal{M}, Q)$ is the smallest Spector 1-class closed under Q and $\breve{Q}$.

Let us mention now some examples of quantifiers.

1) $Q = \exists$, $\breve{Q} = \forall$

2) $\mathcal{S}$ (Souslin quantifier), $\breve{\mathcal{S}} = \mathcal{A}$. Here $\mathcal{S}xP(x) \Leftrightarrow \forall x_0\, \forall x_1 \ldots \exists n P(\langle x_0 \ldots x_n \rangle)$.

3) $\mathcal{G}_0$ (open game quantifier), $\breve{\mathcal{G}}_0 = \mathcal{G}_c$ (= closed game quantifier). Here $\mathcal{G}_0 xP(x) \Leftrightarrow \exists x_0\, \forall x_1\, \exists x_2\, \forall x_3 \ldots \exists n\, P(\langle x_0 \ldots x_n \rangle)$.

4) (Aczel [Ac 2]; see also Hinman [Hi]). For each quantifier Q we can define a new quantifier Q^+ as follows

$$Q^+ xP(x) \Leftrightarrow Qx_0 \breve{Q}x_1\, \exists x_2\, \forall x_3\, Qx_4\, \breve{Q}x_5\, \exists x_6\, \forall x_7 \ldots \exists n\, P(\langle x_0 \ldots x_n \rangle).$$

Here an infinite string of quantifiers

$$Q_0 x_0 Q_1 x_1 \ldots R(x_0, x_1, \ldots)$$

is interpreted as follows: Consider the game

I	II	
X_0		I plays $X_0 \in Q_0$, II plays
	x_0	$x_0 \in X_0$; I plays $X_1 \in Q_1$, II
X_1		plays $x_1 \in X_1$, ... and I
	x_1	wins iff $R(x_0, x_1, \ldots)$.
⋮		

Now by definition,

$$Q_0 x_0 Q_1 x_1 \ldots R(x_0, x_1, \ldots) \Leftrightarrow \text{I has a winning strategy in this game.}$$

The motivation comes from the following simple equivalence:

$$QxP(x) \Leftrightarrow \exists\, X \in Q\, \forall x \in X\, P(x).$$

Clearly $\exists^+ = \mathcal{G}_0$. The results below generalize results of Moschovakis [Mos 1] from $\exists$ to arbitrary Q.

<u>Theorem</u> (Aczel [Ac 2]). Let Γ be a Spector (1- or 2-) class. Then Γ is closed under Q and $\breve{Q}$ iff Γ is closed under Q^+.

In particular, ${}_2\mathrm{IND}(\mathcal{M}, Q)$ is closed under Q^+. Conversely (Aczel [Ac 2]), every θ in ${}_2\mathrm{IND}(\mathcal{M}, Q)$ can be written as $Q^+ y\, \mathcal{R}(\bar{x}, y, \bar{Y})$ with $\mathcal{R}$ elementary in the language of the structure $\langle \mathcal{M}, \mathrm{Seq}, \ell h, (x)_i \rangle$ enlarged by the quantifiers $Q, \breve{Q}$, so that

${}_2\mathrm{IND}(\mathfrak{M},Q)$ is not closed under $\overline{Q^+}$.

§5. Non-monotone induction.

Other interesting examples of Spector 2-classes arise in the theory of nonmonotone induction originally developed by Richter and Aczel and later by Moschovakis in the abstract spirit to be presented below.

5.1. Definition (Moschovakis [Mos 2]. Let $\mathcal{F}$ be a 2-class (viewed here as a collection of - nonmonotone - operators). Call $\mathcal{F}$ typical, nonmonotone on $\mathfrak{M}$ if

i) All second order relations definable in $\mathfrak{M}$ by existential or universal (lightface!) formulas are in $\mathcal{F}$.

ii) $\mathcal{F}$ is closed under $\wedge, \vee$.

iii) $\mathcal{F}$ is closed under substitutions by trivial functions. Here we call $f(\overline{x},\overline{Y})$ with values in M trivial if its graph is quantifier-free definable in $\mathfrak{M}$, possibly with parameters from ω (but not arbitrary ones from M). If $f(\overline{x},\overline{Y})$ takes values n-ary relations we call f trivial if $\overline{z} \in f(\overline{x},\overline{Y})$ is quantifier-free definable, possibly with parameters from ω.

iv) $\mathcal{F}$ is ω-parametrized.

Here are some examples of typical, nonmonotone $\mathcal{F}$:

1) $\mathcal{F} = \Sigma^i_k(\mathfrak{M})$ or $\mathcal{F} = \Pi^i_k(\mathfrak{M})$, $i \geq 1$, $k \geq 1$.

2) $\mathcal{F} = \Sigma^0_k(\mathfrak{M})$ or $\mathcal{F} = \Pi^0_k(\mathfrak{M})$; $k \geq 2$.

Here Σ^0_k is understood to have its matrix built up from atomic formulas in $\langle \mathfrak{M}, \mathrm{Seq}, \ell h, (x)_i \rangle$, containing possibly parameters only from ω, by using the bounded quantifiers $\exists i \leq j$, $\forall i \leq j$.

We now have

5.2. Theorem (Moschovakis [Mos 2]; see also Aczel [Ac 1] and Richter-Aczel [R-A]). Let $\mathcal{F}$ be typical, nonmonotone on $\mathfrak{M}$. Then ${}_2\mathrm{IND}(\mathcal{F})$ is a Spector 2-class.

Our next task is to present a minimality characterization of this type of Spector 2-class. The key notion required here is the notion of $\mathcal{F}$-compactness. This notion captures in this abstract context the basic connection between nonmonotone induction and reflecting properties of ordinals discovered by Aczel and Richter (see [R-A]).

5.3. Definition (Moschovakis [Mos 2]). Let Γ be a 1-class, $\mathcal{F}$ a 2-class. We say that Γ is $\mathcal{J}$-compact if for every $\bar{p} \in M^n$, every $R \in \Gamma(\bar{p})$, every $R^0 \subseteq R$ in $\Delta(\bar{p})$ and every $\varphi(X)$ in $\mathcal{J}(\bar{p})$:

$$\varphi(R) \Rightarrow \exists R^* \in \Delta(\bar{p})\ (R^0 \subseteq R^* \subseteq R \wedge \varphi(R^*)).$$

Correspondingly, a 2-class Γ is $\mathcal{J}$-compact if for each $\bar{Y}$, each $R \in \Gamma(\bar{Y})$, $R^0 \subseteq R$ in $\Delta(\bar{Y})$ and $\varphi(X)$ in $\mathcal{J}(\bar{Y})$ we have:

$$\varphi(R) \Rightarrow \exists R^* \in \Delta(\bar{Y})\ (R^0 \subseteq R^* \subseteq R \wedge \varphi(R^*)).$$

We now have:

5.4. Theorem (Moschovakis [Mos 2]). Let $\mathcal{J}$ be typical, nonmonotone on $\mathfrak{M}$. Then

i) The 2-class ${}_2\mathrm{IND}(\mathcal{J})$ is the smallest Spector 2-class Γ on $\mathfrak{M}$ such that $\mathcal{J} \subseteq \Delta$ and Γ is $\mathcal{J}$-compact.

ii) The 1-class ${}_1\mathrm{IND}(\mathcal{J})$ is the smallest Spector 1-class Γ such that every $\Theta \in \mathcal{J}$ is Δ on Δ and Γ is $\mathcal{J}$-compact.

Here, if Γ is a 1-class and $\Theta(\bar{x}, Y_1 \ldots Y_k)$ is a second order relation we say that Θ is Γ on Δ if for each Q_i, R_i, $1 \leq i \leq k$, in Γ there is $S \in \Gamma$ such that if $\bar{x}_1 \ldots \bar{x}_k$ are such that

$$\forall i \leq k\ \forall \bar{z}_i\ (Q_i(\bar{x}_i, \bar{z}_i) \Leftrightarrow \neg R_i(\bar{x}_i, \bar{z}_i))$$

then

$$S(\bar{x}, \bar{x}_1 \ldots \bar{x}_k) \Leftrightarrow \Theta(\bar{x}, \{\bar{z}_1 : Q_1(\bar{x}_1, \bar{z}_1)\}, \ldots, \{\bar{z}_k : Q_k(\bar{x}_k, \bar{z}_k)\}).$$

We call Θ Δ on Δ if both Θ, $\neg\Theta$ are Γ on Δ.

Some examples are in order.

1) Theorem (Grilliot [Gr]). Every Spector 2-class is Σ^0_2-compact. Thus ${}_2\mathrm{IND}(\Sigma^0_2) = {}_2\mathrm{IND}$ (we have not explicitly indicated $\mathfrak{M}$ here for simplicity)

2) For $n > 2$, ${}_2\mathrm{IND}(\Sigma^0_{n+1}) = {}_2\mathrm{IND}(\Pi^0_n)$ = the smallest Spector 2-class which is Π^0_n-compact. Also ${}_2\mathrm{IND}(\Pi^0_n) < {}_2\mathrm{IND}(\Pi^0_{n+1})$, where

$$\Gamma < \Gamma' \Leftrightarrow \Gamma \subseteq \Delta'$$

denotes strong inclusion between 1- or 2-classes.

§6. Recursion in type 2 objects.

A type 2 object on M is a set ${}^2F$ of subsets of M i.e. ${}^2F \subseteq p(M)$.

Example. ${}^2E_M = \{X \subseteq M : X \neq \emptyset\}$.

Every type 2 object gives rise to Spector 2-class as follows

6.1. Theorem (Moschovakis [Mos 3]. Let ${}^2F$ be a type 2 object on M. Then there is a smallest Spector 2-class Γ on $\mathcal{M}$ such that ${}^2F \in \Delta$. Moreover ${}_1\Gamma$ is the smallest spector 1-class such that ${}^2F$ is Δ on Δ.

Proof. Put

$$Q_F(X) \Leftrightarrow X_0 \cap X_1 \neq \emptyset \vee (X_0 \cup X_1 = M \wedge X_0 \in F),$$

where $X_i = \{x : \langle i, x\rangle \in X\}$. Clearly Q_F is a quantifier. Take now $\Gamma = {}_2\mathrm{IND}(\mathcal{M}, Q_F)$. ⊣

This class Γ has a description in terms of Kleene recursion in higher type objects originally introduced (for $\mathcal{M} = \underset{\sim}{\omega} = \langle\omega,\leq\rangle$) in Kleene [Kl 1,2]. Indeed, the class Γ defined above can be identified as

$\Gamma =_2 \mathrm{env}(\mathcal{M}, {}^2E^{\#}_M, {}^2F) \equiv^{\mathrm{def}}$ class of all second order relations which are semirecursive in ${}^2F$, ${}^2E^{\#}_M$ (= existential quantifier on M viewed as a type 2 functional) and the characteristic functions of $=, R_1, \ldots, R_\ell$.

This will be explained further in §7 below.

Remark: In case $\mathcal{M} = \underset{\sim}{\omega} = \langle\omega, \leq\rangle$, one can replace ${}^2E^{\#}$ by ${}^2E$. The only reason for using ${}^2E^{\#}_M$ instead of ${}^2E_M$ in general is to make sure that ${}_2\mathrm{env}(\mathcal{M}, {}^2E^{\#}_M, {}^2F)$ is closed under $\exists^M$. For $\mathcal{M} = \underset{\sim}{\omega}$ however this follows from the Gandy Selection Theorem. See his paper [Ga], where the basic properties of ${}_2\mathrm{env}({}^2E, {}^2F)$ are first established.

Theorem 6.1 allows us to define a jump operation on Spector 2-classes:

$j(\Gamma)$ = smallest Spector 2-class Γ' such that $\Gamma < \Gamma'$.

Here of course $j\Gamma$ is obtained by applying Theorem 6.1 to a type 2 object ${}^2F$ which is in Γ and is moreover universal i.e. for each $\theta(\overline{x}, \overline{Y})$ in Γ there is $e \in \omega$ such that $\theta(\overline{x}, \overline{Y}) \Leftrightarrow \langle e, \overline{x}, \overline{Y}\rangle \in {}^2F$. For example, if $\mathcal{M} = \underset{\sim}{\omega}$ and $\Gamma = {}_2\mathrm{IND}$, then $j(\Gamma) = {}_2\mathrm{env}(E_1)$, where E_1 is the Tugué object embodying the notion of wellfoundedness: $E_1 = \{X \subseteq \omega : \forall x_0 \forall x_1 \ldots \exists n (\langle x_0 \ldots x_n\rangle \in X)\} = \mathcal{S}$ (the Souslin quantifier).

§7. Recursion in normal type 2 functionals.

A convenient reference for what follows in this and the next two sections is [K-M], especially in regards to unexplained notation and terminology.

Let $PF(M) = \{f : f \text{ is a partial function from } M \text{ into } \omega\}$. A type 2 (partial, monotone) functional on M is a partial map $\varphi : M^n \times PF(M)^k \to \omega$ which is monotone i.e.

$$\varphi(\bar{x}, f_1 \dots f_k) = w \wedge f_1 \subseteq g_1 \wedge \dots \wedge f_k \subseteq g_k \Rightarrow$$

$$\varphi(\bar{x}, g_1 \dots g_k) = w.$$

Example. $${}^2E^{\#}_M(f) = \begin{cases} 0, & \text{if } \exists x(f(x) = 0) \\ 1, & \text{if } \forall x(f(x) \neq 0). \end{cases}$$

Here $f(\bar{x}) \neq 0$ abbreviates "f(x) is defined and has value $\neq 0$".

Let $\bar{\varphi}$ be a finite list of type 2 functionals on $\mathfrak{M}$. Put

${}_2\text{env}(\mathfrak{M},\bar{\varphi}) \equiv$ the 2-envelope of $\bar{\varphi}$ (on $\mathfrak{M}$) $\equiv^{\text{def}}$ the collection of all second order relations on M which are semirecursive in $\bar{\varphi}$ and the characteristic functions of $=, R_1 \dots R_\ell$.

These turn out to be Spector 2-classes for certain types of well-behaved $\bar{\varphi}$ called normal.

7.1. Definition (Moschovakis [Mos 4]). A type 2 functional $\varphi(f)$ on M is called normal in $\bar{\psi}$ if there is a functional $\Delta_\varphi(f,g)$ recursive in $\bar{\psi}$ such that

i) $\varphi(f \restriction \{\bar{x}: g(\bar{x}) = 0\}) \downarrow \Rightarrow \Delta_\varphi(f,g) = 0$

ii) $g \text{ total} \wedge \{\bar{x}: g(\bar{x}) = 0\} \subseteq \text{dom}(f) \wedge$

$\varphi(f \restriction \{\bar{x}: g(\bar{x}) = 0\}) \uparrow \Rightarrow \Delta_\varphi(f,g) = 1.$

Here $\downarrow$ means "is defined" and $\uparrow$ "is undefined".(A similar definition applies to general functionals $\varphi(\bar{x},\bar{f})$). Call $\bar{\varphi}$ normal on $\mathfrak{M}$ if each $\varphi_i \in \bar{\varphi}$ is normal in $\bar{\varphi}$ and the characteristic functions of $=, R_1 \dots R_\ell$.

7.2. Theorem (Moschovakis [Mos 3,4]). Let $\bar{\varphi}$ be a normal sequence of type 2 functionals on $\mathfrak{M}$. Then ${}_2\text{env}(\mathfrak{M}, {}^2E^{\#}_M, \bar{\varphi})$ is the smallest Spector 2-class Γ on $\mathfrak{M}$ such that each functional in $\bar{\varphi}$ is Γ on Γ. Similarly for ${}_1\text{env}(\mathfrak{M}, {}^2E^{\#}_M, \bar{\varphi})$.

Here a type 2 functional $\varphi(\bar{x}, f_1 \dots f_k)$ is Γ on Γ (where Γ is a 2-class - a similar definition applies to 1-classes), if for each $\mathcal{X}_i(\bar{y}_i, \bar{z}_i, \bar{Y}_i)$, $1 \leq i \leq k$, partial functions with values in ω and graphs in Γ, the partial function

$$\psi(\bar{x}, \bar{z}_1, \bar{Y}_1, \dots, \bar{z}_k, \bar{Y}_k) = \varphi(\bar{x}, \lambda\bar{y}_1 \mathcal{X}_1(\bar{y}_1, \bar{z}_1, \bar{Y}_1), \dots, \lambda\bar{y}_k \mathcal{X}_k(\bar{y}_k, \bar{z}_k, \bar{Y}_k))$$

has also graph in Γ.

Again the key to the proof of Theorem 7.2 is Theorem 3.4. The condition that $\bar{\varphi}$ is Γ on Γ guarantees that the operator defining a universal set in ${}_2\text{env}(\mathfrak{M}, {}^2E^{\#}_M, \bar{\varphi})$, say via Kleene's schemes, is Γ on Γ.

Remark. Again, if $\mathcal{M} = \underset{\sim}{\omega}$, $E^{\#}$ can be dropped in 7.2.

Let us consider now some examples.

1) Every type 2 object ${}^2F \subseteq p(M)$ can be identified with the total function ${}^2F : \omega^M \to \omega$ given by

$$ {}^2F(f) = \begin{cases} 0, & \text{if } f \text{ is total} \wedge \{x : f(x) = 0\} \in {}^2F \\ 1, & \text{if } f \text{ is total} \wedge \{x : f(x) = 0\} \notin {}^2F. \end{cases} $$

By making ${}^2F$ undefined on strictly partial $f : M \to \omega$ we can view ${}^2F$ in a natural way as a type 2 functional. Then it is not hard to check that ${}^2E_M$, ${}^2\overline{F}$ is normal for any finite list ${}^2\overline{F}$ of type 2 objects. Moreover

$$ {}_2\mathrm{env}(\mathcal{M}, {}^2E^{\#}_M, {}^2\overline{F}) = {}_2\mathrm{env}(\mathcal{M}, {}^2E^{\#}_M, {}^2E_M, {}^2\overline{F}) $$

= smallest Spector 2-class Γ such that each object in ${}^2\overline{F}$ is in Δ, so that we recover the example in Section 6.

2) (Hinman [Hi]). Let Q be a quantifier on M. We attach to Q the type 2 object ${}^2F_Q \overset{\mathrm{def}}{\equiv} Q$ i.e.

$$ {}^2F_Q(f) = \begin{cases} 0, & \text{if } f \text{ is total} \wedge Qx\, f(x) = 0 \\ 1, & \text{if } f \text{ is total} \wedge \breve{Q}x\, f(x) \neq 0 \end{cases} $$

and the type 2 functional

$$ {}^2F^{\#}_Q(f) = \begin{cases} 0, & \text{if } Qx\, f(x) = 0 \\ 1, & \text{if } \breve{Q}x\, f(x) \neq 0 \\ \uparrow, & \text{otherwise.} \end{cases} $$

thus ${}^2F_{\exists M} = {}^2E_M$, ${}^2E^{\#}_{\exists M} = {}^2E^{\#}_M$. It turns out again (see [K-M]) that ${}^2F^{\#}_Q$ is normal and moreover by the various minimality characterizations mentioned before one can see that

$$ {}_2\mathrm{env}(\mathcal{M}, {}^2E^{\#}_M, {}^2F^{\#}_Q) = {}_2\mathrm{IND}(\mathcal{M}, Q) = $$

the smallest Spector 2-class Γ on $\mathcal{M}$ closed under both $Q, \breve{Q}$. On the other hand

$$ {}_2\mathrm{env}(\mathcal{M}, {}^2E^{\#}_M, {}^2F_Q) = \text{smallest Spector 2-class } \Gamma \text{ on } \mathcal{M} \text{ such that} $$

$Q \in \Delta$ = smallest Spector 2-class Γ such that Δ is uniformly closed under both $Q, \breve{Q}$ i.e. if $\mathcal{P}, \mathcal{Q}$ are in Γ then there are $\mathcal{S}, \mathcal{S}'$ in Γ such that for each $\overline{x}, \overline{Y}$:

If $\forall z(\mathcal{P}(z,\overline{x},\overline{Y}) \Leftrightarrow \neg \mathcal{Q}(z,\overline{x},\overline{Y}))$, then $\mathcal{S}(\overline{x},\overline{Y}) \Leftrightarrow Qz\, \mathcal{P}(z,\overline{x},\overline{Y})$, $\mathcal{S}'(\overline{x},\overline{Y}) \Leftrightarrow Qz\, \mathcal{P}(z,\overline{x},\overline{Y})$.

This is usually also expressed by saying that Γ is closed under the "deterministic" Q and $\breve{Q}$.

In general, ${}_2\mathrm{env}(\mathfrak{M}, {}^2E^{\#}_M, {}^2F_Q) < {}_2\mathrm{env}(\mathfrak{M}, {}^2E^{\#}_M, {}^2F^{\#}_Q)$ - an exception is $Q = \exists$). For example, if $\mathfrak{M} = \underset{\sim}{\omega}$ and $Q = \mathcal{S}$ then ${}_2\mathrm{env}({}^2E, {}^2F_{\mathcal{S}}) = {}_2\mathrm{env}(E_1)$, while ${}_2\mathrm{env}({}^2E, {}^2F^{\#}_{\mathcal{S}}) = {}_2\mathrm{IND}(\mathcal{S}) = {}_2\mathrm{IND}(\Sigma^1_1)$ (by a theorem of Grilliot). It is of course well-known that ${}_2\mathrm{IND}(\Sigma^1_1) \gg {}_2\mathrm{env}(E_1)$.

§8. The second order hierarchy.

The 2-classes $\Sigma^1_n(\mathfrak{M})$, $\Pi^1_n(\mathfrak{M})$ are defined as usual:

$$\Sigma^1_1(\mathfrak{M}) = \{\exists\overline{X}\,\theta(\overline{x},\overline{Y},\overline{X}) \;:\; \theta \text{ (lightface!) elementary on } \mathfrak{M}\}$$

$$\Pi^1_1(\mathfrak{M}) = \smile\Sigma^1_1(\mathfrak{M})$$

$$\Sigma^1_{n+2}(\mathfrak{M}) = \{\exists\overline{X}\,\theta(\overline{x},\overline{Y},\overline{X}) \;:\; \theta \in \Pi^1_{n+1}(\mathfrak{M})\}$$

$$\Pi^1_{n+2}(\mathfrak{M}) = \smile\Sigma^1_{n+2}(\mathfrak{M}),$$

for any $n \geq 0$.

Various 2-classes in this hierarchy give rise to Spector 2-classes in certain cases as the next two theorems show.

8.1. Theorem.

i) (Kleene [Kl 3] for $\mathfrak{M} = \underset{\sim}{\omega}$; Barwise-Gandy-Moschovakis [B-M-G] in general). For each countable $\mathfrak{M}$, $\Pi^1_1(\mathfrak{M}) = {}_2\mathrm{IND}(\mathfrak{M})$.

ii) (Moschovakis [Mos 1]). For each countable $\mathfrak{M}$, $\Sigma^1_2(\mathfrak{M})$ is a Spector 2-class.

Theorem 8.1 holds also for certain uncountable $\mathfrak{M}$'s of "strong cofinality ω" e.g. $\langle V_\lambda, \epsilon\rangle$, where $\mathrm{cof}(\lambda) = \omega$ (Chang-Moschovakis; see [Mos 1]).

8.2. Theorem.

i) (Martin [Ma], Moschovakis [A-M]). Assume Projective Determinacy. If $\mathfrak{M}$ is countable, then all $\Pi^1_{2n+1}(\mathfrak{M})$, $\Sigma^1_{2n+2}(\mathfrak{M})$ for $n \geq 0$ are Spector 2-classes.

ii) (Addison [Ad]). Assume V = L. Then for any $\mathfrak{M}$, all $\Sigma^1_{2n+2}(\mathfrak{M})$, where $n \geq 0$, are Spector 2-classes. If moreover $\mathfrak{M}$ is essentially uncountable i.e. $\mathcal{W}\mathcal{F} = \{X \subseteq M: \neg\exists x_0\, \exists x_1 \ldots \forall i\, \langle x_i, x_{i+1}\rangle \in X\}$ (= the notion of wellfoundedness) is in $\Delta^1_1(\mathfrak{M})$, then $\Sigma^1_1(\mathfrak{M})$ is also a Spector 2-class.

It is interesting to note here that if one assumes AD, on top of ZF + DC, then on the one hand for $\mathfrak{M} = \mathbb{R}$ = structure of the reals, we have that $\Sigma^1_1(\mathbb{R})$ is a Spector

2-class (Martin, Solovay, Kechris, ...) while on the other if $\mathcal{M} = \underset{\sim}{\omega}_1 = \langle\omega_1, \leq\rangle$, $\Pi^1_1(\underset{\sim}{\omega}_1)$ is a Spector 2-class (Kechris [Ke 1]). Harrington has recently shown both that

$$\mathrm{Con(ZFC)} \Rightarrow \mathrm{Con(ZFC} + \Pi^1_1(\underset{\sim}{\omega}_1) \text{ is a Spector 2-class})$$

and also that

$$\mathrm{Con(ZFC)} \Rightarrow \mathrm{Con(ZFC} + \Pi^1_1(\mathbb{R}) \text{ is a Spector 2-class}).$$

§ 9. Recursion in type 3 objects.

A type 3 object on M is a collection ${}^3F$ of subsets of p(M) i.e. ${}^3F \subset p(p(M))$.

Example. ${}^3E_M = \{\mathcal{S} \subseteq p(M) : \mathcal{S} \neq \emptyset\}$.

We let again

$${}_2\mathrm{env}(\mathcal{M}, {}^3\overline{F}) \equiv^{\mathrm{def}} \text{all second order relations which are semirecursive}$$

in ${}^3\overline{F}$ and the characteristic functions of $=$, $R_1 \ldots R_\ell$.

9.1. Theorem. Let ${}^3\overline{F}$ be a finite list of type 3 objects on M. Then ${}_2\mathrm{env}(\mathcal{M}, {}^3E_M, {}^3\overline{F})$ is a Spector 2-class. Moreover it is closed under $\forall^{p(M)} \equiv \forall^1$ and the deterministic $\exists^{p(M)} \equiv \exists^1$.

Here we say that a 2-class Γ is closed under the deterministic $\exists^1$ if for each $\mathcal{P}, \mathcal{Q}$ in Γ there is $\mathcal{R} \in \Gamma$ such that for each $\overline{x}, \overline{Y}$:

If $\forall Z(\mathcal{P}(\overline{x},\overline{Y},Z) \Leftrightarrow \neg\mathcal{Q}(\overline{x},\overline{Y},Z))$, then $\mathcal{R}(\overline{x},\overline{Y}) \Leftrightarrow \exists Z\ \mathcal{P}(\overline{x},\overline{Y},Z)$.

We will see later that ${}_2\mathrm{env}(\mathcal{M}, {}^3E_M, {}^3\overline{F})$ is never closed under $\exists^1$.

The two crucial steps in proving Theorem 9.1. are first the verification that ${}_2\mathrm{env}(\mathcal{M}, {}^3E_M, {}^3\overline{F})$ is normed and this is due to Moschovakis [Mos 5] and second the verification that ${}_2\mathrm{env}(\mathcal{M}, {}^3E_M, {}^3\overline{F})$ is closed under $\exists^M$ and this is the Grilliot-Harrington-MacQueen Theorem (see Harrington-MacQueen [H-Ma]).

Our next goal is to provide again a minimality characterization of these envelopes. For that it will be convenient to separate the role of ${}^3E_M$ in the list ${}^3E_M, {}^3\overline{F}$.

9.2. Definition. A Spector 2-class on $\mathcal{M}$ which is closed under $\forall^1$ and the deterministic $\exists^1$ will be called an E-Spector 2-class.

Recall also that if Γ is a 2-class and ${}^3F$ a type 3 object then we say that ${}^3F$ is Γ on Δ if for each $\mathcal{P}, \mathcal{Q}$ in Γ there is $\mathcal{R} \in \Gamma$ such that for each $\overline{x}, \overline{Y}$:

If $\forall Z(\mathcal{P}(\overline{x},\overline{Y},Z) \Leftrightarrow \neg\mathcal{Q}(\overline{x},\overline{Y},Z))$, then $\mathcal{R}(\overline{x},\overline{Y}) \Leftrightarrow \{Z : \mathcal{P}(\overline{x},\overline{Y},Z)\} \in {}^3F$. Call also

${}^3F$ $\underline{\Delta \text{ on } \Delta}$ if both ${}^3F$, $\neg\, {}^3F$ are Δ on Δ. Thus for example ${}^3E_M$ is Δ on Δ iff Γ is closed under the deterministic $\exists^1$, $\forall^1$.

We are now ready to state

9.3. Theorem (Moschovakis [Mos 3]. Let ${}^3\overline{F}$ be a list of type 3 objects on M. Then ${}_2\text{env}(\mathcal{M}, {}^3E_M, {}^3\overline{F})$ is the smallest E-Spector 2-class Γ on $\mathcal{M}$ such that each object in ${}^3\overline{F}$ is Δ on Δ.

The key to the proof is the full First Recursion Theorem for Spector 2-classes - a basic closure property of such classes under appropriate inductive definitions operating on p(M) instead of M this time. One of course needs to consider such inductive definitions since universal sets in ${}_2\text{env}(\mathcal{M}, {}^3E_M, {}^3\overline{F})$ are defined by operators which act essentially on p(M) as opposed to those used in Sections 4-7 which operate on M and have the members of p(M) carried through as parameters.

Let $\Phi(\overline{x},\overline{Y},\mathcal{S})$ be an operative third order relation i.e. $\mathcal{S}$ varies over second order relations of the appropriate signature so that $\mathcal{S}(\overline{x},\overline{Y})$ makes sense. For such a Φ we let again

$$\Phi^{\xi}(\overline{x},\overline{Y}) \Leftrightarrow \Phi^{<\xi}(\overline{x},\overline{Y}) \vee \Phi(\overline{x},\overline{Y},\Phi^{<\xi})$$

and

$$\Phi^{\infty}(\overline{x},\overline{Y}) \Leftrightarrow \exists\xi\ \Phi^{\xi}(\overline{x},\overline{Y}).$$

If Γ is a 2-class then we say that Φ is $\underline{\Gamma \text{ on } \Gamma}$ if for each $\mathcal{P} \in \Gamma$ the second order relation

$$\mathcal{S}(\overline{x},\ \overline{Y},\ \overline{z},\ \overline{W}) \Leftrightarrow \Phi(\overline{x},\overline{Y},\ \{(\overline{x}',\overline{Y}'):\ \mathcal{P}(\overline{z},\ \overline{W},\ \overline{x}',\ \overline{Y}')\}$$

is also in Γ.

Now we have

9.4. Theorem (First Recursion Theorem - Moschovakis [Mos 3]). Let Γ be a Spector 2-class and let $\Phi(\overline{x},\overline{Y},\mathcal{S})$ be operative and monotone on $\mathcal{S}$. Then if Φ is Γ on Γ, $\Phi^{\infty} \in \Gamma$.

§10. Recursion in normal type 3 functionals.

The results of Section 9 can be generalized to normal type 3 functionals.

Let $PF(p(M)) = \{f^2 : f^2$ is a partial function from p(M) into $\omega\}$. A type 3 (partial, monotone) functional on M is a partial map $\Phi : p(M)^n \times PF(p(M))^k \to \omega$ which is monotone i.e.

$$\Phi(\overline{X},\ f_1^2 \ldots f_k^2) = w \wedge f_1^2 \subseteq g_1^2 \wedge \ldots \wedge f_k^2 \subseteq g_k^2 \Rightarrow$$

$$\Phi(\overline{X}, g_1^2 \ldots g_k^2) = w.$$

Example. $${}^3E_M^{\#}(f^2) = \begin{cases} 0, & \text{if } \exists X\ f^2(X) = 0 \\ 1, & \text{if } \forall X\ f^2(X) \neq 0. \end{cases}$$

Let $\overline{\Phi}$ be a finite list of type 3 functionals on M. Put again ${}_2\text{env}(\mathcal{M},\overline{\Phi}) \equiv^{def}$ the class of second order relations on M which are semirecursive in $\overline{\Phi}$ and the characteristic functions of =, $R_1 \ldots R_\ell$.

A list $\overline{\Phi}$ is normal on $\mathcal{M}$ if it satisfies the obvious modification of the definition given in 7.1. We now have

10.1. Theorem (Moschovakis [Mos 3,4]). Let $\overline{\Phi}$ be a normal sequence of type 3 functionals on $\mathcal{M}$. Then ${}_2\text{env}(\mathcal{M}, {}^2E_M^{\#}, \overline{\Phi})$ is the smallest Spector 2-class Γ such that each functional in $\overline{\Phi}$ is Γ on Γ.

Here we say as usual that a type 3 functional $\Phi(\overline{X},f^2)$ is Γ on Γ if for each $\mathcal{X}(\overline{x},\overline{Y},Z)$ a partial function with values in ω and graph in Γ, the partial function

$$\psi(\overline{x},\overline{X},\overline{Y}) = \Phi(\overline{X}, \lambda Z\, \mathcal{X}(\overline{x},\overline{Y},Z))$$

has also graph in Γ.

Remark: Again ${}^2E_M^{\#}$ is added to ensure that ${}_2\text{env}(\mathcal{M}, {}^2E_M^{\#}, \overline{\Phi})$ is closed under $\exists^M$ and it can be dropped if $\mathcal{M} = \underset{\sim}{\omega}$.

Let us mention now some examples in analogy with those discussed in Section 7.

1) As with type 2 objects each type 3 object ${}^3F$ can be naturally viewed as a type 3 functional. Then for each finite list ${}^3\overline{F}$ one can easily check again that the extended list ${}^3E_M$, ${}^3\overline{F}$ is normal. Moreover by the Grilliot-Harrington-MacQueen Theorem

$${}_2\text{env}(\mathcal{M}, {}^3E_M, {}^3\overline{F}) = {}_2\text{env}(\mathcal{M}, {}^2E_M^{\#}, {}^3E_M, {}^3\overline{F})$$

so we come back to the example in Section 9.

2) Each quantifier Q on p(M) gives rise as before to a type 3 object ${}^3F_Q$ and a type 3 functional ${}^3F_Q^{\#}$. For example ${}^3F_{\exists^1} = {}^3E_M$, and ${}^3F_{\exists^1}^{\#} = {}^3E_M^{\#}$. Again ${}^3F_Q^{\#}$ is normal. Now it is not hard to check that

$${}_2\text{env}(\mathcal{M}, {}^3E_M, {}^3F_Q) = \text{smallest E-Spector}$$

2-class on $\mathcal{M}$ closed under the deterministic $Q, \breve{Q}$, while

$${}_2\text{env}(\mathcal{M}, {}^3E_M^{\#}, {}^3F_Q^{\#}) = \text{smallest Spector}$$

2-class on $\mathfrak{M}$ closed under $\exists^1$, $\forall^1$, Q and $\breve{Q} = {}_1\mathrm{IND}(\mathfrak{M}^1, Q)$, where $\mathfrak{M}^1 = \langle M \cup p(M), \ldots \rangle$ is the type 1 domain associated with M. In particular,

$$ {}_2\mathrm{env}(\mathfrak{M}, {}^3E_M) < {}_2\mathrm{env}(\mathfrak{M}, {}^3E_M^{\#}), $$

since the negatism of a second order relation in ${}_2\mathrm{env}(\mathfrak{M}, {}^3E_M)$ can be obtained by existential over p(M) quantification applied to a second order relation in ${}_2\mathrm{env}(\mathfrak{M}, {}^3E_M)$ (see Section 18 here).

§11. Picture of some examples for $\mathfrak{M} = \underset{\sim}{\omega}$.

$\Pi^1_1 = {}_2\mathrm{IND} < {}_2\mathrm{env}(E_1) < {}_2\mathrm{IND}(\Sigma^0_3) < \ldots < {}_2\mathrm{IND}(\Pi^1_1) < {}_2\mathrm{env}(E_1^{\#} \equiv^{\mathrm{def}} {}^2F_{\mathcal{S}}^{\#}) = {}_2\mathrm{IND}(\mathcal{S}) =$
${}_2\mathrm{IND}(\Sigma^1_1) < \underbrace{\Sigma^1_2 \; (< \Pi^1_3 < \ldots)}_{\mathrm{PD}} < {}_2\mathrm{env}({}^3E) < {}_1\mathrm{IND}(\mathbb{R}).$

PART II
REFLECTING SPECTOR 2-CLASSES

A. The basic notions

§12. Reflecting and rigid Spector 2-classes.

There is a basic classification of Spector 2-classes into two major categories the distinguishing characteristic being that of reflection.

12.1. Definition. Let Γ be a Spector 2-class on $\mathfrak{M}$. We call Γ self-compact or reflecting if Γ is Γ-compact. Explicitly this means any one of the equivalent formulations below

i) For each $\overline{Y}$, each $\varphi(X)$ in $\Gamma(\overline{Y})$ and each R in $\Gamma(\overline{Y})$ and $R^0 \subseteq R$ in $\Delta(\overline{Y})$ we have:

$$ \varphi(R) \Rightarrow \exists R^* \in \Delta(\overline{Y})(R^0 \subseteq R^* \subseteq R \wedge \varphi(R^*)). $$

ii) As in i) dropping R^0.

iii) As in ii) but not requiring in the conclusion that $R^* \subseteq R$ i.e. for each $\overline{Y}$, each $\varphi(X)$ in $\Gamma(\overline{Y})$ and each R in $\Gamma(\overline{Y})$:

$$ \varphi(R) \Rightarrow \exists R^* \in \Delta(\overline{Y}) \; \varphi(R^*). $$

iv) For each $\overline{Y}$, no $W \subseteq M^n$, $W \in \Gamma(\overline{Y}) - \underset{\sim}{\Delta}(\overline{Y})$ can be a $\Gamma(\overline{Y})$-singleton (Here W is a $\Gamma(\overline{Y})$ singleton iff $\{W\} \in \Gamma(\overline{Y})$).

Of these equivalences only the direction iv) $\Rightarrow$ iii) needs possibly any comment. Assume iv) holds but, towards a contradiction, for some $\overline{Y}$, some $\varphi(X)$ in $\Gamma(\overline{Y})$ and some $R \in \Gamma(\overline{Y})$, $\varphi(R)$ holds but for no $R^* \in \Delta(\overline{Y})$ $\varphi(R^*)$ holds. Then using that Γ is a Spector 2-class it is not hard to check that actually for no $R^* \in \underset{\sim}{\Delta}(Y)$ $\varphi(R^*)$ holds. Let now σ be a $\Gamma(\overline{Y})$-norm on R and note that the following shows that $\{R\} \in \Gamma(\overline{Y})$ contradicting iv):

$$R' = R \Leftrightarrow R' \subseteq R \wedge \forall\overline{x}\, \forall\overline{y}$$
$$(\overline{x} \in R' \wedge \sigma(\overline{y}) \leq \sigma(\overline{x}) \Rightarrow \overline{y} \in R') \wedge \varphi(R').$$

The last condition v) mentioned above leads to the opposite to reflection notion of rigidity.

12.2. Definition. A Spector 2-class Γ is called uniformly non-reflecting or rigid if for any $\mathcal{H}(\overline{x},\overline{Y})$ in Γ such that for all $\overline{Y}$,

$$\mathcal{H}^{\overline{Y}} \equiv^{\text{def}} \{\overline{x} : \mathcal{H}(\overline{x},\overline{Y})\} \in \Gamma(\overline{Y}) - \underset{\sim}{\Delta}(\overline{Y})$$

the second order relation

$$\mathsf{R}_{\mathcal{H}}(W,\overline{Y}) \Leftrightarrow W = \mathcal{H}^{\overline{Y}}$$

is in Γ i.e. each $\mathcal{H}^{\overline{Y}}$ is a $\Gamma(\overline{Y})$-singleton uniformly in $\overline{Y}$.

Since rigidity is not the literal but rather the uniform negation of reflection one can find pathological examples of Spector 2-classes which cannot be classified in one of these two categories. In practice however all natural examples of Spector 2-classes belong to one or the other of these two groups.

Examples. We shall see later in Sections 13-17 that the examples in Sections 4-7 are all rigid while the ones in Sections 8 (except for Π^1_1), 9 and 10 (mostly) are reflecting.

The notion of reflection described above was discovered by Harrington [Ha 1] (who in the context of recursion in higher types proved that ${}_2\text{env}(\mathcal{M}, {}^3E_M, {}^3\overline{F})$ is reflecting) and independently by Kechris [Ke 2] (who in the context of descriptive set theory proved that any Spector 2-class Γ closed under $\forall^1$ for which $\mathcal{H}\,\mathcal{F} \in \Delta$-like e.g. Π^1_{2n+1}, for $n > 0$, from PD - is reflecting).

B. Rigid Spector 2-classes

§13. General results and examples.

The structure of rigid Spector 2-classes is in general better understood than that of the reflecting ones. This will become apparent as we proceed. As a first step we shall state now a number of equivalent but quite different looking versions of the notion of rigidity, which will hopefully illuminate its character.

13.1. Theorem. Let Γ be a Spector 2-class on $\mathcal{M}$. Then the following are equivalent:

i) Γ is rigid

ii) There is a universal $\mathcal{U}(x,Y)$ in Γ such that the second order relation:

$$\mathcal{R}_{\mathcal{U}}(W,Y) \Leftrightarrow W = \mathcal{U}^Y$$

is in Γ.

(Recall here that $\mathcal{U}(x,Y) \in \Gamma$ is universal if for each $\mathcal{P}(x,Y)$ in Γ there is $e \in \omega$ such that $\mathcal{P}(x,Y) \Leftrightarrow \mathcal{U}(\langle e,x\rangle, Y)$.

iii) Γ admits Γ-norms without gaps i.e. there is $\mathcal{U}$ universal in Γ and a Γ-norm $\sigma : \mathcal{U} \to$ Ordinals such that

$$\mathcal{U}(x,Y) \wedge \sigma(x,Y) > \xi \Rightarrow \exists x^*(\mathcal{U}(x^*,Y) \wedge \xi = \sigma(x^*,Y))$$

iv) Let $\underset{\sim}{\delta}^Y = \sup\{\xi : \xi$ is the length of a $\underset{\sim}{\Delta}(Y)$ $(= \underset{\sim}{\Gamma}(Y) \cap \underset{\sim}{\check{\Gamma}}(Y))$ prewellordering of $M\}$. Then the second order relation

$$\mathcal{T}(X,Y) \Leftrightarrow \underset{\sim}{\delta}^X \leq \underset{\sim}{\delta}^{\bar{Y}}$$

is in $\check{\Gamma}$.

v) (Harrington) The second order relation

$$\mathcal{P}(X,Y) \Leftrightarrow X \in \Gamma(Y) - \underset{\sim}{\Delta}(Y)$$

is in Γ.

Before we proceed to the proofs of these equivalences let us comment on them. ii) is just a convenient technical reformulation. iii) makes precise and clear the intuitive but vague idea that reflection has something to do "with gaps in hierarchies (induced by norms on second order relations)". Clearly any Spector 2-class which is inductively generated in a reasonable fashion by operations on M and in which elements of p(M) are just carried through as parameters, admits norms without gaps. Hence,

13.2. <u>Corollary</u>. The examples in Sections 4-7 and $\Pi_1^1(\mathfrak{M})$, for countable $\mathfrak{M}$, from Section 8, are rigid Spector 2-classes.

See now 14.1.ii) here.

Finally, in relation to iv) the reader should recall the very useful fact in the theory of Π_1^1 sets on $\underset{\sim}{\omega}$ that

$$\omega_1^X \leq \omega_1^Y$$

is Σ_1^1. iv) shows that the Spector 2-classes which share the abstract formulation of this property are exactly the rigid ones.

We proceed now to the

Proof of Theorem 13.1. Clearly i) ⇒ ii). To show ii) ⇒ iii) let $\mathcal{U}$ be as in ii), let σ' be a Γ-norm on $\mathcal{U}$ and let σ be its corresponding collapsed norm i.e. for each $(x,Y) \in \mathcal{U}$

$\sigma(x,Y)$ = order type of $\{(x',Y) : \sigma'(x',Y) \leq \sigma'(x,Y)\}$ under the prewellordering on $\mathcal{U}$ induced by σ'.

Clearly σ has no gaps. It is enough therefore to prove it is a Γ-norm. Clearly if $(x,Y) \in \mathcal{U}$, then $(x',Y') \in \mathcal{U} \wedge \sigma(x',Y') \leq \sigma(x,Y)$ is in Γ. We will now show that it is also in $\breve{\Gamma}$ which will complete the proof.

Note that $\underset{\sim}{\delta}^{Y'} = \sup\{\sigma(x,Y') : (x,Y') \in \mathcal{U}\}$. Thus for $(x,Y) \in \mathcal{U}$, the relation $\sigma(x,Y) \geq \underset{\sim}{\delta}^{Y'}$ is in Δ. Indeed it is clearly in $\breve{\Gamma}$ and to see that it is also in Γ note that by the Spector Criterion, if $\sigma(x,Y) \geq \underset{\sim}{\delta}^{Y'}$ then $\mathcal{U}^{Y'} \in \underset{\sim}{\Delta}(Y,Y')$ so

$$\sigma(x,Y) \geq \underset{\sim}{\delta}^{Y'} \Leftrightarrow \exists\, W \in \underset{\sim}{\Delta}(Y,Y')(W = \mathcal{U}^{Y'} \wedge \forall\, z \in W(\sigma(z,Y') \leq \sigma(x,Y)).$$

From the above we conclude now that if $(x,Y) \in \mathcal{U}$ then:

$$\begin{aligned}&(x',Y') \in \mathcal{U} \wedge \sigma(x',Y') \leq \sigma(x,Y) \Leftrightarrow\\ &[\sigma(x,Y) \geq \underset{\sim}{\delta}^{Y'} \wedge \forall\, W \in \underset{\sim}{\Delta}(Y,Y')(W = \mathcal{U}^{Y'} \Rightarrow x' \in W)] \vee\\ &[\sigma(x,Y) < \underset{\sim}{\delta}^{Y'} \wedge \forall\, x^*((x^*,Y') \in \mathcal{U} \wedge \sigma(x^*,Y') =\\ &\sigma(x,Y) \Rightarrow \sigma'(x^*,Y') \geq \sigma'(x',Y'))],\end{aligned}$$

which completes the proof.

To prove that iii) ⇒ iv) let $\mathcal{U}$, σ be as in iii). Then

$$\underset{\sim}{\delta}^X \leq \underset{\sim}{\delta}^Y \Leftrightarrow \forall x[(x,X) \in \mathcal{U} \Rightarrow \exists y(\sigma(y,Y) = \sigma(x,X))].$$

For the direction iv) ⇒ i) let $\mathcal{U}(x,Y)$ be such that for each $Y, \mathcal{U}^Y \in \Gamma(Y) - \underset{\sim}{\Delta}(Y)$. Let σ be a Γ-norm on $\mathcal{U}$. Then

$$W = \mathcal{U}^Y \Leftrightarrow W \subseteq \mathcal{U}^Y \wedge \forall x \in W\, \forall x'(\sigma(x',Y) \leq \sigma(x,Y) \Rightarrow x' \in W) \wedge \underset{\sim}{\delta}^{Y,W} > \underset{\sim}{\delta}^Y.$$

Replacing $\underset{\sim}{\delta}^{Y,W} > \underset{\sim}{\delta}^Y$ by $W \in \Gamma(Y) - \underset{\sim}{\Delta}(Y)$ we obtain a proof that v) ⇒ i).

Finally to prove that ii) ⇒ v) let $\mathcal{U}(x,Y)$ be universal. Then

$$X \in \Gamma(Y) - \underset{\sim}{\Delta}(Y) \Leftrightarrow \exists\, W \in \underset{\sim}{\Delta}(X,Y)(W = \mathcal{U}^Y \wedge \wedge\, \exists\, e \in \omega(W_e = X) \wedge \underset{\sim}{\delta}^{X,Y} > \underset{\sim}{\delta}^X).$$

⊣

Our next goal is to bring forth some aspects of the theory of rigid Spector 2-classes related to the comparison between monotone and nonmonotone inductive definability.

If $\mathcal{F}$ is a 2-class, by $\mathcal{F}^{\mathrm{mon}}$ we denote the collection of operative $\varphi(\overline{x},\overline{Y},S)$ in $\mathcal{F}$ which are monotone in S. A well-known result of Grilliot asserts that for $\mathfrak{M} = \underset{\sim}{\omega}$, ${}_2\mathrm{IND}(\mathbf{\Sigma}_1^1) = {}_2\mathrm{IND}(\Sigma_1^{1,\mathrm{mon}})$. The next result provides the proper content that explains this theorem. It was proved originally for most interesting rigid Γ's as a straightforward combination of a result of Harrington (14.1. below) and a result of Harrington-Moschovakis (14.2. below). The general version given next (and a different proof) is due to Harrington-Kechris [H-K 1].

13.3. Theorem. Let Γ be a Spector 2-class on $\mathfrak{M}$. If Γ is rigid, then ${}_2\mathrm{IND}(\breve{\Gamma}) = {}_2\mathrm{IND}((\breve{\Gamma})^{\mathrm{mon}})$.

It is an open problem if the converse holds as well.

A corresponding result holds for inductions in Γ as well (for most Γ's).

13.4. Theorem (Harrington-Kechris [H-K 1]). Let Γ be a Spector 2-class on $\mathfrak{M}$. If $\mathcal{U}\mathcal{F} \in \Delta$ and Γ is rigid, then ${}_2\mathrm{IND}(\Gamma) = {}_2\mathrm{IND}(\Gamma^{\mathrm{mon}})$.

The hypothesis $\mathcal{U}\mathcal{F} \in \Delta$ is needed as the counterexample $\Gamma = {}_2\mathrm{IND}(\mathfrak{M})$ on a countable $\mathfrak{M}$ shows.

It seems also relevant to mention here the following

13.5. Theorem (Aanderaa [Aa]). If $\mathcal{F}$ is typical, nonmonotone on $\mathfrak{M}$ and if $\mathcal{F}$ is normed and closed under $\exists^M$ then ${}_2\mathrm{IND}(\mathcal{F}) < {}_2\mathrm{IND}(\breve{\mathcal{F}})$.

In particular, for any Spector 2-class Γ, ${}_2\mathrm{IND}(\Gamma) < {}_2\mathrm{IND}(\breve{\Gamma})$.

§14. The Harrington Representation Theorem

The structure of rigid Spector 2-classes is made transparent by the following result of Harrington.

14.1. Theorem (Harrington [Ha 2]).

i) Let Γ be a Spector 1-class on $\mathfrak{M}$. Then there is a quantifier Q on $\mathfrak{M}$ such that $\Gamma = {}_1\mathrm{IND}(\mathfrak{M}, Q)$.

ii) Let Γ be a regular Spector 2-class on $\mathfrak{M}$ i.e. $\Gamma = {}_2\mathrm{IND}(\mathfrak{M})$ or else $\mathcal{U}\mathcal{F} \in \Delta$. If Γ is rigid, then $\Gamma = {}_2\mathrm{IND}(\mathfrak{M}, Q)$ for some quantifier Q on M.

The reader should go back and recall 13.1. iii) in connection with ii) above.

The restriction of regularity is needed above as a counterexample of Harrington [Ha 2] shows. Since however any natural Spector 2-class Γ which is not equal to ${}_2\mathrm{IND}(\mathfrak{M})$ strongly contains this minimal Spector 2-class, a fact which implies that $\mathcal{U}\,\mathcal{J} \in \Delta$, this condition of regularity is always met in practice, modulo pathological counterexamples.

By combining i) and ii) we conclude also that every Spector 1-class can be extended to a rigid Spector 2-class. We will see later on that some Spector 1-classes cannot be extended to reflecting Spector 2-classes.

For regular rigid Spector 2-classes one has the following stronger version of 13.3, which was actually proved before that result.

14.2. Theorem (Harrington-Moschovakis [H-Mo]). Let Q be a quantifier on M. Let $\Gamma = {}_2\mathrm{IND}(\mathfrak{M},Q)$. Then

$${}_2\mathrm{IND}(\breve{\Gamma}) = {}_2\mathrm{IND}((\breve{\Gamma})^{\mathrm{pos}}) = {}_2\mathrm{IND}(\mathfrak{M},Q^+).$$

Here, for $\Gamma = {}_2\mathrm{IND}(\mathfrak{M},Q)$,

$(\breve{\Gamma})^{\mathrm{pos}}$ = all operative $\varphi(\bar{x},\bar{Y},S)$ of the form $Q^+ u\, \psi(u, \bar{x}, \bar{Y}, S_+)$, where ψ is elementary in $\langle \mathfrak{M}, \mathrm{Seq}, \ell h, (x)_i \rangle$ and S occurs positively in ψ.

(Recall here the remarks at the end of Section 4).

In view of 14.2 and 13.3 one can define a strong jump of rigid Spector classes:

$$j^+(\Gamma) = {}_2\mathrm{IND}(\breve{\Gamma}).$$

If $\Gamma = {}_2\mathrm{IND}(\mathfrak{M},Q)$, then $j^+(\Gamma) = {}_2\mathrm{IND}(\mathfrak{M},Q^+)$.

§15. Characterization and Classification problems.

An interesting general problem in the theory of Spector (1- or 2-) classes is to provide reasonable abstract characterizations of various significant collections of examples which arise in inductive definability, higher type recursion etc. and thereby gain a better understanding of the role of these examples in the general theory and hopefully move towards some sort of classification theory of Spector (1- or 2-)classes. The Harrington Representation Theorem clearly settles these questions for the case of positive elementary induction in a quantifier. We shall concentrate here on another important example: recursion in a type 2 object.

For simplicity we shall state the next result only for boldface Spector 1-classes.

15.1. <u>Theorem</u> (Harrington-Kechris [H-K 2], Simpson). Let Γ be a boldface Spector 1-class on $\mathcal{M}$. Then the following are equivalent:

i) $\Gamma = {}_1\underset{\sim}{\text{env}}(\mathcal{M}, {}^2E^{\#}_M, {}^2F)$, for some type 2 object ${}^2F$ on M.

ii) The companion of Γ (in the sense of Moschovakis [Mos 1]) is not Mahlo.

Here the companion $\mathcal{A} = \langle \mathcal{M}; A, \varepsilon, R\rangle$ (an admissible structure above $\mathcal{M}$ - see Barwise [Ba]) is called <u>Mahlo</u> if for each $X \subseteq A \cup M$ and $x \in A \cup M$, if $X \in \Delta_1(\mathcal{A})$, then there is $\mathcal{B} = \langle \mathcal{M}; B, \varepsilon, R \cap (M \cup B)\rangle \in A$ such that $x \in M \cup B$ and also $\langle \mathcal{B}, X \cap (B \cup M)\rangle$ is admissible above $\mathcal{M}$.

We immediately now have

15.2. <u>Corollary</u>. If Γ, a Spector 1-class on $\mathcal{M}$, is Π^0_2-compact then $\Gamma \neq_1 \text{env}(\mathcal{M}, {}^2E^{\#}_M, {}^2F)$ for any type 2 object ${}^2F$ on M.

<u>Proof</u>. By [Mos 2], the companion of $\underset{\sim}{\Gamma}$ is Π_3-reflecting, therefore Mahlo. $\dashv$

15.3. <u>Corollary</u>. For each type 2 object ${}^2F$ on M, ${}_1\text{env}(\mathcal{M}, {}^2E^{\#}_M, {}^2F)$ is never equal to ${}_1\text{IND}(\mathcal{F})$ for any typical, nonmonotone $\mathcal{F} \supseteq \Pi^0_2$. It is also never equal to ${}_1\Gamma$ for any reflecting Spector 2-class Γ.

It would be nice to have also a characterization of the Spector 1-classes of the form ${}_1\text{IND}(\mathcal{F})$, with $\mathcal{F}$ typical, nonmonotone (possibly also satisfying other reasonable properties) and containing Π^0_2 to avoid trivialities. Every such class should be Π^0_2-compact but we do not know if this is enough.

15.4. <u>On the superjump</u> (Harrington [Ha 3,4]). Let us take now $\mathcal{M} = \underset{\sim}{\omega}$. In concluding this section we shall illustrate how the results of Harrington on the superjump fit nicely in this context. In [Ha 4] Harrington introduces a type 2 functional ${}^2S^{\#}$ such that ${}^2S^{\#}$ is normal and for any Spector 1-class Γ the following are equivalent:

i) Γ is Mahlo (i.e. its companion is Mahlo)

ii) ${}^2S^{\#}$ is Γ on Γ.

Therefore, the smallest Mahlo Spector 1-class is equal to ${}_1\text{env}({}^2S^{\#})$. By the general companion theory it follows now that for $P \subseteq \omega^n$:

$$P \in {}_1\text{env}({}^2S^{\#}) \Leftrightarrow P \in \Sigma_1(L_{\rho_0}),$$

where ρ_0 = least recursively Mahlo ordinal.

Roughly speaking ${}^2S^{\#}$ diagonalizes "over the type" of inductive definition on ω used to define recursion in type 2 objects. Moreover, as Harrington shows in [Ha 4] ${}^2S^{\#}$ is recursive in ${}^3S$ (=type 3 superjump), while for each type 2 object ${}^2F$, the superjump of ${}^2F$ is uniformly recursive in ${}^2S^{\#}$, ${}^2F$. Thus

$$_1sc({}^2S^{\#}) = p(\omega) \cap L_{\rho_0} = {}_1sc({}^3S),$$

where $_1sc(*) \equiv$ 1-section of $* \equiv^{def}$ the subsets of ω Kleene recursive in $*$.

C. Reflecting Spector 2-classes

§16. General theory and examples.

By analogy with 13.1. we shall provide first a list of equivalent reformulations of the notion of reflection.

16.1. Theorem. Let Γ be a Spector 2-class on $\mathcal{M}$. Thenthe following are equilent:

i) Γ is reflecting.

ii) For each $\mathcal{P}$ in Γ, $\mathcal{S}(\bar{x},\bar{Y}) \Leftrightarrow \exists Z \in \Gamma(\bar{x},\bar{Y})\mathcal{P}(\bar{x},\bar{Y},Z)$ is also in Γ.

iii) $\mathcal{U}\mathcal{F} \in \Delta$ and Γ is monotone reflecting i.e. for each $\bar{Y}$, each $\varphi(X)$in $\Gamma(\bar{Y})$ which is monotone and each $R \in \Gamma(\bar{Y})$

$$\varphi(R) \Rightarrow \exists R^* \in \Delta(\bar{Y})[R^* \subseteq R \wedge \varphi(R^*)]$$

iv) As in iii) with $\varphi \in \Delta(\bar{Y})$.

v) $\mathcal{U}\mathcal{F} \in \Delta$ and for each quantifier $Q \in \Gamma(\bar{Y})$ (or even $\Delta(\bar{Y})$) $_1\Gamma(\bar{Y})$ is closed under Q.

vi) $\mathcal{U}\mathcal{F} \in \Delta$ and for each $\bar{Y}$, $_2IND(\Gamma(\bar{Y})^{mon}) = \Gamma(\bar{Y})$.

vii) As in vi) with $_2IND(\ldots) = \Gamma(\ldots)$ replaced by $_1IND(\ldots) = {}_1\Gamma(\ldots)$.

Before we proceed to the proofs of these equivalences let us make some comments again. ii) should be compared with the fact that every Spector 2-class is closed under $\exists Z \in \Delta(\bar{x},\bar{Y})$. Strengthening $\Delta(\ldots)$ to $\Gamma(\ldots)$ leads to reflection. iii) is a key reformulation since it reduces reflection to its monotone version provided $\mathcal{U}\mathcal{F} \in \Delta$. This extra hypothesis is needed since $\Pi^1_1(\mathcal{M})$, for $\mathcal{M}$ countable, is monotone reflecting, but of course rigid. From this point of view reflection appears to be triggered by the fact that $\mathcal{U}\mathcal{F} \in \Delta$. iv) is a technical strengthening of iii), while v), vi), vii) are essentially convenient different ways of looking at iii). For example, from v) we immediately now obtain

Corollary (Kechris [Ke 2]). Let Γ be a Spector 2-class on $\mathcal{M}$ such that $\mathcal{U}\mathcal{F} \in \Delta$ and Γ is closed under either $\exists^1$ or $\forall^1$. Then Γ is reflecting.

Proof. We verify v). If $R(x, -)$ is in ${}_1\Gamma(\overline{Y})$ and $Q \in \Gamma(\overline{Y})$ then by standard tricks

$$Q\, x\, R(x, -) \Leftrightarrow \exists X\, [X \in Q \wedge \forall x \in X\, R(x, -)]$$
$$\Leftrightarrow \forall X\, [\forall x\, (R(x, -) \Rightarrow x \in X) \Rightarrow X \in Q].$$
⊣

Corollary i) (Kechris [Ke 2]). The examples in Section B except for $\Pi_1^1(\mathcal{M})$, for countable $\mathcal{M}$, are reflecting.

ii) (Harrington [Ha 1]). The example in Section 9 - recursion in type 3 objects - is reflecting.

We give now the

Proof of Theorem 16.1. i) ⇒ ii) follows from the fact that if Γ is reflecting then for $\theta \in \Gamma$

$$\exists Z \in \Gamma(\overline{x},\overline{Y})\theta(\overline{x},\overline{Y},Z) \Leftrightarrow \exists Z \in \Delta(\overline{x},\overline{Y})\theta(\overline{x},\overline{Y},Z).$$

To prove that ii) ⇒ i) let $W \in \Gamma(\overline{Y}) - \underset{\sim}{\Delta}(\overline{Y})$ be a $\Gamma(\overline{Y})$ singleton towards a contradiction. Then

$$\overline{x} \notin W \Leftrightarrow \exists U \in \Gamma(\overline{Y})(U \in \{W\} \wedge x \notin U)$$

so $W \in \Delta(Y)$, a contradiction.

To see that i) ⇒ iii) it is clearly enough to show that $\mathcal{WF} \in \Delta$, if Γ is reflecting. But noting that for $X \subseteq M$ the wellfounded part of $<_X \equiv^{\text{def}} \{(x,y) : \langle x,y\rangle \in X\}$ is in ${}_2\text{IND}(\langle \mathcal{M}, X\rangle)$, therefore in $\Gamma(X)$, we have

$$\neg\mathcal{WF}(X) \Leftrightarrow \exists Z \in \Gamma(X)(\text{Field}(<_X) - Z \neq \emptyset \wedge \text{Field}(<_X) - Z \text{ has no } <_X\text{-minimal element}]$$

so by ii) $\mathcal{WF} \in \Delta$.

Obviously iii) ⇒ iv), so let us prove iv) ⇒ i): Let $\varphi(X)$ be in $\Gamma(\overline{Y})$ and assume $R \in \Gamma(\overline{Y}) - \underset{\sim}{\Delta}(\overline{Y})$ is the unique solution of φ, towards a contradiction. Let σ be a $\Gamma(\overline{Y})$-norm on R. Put

$\psi(\leq) \Leftrightarrow \leq$ is a prewellordering which is an initial segment (not necessarily proper) of the prewellordering $\leq^\sigma$ induced on R by σ.

Then easily $\psi \in \Gamma(\overline{Y})$. Also if

$$\psi'(\leq) \Leftrightarrow \psi(\leq) \wedge \varphi(\text{Field }(\leq))$$

then ψ' has the unique solution $\leq^\sigma$. So by Spector's Criterion if

$$\theta(X) \Leftrightarrow X \in \mathcal{WF} \wedge |X| \geq \underset{\sim}{\delta}^{\overline{Y}}$$

then

$$\mathcal{O}(X) \Leftrightarrow X \in \mathcal{W}\mathcal{F} \wedge \exists \leq \in \underset{\sim}{\Delta}(x,\overline{Y})$$
$$[\psi'(\leq) \wedge |\leq| \leq |X|]$$

so $\mathcal{O} \in \Delta(\overline{Y})$. If now

$$\mathcal{H}(X) \Leftrightarrow X \notin \mathcal{W}\mathcal{F} \vee \mathcal{O}(X)$$

then $\mathcal{H} \in \Delta(\overline{Y})$ is monotone and is satisfied by $X = \{\langle x,y\rangle : x, y \in W \wedge \sigma(x) < \sigma(Y)\}$ so by monotone reflection it is satisfied by some $Z \subseteq X$, $Z \in \Delta(\overline{Y})$. Clearly $Z \in \mathcal{W}\mathcal{F}$, therefore $\mathcal{O}(Z)$ i.e. $|Z| \geq \underset{\sim}{\delta}^{\overline{Y}}$, a contradiction.

To prove that iii) $\Rightarrow$ v) let $Q \in \Gamma(\overline{Y})$ be a quantifier and let $P(x,y)$ be in ${}_1\Gamma(\overline{Y})$. Then

$$Qy\ P(\overline{x},y) \Leftrightarrow \{y : P(\overline{x},y)\} \in Q$$
$$\Leftrightarrow \exists Z \in \Delta(\overline{x},\overline{Y})[Z \subseteq \{y: P(\overline{x},y)\} \wedge Z \in Q],$$

by monotone reflection and we are done. For the converse we prove that v) $\Rightarrow$ iv). So let $\varphi(X)$ be in $\Delta(\overline{Y})$ and monotone and assume $\varphi(R)$ holds, where $R \in \Gamma(\overline{Y})$. We can clearly assume that φ is a quantifier (i.e. $\varphi \neq \emptyset$ and $\varphi \neq p(M)$) and towards a contradiction that for no $R' \in \underset{\sim}{\Delta}(\overline{Y})$, $R' \subseteq R$, $\varphi(R')$ holds. Then if σ is a $\Gamma(\overline{Y})$-norm on R we have

$$\overline{x} \notin W \Leftrightarrow \varphi(\{\overline{y} : \overline{y} <^*_\sigma \overline{x}\}),$$

so $W \in \Delta(\overline{Y})$, a contradiction. Thus there is some $R' \in \underset{\sim}{\Delta}(\overline{Y})$ such that $\varphi(R')$ holds. From this by standard arguments on Spector 2-classes it follows that there is $R^* \in \Delta(\overline{Y})$ such that $\varphi(R^*)$.

To see that i) $\Rightarrow$ vi) let $\varphi(\overline{x},\overline{Z},S)$ be in $\Gamma(\overline{Y})^{mon}$. To prove that ${}_2\mathrm{IND}(\Gamma(\overline{Y})^{mon}) \subseteq \Gamma(\overline{Y})$ it is enough by Theorem 3.4. to show that φ is $\Gamma(\overline{Y})$ on $\Gamma(\overline{Y})$. But if $\mathcal{O}(\overline{x}',\overline{u},\overline{V})$ is in $\Gamma(\overline{Y})$ then by reflection

$$\varphi(\overline{x},\overline{Z}, \{\overline{x}' : \mathcal{O}(\overline{x}',\overline{u},\overline{V})\}) \Leftrightarrow$$
$$\exists S \in \Delta(\overline{x},\overline{Z},\overline{u},\overline{V},\overline{Y})[\varphi(\overline{x},\overline{Z},S) \wedge S \subseteq \{\overline{x}' : \mathcal{O}(\overline{x}',\overline{u},\overline{V})\}] .$$

Since obviously vi) $\Rightarrow$ vii) we can complete the proof of the theorem by proving vii) $\Rightarrow$ v). So let $Q \in \Gamma(\overline{Y})$ and $P(\overline{x},y) \in \Gamma(\overline{Y})$. We want to show that $Qy\ P(\overline{x},y)$ is also in $\Gamma(\overline{Y})$. For that consider the following system of simultaneous inductions

$$\varphi_1(\overline{x},y,S,U) \Leftrightarrow P(\overline{x},y)$$
$$\varphi_2(\overline{x},S,U) \Leftrightarrow Qy\ S(\overline{x},y)$$

Clearly they are in $\Gamma(\overline{Y})$ and monotone. Moreover $\varphi_1^\infty(\overline{x},y) \Leftrightarrow P(\overline{x},y)$ and $\varphi_2^\infty(\overline{x}) \Leftrightarrow Qy\ \varphi_1^\infty(\overline{x},y) \Leftrightarrow Qy\ P(\overline{x},y)$. Since $\varphi_2^\infty \in \Gamma(\overline{Y})$ we have that $Qy\ P(\overline{x},y) \in \Gamma(\overline{Y})$ and we are done. $\dashv$

In view of vi) we have that for any reflecting Spector 2-class Γ on $\mathfrak{M}$

(*) $$_2\mathrm{IND}(\Gamma^{\mathrm{mon}}) = \Gamma.$$

This should be contrasted with 13.4. (*) seems to put in a proper context standard results like (say for $\mathfrak{M} = \underset{\sim}{\omega}$) $_2\mathrm{IND}(\Sigma_2^{1,\mathrm{mon}}) = \Sigma_2^1$ etc.

What about $_2\mathrm{IND}((\breve{\Gamma})^{\mathrm{mon}})$? Recall that 13.3 says that if Γ is rigid then $_2\mathrm{IND}((\breve{\Gamma})) = {}_2\mathrm{IND}(\breve{\Gamma})$. One should suspect by analogy with the preceding fact (*) that if Γ is reflecting then

(**) $$_2\mathrm{IND}((\breve{\Gamma})^{\mathrm{mon}}) = \breve{\Gamma}.$$

This is however an open problem. We can however establish the converse of (**) easily from 16.1.

<u>Corollary</u>. Let Γ be a Spector 2-class on $\mathfrak{M}$. If for all $\overline{Y}$, $_1\mathrm{IND}(\breve{\Gamma}(\overline{Y})^{\mathrm{mon}}) = {}_1\breve{\Gamma}(\overline{Y})$ then Γ is reflecting.

<u>Proof</u>. Assume the hypothesis. We prove 16.1 v). Let Q be a quantifier in $\Gamma(\overline{Y})$ and let $P(\overline{x},y) \in \Gamma(\overline{x})$. We have now

$$\begin{aligned} Qy\ P(\overline{x},y) &\Leftrightarrow \neg Qy \neg P(\overline{x},y) \\ &\Leftrightarrow \neg \breve{Q}y\ S(\overline{x},y) \end{aligned}$$

where $S = \neg P \in \breve{\Gamma}(\overline{Y})$. So it is enough to show that $\breve{Q}yS(\overline{x},y)$ is in $\breve{\Gamma}(\overline{Y})$. But clearly $\breve{Q}$ is a quantifier in $\breve{\Gamma}(\overline{Y})$ so, as in the proof that 16.1 vii) $\Rightarrow$ v), $\breve{Q}y\ S(\overline{x},y) \in \breve{\Gamma}(\overline{Y})$. ⊣

§17. <u>The smallest reflecting Spector 2-class</u>.

We shall present here Harrington's construction of the smallest reflecting Spector 2-class. For simplicity we shall take $\mathfrak{M} = \underset{\sim}{\omega}$ here.

17.1. <u>Definition</u>. Let for each $X \subseteq \omega$,

$$\pi_0(X) = \text{least non-projectible ordinal relative to } X.$$

Let also Π_0 be the 2-class defined as follows:

$$\mathcal{P}(X) \in \Pi_0 \Leftrightarrow \text{there is a } \Sigma_1 \text{ formula, with no parameters, such that } \mathcal{P}(X) \Leftrightarrow L_{\pi_0(X)}[X] \models \varphi(X).$$

We now have

17.2. <u>Theorem</u> (Harrington [Ha 5]). The 2-class Π_0 is the smallest reflecting Spector 2-class on $\underset{\sim}{\omega}$.

Harrington [Ha 5] constructs also a type 3 functional ${}^3CL$ which is essentially the diagonalization operator for arbitrary inductive definitions on ω. He shows that ${}^2E$, ${}^3CL$ is normal and moreover that for any Spector 2-class Γ:

Γ is reflecting iff ${}^3CL$ is Γ on Γ. Thus

$$\Pi_0 = {}_2\mathrm{env}({}^2E, {}^3CL).$$

Furthermore, he shows that if ${}^3R^*$ is the Liapunov operator (i.e. ${}^3R^*({}^2F_Q) = {}^2F_{\hat{Q}}$, where $\hat{Q}x\, P(x) \Leftrightarrow Qx_0\, \breve{Q}x_1 \ldots \forall n\, P(\langle x_0 \ldots x_n\rangle))$ then ${}^3CL$ is recursive in ${}^3R^*$, ${}^2E$ while ${}^2F_{\hat{Q}}$ is uniformly recursive in ${}^3CL$, ${}^2E$ and ${}^2F_Q$. Thus

$$ {}_1sc({}^3R^{*2}E) = {}_1sc({}^3CL, {}^2E) = p(\omega) \cap L_{\sigma_0(\pi_0)},$$

where $\sigma_0(\pi_0)$ = least ordinal stable in π_0.

And we conclude this part of the paper with two open problems:

1) Can we represent every reflecting Spector 2-class as ${}_2\mathrm{env}({}^3CL, \Phi)$, for some type 3 functional Φ so that ${}^3CL, \Phi$ is normal ?

2) Find a reasonable abstract characterization of the collection of all ${}_1\Gamma$ for reflecting Spector 2-classes Γ. Is for example Π_0-compactness enough?

PART III. REFLECTION IN RECURSION IN HIGHER TYPES

§18. Inductive analysis of the 2-envelope of a type 3 object.

The farther theory of reflection depends usually on the special properties of the Spector 2-classes on which it is applied. For example, in the context of descriptive set theory under PD reflection mixes with the uniformization and basis properties of the classes Π^1_{2n+1} for $n > 0$ to produce what is known as Q-theory (see Martin-Solovay [M-S] and Kechris [Ke 2]). Our purpose in the last part of this paper will be to present an exposition of the form that reflection takes in the context of recursion in type 3 objects. Before we do that however it will be useful to review some important facts about 2-envelopes of type 3 objects.

Recall first that an E-Spector 2-class on $\mathcal{M}$ is a Spector 2-class Γ on $\mathcal{M}$ such that

i) Γ is closed under $\forall^1$

ii) Γ is closed under the deterministic $\exists^1$.

Recall also that ${}_2\mathrm{env}(\mathfrak{M}, {}^3E_M, {}^3F)$ is the smallest E-Spector 2-class Γ such that ${}^3F$ is Δ on Δ. In particular ${}_2\mathrm{env}(\mathfrak{M}, {}^3E_M)$ is the smallest E-Spector 2-class on $\mathfrak{M}$.

It will be convenient in the following to let (following Sacks)

$$I = p(M) = \text{set of } \underline{\text{individuals}}$$

(and occasionally call M = set of <u>subindividuals</u>). If Γ is a 2-class we call as usual $\mathcal{U} \in \Gamma$, $\mathcal{U} \subseteq I$ universal if for each $\mathcal{P} \subseteq I$ in Γ, there is $e \in \omega$ such that $\mathcal{P}(X) \Leftrightarrow \langle e,X\rangle \in \mathcal{U}$.

As it turns out one can single out the 2-envelopes of type 3 objects among all E-Spector 2-classes by an important structural property which we now isolate.

18.1. <u>Definition</u>. An E-Spector 2-class Γ is <u>Π_1-like</u> <u>inductive</u> if there is $\mathcal{R} \subseteq I \times I$ such that

i) $\mathcal{R} \in \Gamma$

ii) $\mathcal{R}$ is <u>piecewise Δ</u> i.e. for each $X \in I$, $\mathcal{R}_X = \{Y : \mathcal{R}(X,Y)\} \in \Delta(X)$,

and if we define the operator

$$\Phi(X,\mathcal{S}) \Leftrightarrow \forall Y(\mathcal{R}(X,Y) \Rightarrow Y \in \mathcal{S})$$

then $\Phi^\infty (\subseteq I)$ is universal for Γ.

We have now

18.2. <u>Theorem</u>

i) (Moschovakis [Mos 5]). For every type 3 object ${}^3F$, ${}_2\mathrm{env}(\mathfrak{M}, {}^3E_M, {}^3F)$ is Π_1-like inductive.

ii) (Harrington). If Γ is an E-Spector 2-class which is Π_1-like inductive then there is a type 3 object ${}^3F$ such that $\Gamma = {}_2\mathrm{env}(\mathfrak{M}, {}^3E_M, {}^3F)$.

<u>Corollary</u> (Moschovakis [Mos 5]). The 2-class ${}_2\mathrm{env}(\mathfrak{M}, {}^3E_M, {}^3F)$ is never closed under $\exists^1$.

<u>Proof</u>. If $\mathcal{R}$ is as above then

$$X \notin \Phi^\infty \Leftrightarrow \exists X_0, X_1, \ldots (X_0 = X \wedge \forall n\, \mathcal{R}(X_n, X_{n+1})). \qquad \dashv$$

For the proof of 18.1 i) one takes, $\mathcal{R}(X,Y)$ to be "Y is a subcomputation of X", where subcomputation is interpreted say in terms of the Kleene schemata. For details see for example [Ke 3].

§19. Reflecting ordinals.

In this section, Γ will denote a Spector 2-class of the form ${}_2\mathrm{env}(\mathcal{M}, {}^3E_M, {}^3F)$ or equivalently an E-Spector 2-class which is Π_1-like inductive. Many of the results and definitions will be true however for any E-Spector 2-class - see [Ke 3] for more details.

Let us first define a number of important ordinal invariants associated with Γ:

For each individual X let $\kappa_0^{\Gamma,X} \equiv \kappa_0^X \equiv^{\mathrm{def}} \sup\{\xi : \xi$ is the length of a $\Delta(X)$ prewell-ordering of $I\} \equiv$ sup of the "$\Delta(X)$-ordinals". If σ is a Γ-norm on $\mathcal{U}$, a universal set of individuals (σ mapping $\mathcal{U}$ onto an ordinal), then

$$\kappa_0^X = \sup\{\sigma(\langle e,X\rangle) : \langle e,X\rangle \in \mathcal{U}\}$$

$$= \sup\{\sigma(Y): Y \in \mathcal{U},\ Y \in \Delta(X)\}$$

$\equiv$ sup of the ordinals having integer "modulo X" notations i.e. the "X-constructive" ordinals.

Put also

$$\kappa^\Gamma = \sup\{\kappa_0^{\Gamma,X}: X \in I\} = \sup\{\sigma(X) : X \in \mathcal{U}\}.$$

In the present context Sacks and Harrington have also interpreted reflection in terms of ordinals.

19.1. Definition. An ordinal $\xi \leq \kappa^\Gamma$ is called X-reflecting if for each $e \in \omega$

$$\exists\, Y(\sigma(\langle e,X,Y\rangle) < \xi) \Rightarrow \exists\, Y(\sigma(\langle e,X,Y\rangle) < \kappa_0^X).$$

(This depends, a priori, on $\mathcal{U},\sigma$).

That there are nontrivial X-reflecting ordinals (i.e. $> \kappa_0^X$) follows from the next result, which in this context is essentially an equivalent way of saying that Γ is reflecting.

19.2. Theorem (Harrington [Ha 1]). If $P^{\Gamma,X} \equiv P^X \subseteq M$ is a $\Gamma(X)$-universal set of subindividuals then $\kappa_0^{P^X} > \kappa_0^X$ and $\kappa_0^{P^X}$ is X-reflecting.

Sacks [Sa 2] had earlier shown that $\kappa_M^X = \sup\{\sigma(\langle y,X\rangle): y \in M, \langle y,X\rangle \in \mathcal{U}\}$ is X-reflecting. For many M, like e.g. $M = p(\omega)$, $\kappa_M^X > \kappa_0^X$ but if $M = \omega$, then $\kappa_M^X = \kappa_0^X$.

On the other hand by the non-closure of Γ under $\exists^1$ we easily have that:

κ^Γ is not X-reflecting.

This leads to the

19.3. Definition. Let

$$\lambda^{\Gamma,X}_{\mathcal{U},\sigma} \equiv \lambda^X \equiv^{\text{def}} \sup\{\xi : \xi \text{ is X-reflecting}\}$$

and

$$\kappa^{\Gamma,X}_r \equiv \kappa^X_r \equiv^{\text{def}} \sup\{\kappa^{X,Y}_0 : \kappa^{X,Y}_0 \leq \lambda^X\}.$$

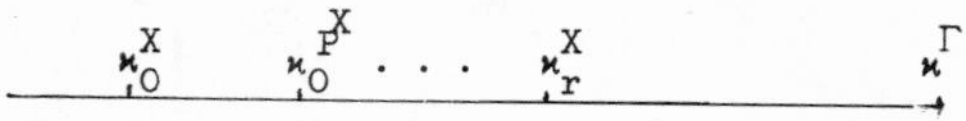

Although, as suggested by our notation, λ^X may depend on $\mathcal{U},\sigma$ $\kappa^{\Gamma,X}_r$ is an invariant of the 2-class Γ (and X). This can be seen as follows: Let

$$\mathcal{S}^{\Gamma,X} \equiv \mathcal{S}^X \equiv^{\text{def}} \{Y : \kappa^{X,Y}_0 \leq \kappa^X_r\}.$$

Since $\kappa^X_r = \sup\{\kappa^{X,Y}_0 : Y \in \mathcal{S}^X\}$ it is enough to prove that $\mathcal{S}^X$ is an invariant. This is an immediate consequence of

19.4. Theorem (Kechris [Ke 3]). For each individual X, $\mathcal{S}^X$ is the largest set $\mathcal{S} \subseteq I$ with the following property:

For each $\mathcal{P}(n,Y)$ in $\Gamma(X)$ the relation $\exists Y \in S\, \mathcal{P}(n,Y)$ is also in $\Gamma(X)$.

Before we proceed to list several properties of the ordinal invariant κ^X_r we shall give a set theoretical interpretation of the reflecting ordinals due to Harrington [Ha 1].

Let ${}^3F$ be a type 3 object such that $\Gamma = {}_2\text{env}(\mathcal{M}, {}^3E_M, {}^3F)$. Let $L_\xi[I; {}^3F]$ be the usual constructibility hierarchy built on the basis of I as a set of urelements (with a more or less obvious structure attached to it) and ${}^3F$ as an extra predicate (see Barwise [Ba]). Roughly speaking, for $\xi < \kappa^\Gamma$ one can view $L_\xi[I; {}^3F]$ as the set theoretic counterpart of $\mathcal{U}_\xi = \{Y : Y \in \mathcal{U} \wedge \sigma(Y) < \xi\}$. In fact it turns out (see [Ha 1])that $\mathcal{S} \subseteq I$ is in Γ iff there is a Σ_1 formula φ such that

$$X \in \mathcal{S} \Leftrightarrow L_{\kappa^X_0}[I; {}^3F] \models \varphi(X).$$

Call now $\xi \leq \kappa^\Gamma$ X-reflecting' iff for all Σ_1 formulas φ

$$L_\xi[I; {}^3F] \models \varphi(X) \Rightarrow L_{\kappa^X_0}[I; {}^3F] \models \varphi(X).$$

Then,

$$\kappa^X_r = \sup\{\xi : \xi \text{ is X-reflecting}'\}.$$

Terminology. κ^X_r is called the largest X-reflecting ordinal.

Here are some of its basic properties

19.5. Theorem

i) (Harrington [Ha 1]). For all X, Y : $\kappa_r^X \leq \kappa_r^{X,Y}$.

ii) For all X, Y : $\kappa_r^{X,Y} = \kappa_r^X \Leftrightarrow \kappa_0^{X,Y} < \kappa_r^X$. (See [Ke 3])

iii) For all X, κ_r^X is not attained i.e. for no Y, $\kappa_r^X = \kappa_0^{X,Y}$. (See [Ke 3]).

iv) (Harrington [Ha 1]). For all X, cofinality $(\kappa_r^X) = \omega$.

v) (Harrington). The second order relation $\kappa_r^X \leq \kappa_r^Y$ is in $\breve{\Gamma}$.

vi) For all $y \in M$ (viewed here as a subset of I) $\kappa_r^{X,y} = \kappa_r^X$.

We have introduced before the set

$$\mathcal{S}^X = \{Y : \kappa_0^{X,Y} < \kappa_r^X\} = \{Y : \kappa_r^{X,Y} = \kappa_r^X\}.$$

In view of 19.5.vi) $M \subseteq \mathcal{S}^X$. This set $\mathcal{S}^X$ turns out to play a significant role in the structure theory of Γ, as the next two results show.

19.6. Theorem (Kechris [Ke 3]). For each X, $\mathcal{S}^X$ is a basis for $\breve{\Gamma}(X)$ sets of individuals i.e. for each $\mathcal{P} \neq \emptyset$, $\mathcal{P} \in \breve{\Gamma}(X)$ there is $Y \in \mathcal{P}$ such that $\kappa_r^{X,Y} = \kappa_r^X$.

This is of course the analog of the well-known Gandy Basis Theorem in the theory of Π_1^1 sets of reals which asserts that every nonempty Σ_1^1 set of reals contains a real X with $\omega_1^X = \omega_1$. In fact, the ordinal invariant κ_r^X plays in the present context very closely the role that ω_1^X plays in the theory of Π_1^1 sets. For example, note that 19.5(v) above is the analog of the fact that $\omega_1^X \leq \omega_1^Y$ is Σ_1^1.

Our second result about $\mathcal{S}^X$ is related to the Grilliot-Harrington-MacQueen Theorem. First let us reformulate this result as a basis theorem, as it is customarily done in this context:

If $\mathcal{P} \subseteq I$ is in $\Gamma(X)$ and $\mathcal{P} \cap M \neq \emptyset$, then $\mathcal{P}$ has a nonempty $\Delta(X)$ subset.

In view of this fact let us call a set $\mathcal{S} \subseteq I$ of individuals X-good if for all $\mathcal{P} \in \Gamma(X)$ if $\mathcal{P} \cap \mathcal{S} \neq \emptyset$ then $\mathcal{P}$ contains a nonempty $\Delta(X)$ subset. Thus by the Grilliot-Harrington-MacQueen Theorem M is good while by the Moschovakis Theorem I is not.

19.7. Theorem (Kechris [Ke 3]). For each X, $\mathcal{S}^X$ is the largest X-good set. In fact, $\mathcal{P} \in \Gamma(X)$ contains a $\Delta(X)$ nonempty subset iff $\mathcal{P} \cap \mathcal{S}^X \neq \emptyset$.

Thus since $M \subsetneqq \mathcal{S}^X \subsetneqq I$, $\mathcal{S}^X$ measures exactly the obstruction to the desirable but false property that every nonempty $\Gamma(X)$ set should contain a nonempty $\Delta(X)$ subset.

§20. Applications

The concept of reflection has been used to establish a number of important results in the theory of envelopes of higher type objects. We shall concentrate here on those that are related to the general problem of the effect of the type of an object on the structure of its envelopes and sections. To start with we have

20.1. Theorem (Moschovakis [Mos 3]). If ${}^2F$ is a type 2 object on M then ${}_1\mathrm{env}(\mathfrak{M}, {}^2E^{\#}_M, {}^2F)$ is never equal to ${}_1\mathrm{env}(\mathfrak{M}, {}^3E_M, {}^3F)$ for any type 3 object ${}^3F$. Similarly for higher than 3 type objects.

Thus the 1-envelope of a type 2 object encodes its type. On the other hand one has

20.2. Theorem (The Plus 2 Theorem - Harrington [Ha 1]). The following are equivalent for any Spector 1-class γ on $\mathfrak{M}$:

i) $\gamma = {}_1\Gamma$ for an E-Spector 2-class Γ.

ii) $\gamma = {}_1\mathrm{env}(\mathfrak{M}, {}^3E_M, {}^3F)$ for some type 3 object ${}^3F$.

In particular (and this gives its name to the theorem), the 1-envelope of a higher than 3 object ${}^jF$-together with ${}^jE_M$ of course - is also the 1-envelope of a type 3 object ${}^3F$(and ${}^3E_M$).

Remark: Harrington [Ha 6] extends the preceding result to obtain a nice characterization of the collection of ${}_1\mathrm{env}(\mathfrak{M}, {}^3E_M, {}^3F)$ for type 3 objects ${}^3F$, which roughly speaking reads as follows:

A Spector 1-class Γ on $\mathfrak{M}$ is of the form ${}_1\mathrm{env}(\mathfrak{M}, {}^3E_M, {}^3F)$ for some type 3 object ${}^3F$ iff every second order relation in ${}_2\mathrm{env}(\mathfrak{M}, {}^3E_M)$ is Γ on Δ and Γ is "strongly ${}_2\mathrm{env}(\mathfrak{M}, {}^3E_M)$-reflecting". This last notion is a stronger version of ${}_2\mathrm{env}(\mathfrak{M}, {}^3E_M)$-compactness (which Harrington shows is not sufficient).

Of course there are many E-Spector 2-classes which are not equal to ${}_2\mathrm{env}(\mathfrak{M}, {}^3E_M, {}^3F)$ for any type 3 object ${}^3F$, as for example the 2-envelope of a type 4 object ${}^4F$ and ${}^4E_M$. On the other hand letting

$${}_2\mathrm{sc}(\mathfrak{M}, {}^3E_M, {}^3F) \equiv^{\mathrm{def}} \text{the } \underline{\text{2-section}} \text{ of } {}^3E_M, {}^3F$$
$$(\text{i.e. the } \Delta \text{ of } \Gamma = {}_2\mathrm{env}(\mathfrak{M}, {}^3E_M, {}^3F))$$

we have

20.3. Theorem (The Plus One Theorem - Sacks [Sa 2]). The following are equivalent for any 2-class Λ:

i) There is an E-Spector 2-class Γ such that $\Lambda = \Delta$.

ii) $\Lambda = {}_2\mathrm{sc}(\mathcal{M}, {}^3E_M, {}^3F)$, for some type 3 object ${}^3F$.

In particular, the 2-section of a higher than type 3 object ${}^jF$ and ${}^jE_M$ is also equal to the 2-section of a type 3 object ${}^3G$ and ${}^3E_M$.

An exposition of these results can be found in Moldestad [Mo 1].

Remark: For $\mathcal{M} = \underset{\sim}{\omega}$, Sacks [Sa 1] shows that the above theorem is also true if one lowers everywhere "2" to "1" and replaces of course "E-Spector" by just "Spector". In both [Sa 1,2] Sacks also obtains abstract characterizations of sections.

To summarize: Take $\mathcal{M} = \underset{\sim}{\omega}$. Then by Sacks the 1-section of a higher type object ${}^jF$ and ${}^jE$ $(j \geq 2)$ fails to encode its type being always also the 1-section of a type 2 object ${}^2G$ and ${}^2E$. On the other hand by Moschovakis the 1-envelope of ${}^jF$, ${}^jE_M$ encodes the distinction between type $j = 2$ and $j > 2$. Finally, by Harrington this envelope cannot further distinguish the type if it is bigger than 2, since it is always the 1-envelope of a type 3 object. This leads naturally to the

Question: Are there reasonable invariants "living on ω" which will distinguish further the type of the object from which they come?

Moschovakis observed that standard results imply that the following will do: Take ${}^kF$ to be a type $k \geq 2$ object and let $\Gamma_n({}^kF)$ = the collection of all relations on ω of the form $\exists\alpha^n P(x,\alpha^n)$, where P is Kleene semirecursive in ${}^kE$, ${}^kF$ and α^n varies over the type n objects on ω. Then $\vec{\Gamma}({}^kF) = \langle\Gamma_0({}^kF),\Gamma_1({}^kF),\ldots\rangle$ determines the type k of ${}^kF$ since for $k \geq 3$, k = least i such that $\Gamma_{i-2}({}^kF) \neq \Gamma_{i-3}({}^kF)$ by the Grilliot-Harrington-MacQueen Theorem and 18.2.i) (see proof of Corollary), while if $k = 2, \Gamma_0({}^2F) \neq \Gamma_0({}^jF)$ for any $j \geq 3$ by 20.1.

To what extent there is a more revealing answer to the above question remains an open problem.

Addendum. Martin has recently shown that the converse to Theorem 13.3 does not hold. He proves that if Π_0 = least reflecting Spector 2-class on $\underset{\sim}{\omega}$, then $\Pi_0 \subseteq \mathrm{IND}((\breve{\Pi}_0)^{\mathrm{mon}})$ (and therefore ${}_2\mathrm{IND}((\breve{\Pi}_0)^{\mathrm{mon}}) = {}_2\mathrm{IND}(\breve{\Pi}_0))$. This also shows that ${}_2\mathrm{IND}((\breve{\Gamma})^{\mathrm{mon}}) = \breve{\Gamma}$ does not necessarily hold for all reflecting Γ (see the results before the last corollary of §16).

REFERENCES

[Aa] S. Aanderaa, Inductive definitions and their closure ordinals, J. E. Fenstad and P. G. Hinman, Eds. North-Holland, Amsterdam, 1974.

[Ac 1] P. Aczel, Stage comparison theorems and game playing with inductive definitions, preprint.

[Ac 2] P. Aczel, Quantifiers, games and inductive definitions, Proc. of 3rd Scandinavian Logic Symp., S. Kanger, Ed, North Holland, Amsterdam, 1975, 1-14.

[Ad] J. W. Addison, Some consequences of the axiom of constructibility, Fund. Math., 46(1959a), 123-135.

[A-M] J. W. Addison and Y. N. Moschovakis, Some consequences of the axiom of definable determinateness, Proc. of Nat. Acad. of Sc., USA, 59(1968), 708-712.

[Ba] K. J. Barwise, Admissible sets and structures, Springer-Verlag, New York and Heidelberg, 1975.

[B-G-M] K. J. Barwise, R. O. Gandy and Y. N. Moschovakis, The next admissible set, J. Symb. Logic, 36(1971), 108-120.

[Ga] R. O. Gandy, General recursive functionals of finite type and hierarchies of functions, Proc. of Logic Colloq. Clermont Ferrand, 1962, 5-24.

[Gr] T. J. Grilliot, Inductive definitions and computability, Trans. Amer. Math. Soc., 158(1971), 309-317.

[Ha 1] L. A. Harrington, Contributions to recursion theory on higher types, Ph.D. Thesis, M.I.T., 1973.

[Ha 2] L. A. Harrington, Monotone quantifiers and inductive definability, mimeographed notes, April 1975.

[Ha 3] L. A. Harrington, The superjump and the first recursively Mahlo ordinal, J. E. Fenstad and P. G. Hinman, Eds, Generalized Recursion Theory, North-Holland, Amsterdam, 43-52.

[Ha 4] L. A. Harrington, The superjump revisited, mimeographed note, Sept. 1974.

[Ha 5] L. A. Harrington, The Kolmogorov R-operator and the first non-projectible ordinal, mimeographed note, June 1975.

[Ha 6] L. A. Harrington, Abstract 1-envelopes, mimeographed note, Sept. 1974.

[H-K1] L. A. Harrington and A. S. Kechris, On monotone vs. nonmonotone induction, Bull. of Amer. Math. Soc. 82(1976), 888-890.

[H-K2] L. A. Harrington and A. S. Kechris, On characterizing Spector classes, J. Symb. Logic, 40(1975), 19-24.

[H-Ma] L. A. Harrington and D. MacQueen, Selection in abstract recursion theory, J. Symb. Logic, 41(1976), 153-158.

[H-Mo] L. A. Harrington and Y. N. Moschovakis, On positive induction vs. non-monotone induction, mimeographed note, 1974.

[Hi] P. G. Hinman, Hierarchies of effective descriptive set theory, Trans. Amer. Math. Soc., 142(1969),111-140.

[Ke 1] A. S. Kechris, Countable ordinals and the analytical hierarchy II, preprint.

[Ke 2] A. S. Kechris, The theory of countable analytical sets, Trans. Amer.Math. Soc., 202(1975), 259-297.

[Ke 3] A. S. Kechris, The structure of envelopes: A survey of recursion theory in higher types, M.I.T. Logic Seminar Notes, Dec. 1973.

[Ke-M] A. S. Kechris and Y. N. Moschovakis, Recursion in higher types, Handbook of Mathematical Logic, K. J. Barwise, Ed., North-Holland, 1977, 681-737.

[Kl 1] S. C. Kleene, Recursive functionals and quantifiers of finite types I, Trans. Amer. Math. Soc. 91(1959), 1-52.

[Kl 2] S. C. Kleene, Recursive functionals and quantifiers of finite type II, Trans. Amer. Math. Soc., 108(1963), 106-142.

[Kl 3] S. C. Kleene, On the forms of the predicates in the theory of constructive ordinals (second paper), Amer. J. Math. 77(1955), 405-428.

[Ma] D. A. Martin, The axiom of determinateness and reduction principles in the analytical hierarchy, Bull. Amer. Math. Soc., 74(1968), 687-689.

[M-S] D. A. Martin and R. M. Solovay, Basis Theorems for Π^1_{2k} sets of reals, to appear.

[Mol] J. Moldestad, Computations in higher types, Lecture Notes in Math., Springer-Verlag, Vol. 574.

[Mos 1] Y. N. Moschovakis, Elementary Induction on Abstract Structures, North-Holland, 1974.

[Mos 2] Y. N. Moschovakis, On nonmonotone inductive definability, Fund. Math., 82 (1974), 39-83.

[Mos 3] Y. N. Moschovakis, Structural characterizations of classes of relations, J. E. Fenstad and P. G. Hinman, Eds., Generalized Recursion Theory, North-Holland, 53-79.

[Mos 4] Y. N. Moschovakis, On the basic notions in the theory of induction,Logic, Found. of Math. & Comp. Th., R.E. Butts & J. Hintikka, Reidel,1977,207-236.

[Mos 5] Y. N. Moschovakis, Hyperanalytic predicates, Trans. Amer. Math. Soc., 129 (1967), 249-282.

[Mos 6] Y. N. Moschovakis, Descriptive Set Theory, North-Holland, forthcoming.

[R-A] W. Richter and P. Aczel, Inductive definitions and reflecting properties of admissible ordinals, Generalized Recursion Theory, J. E. Fenstad and P. G. Hinman, Eds., North-Holland, 1974, 301-381.

[Sa 1] G. E. Sacks, The 1-section of a type n object, Generalized Recursion Theory, J. E. Fenstad and P. G. Hinman, Eds, North-Holland, 1974, 81-93.

[Sa 2] G. E. Sacks, The k-section of a type n object, Amer. J. of Math., 99 (1977), 901-917.

J.E. Fenstad, R.O. Gandy, G.E. Sacks (Eds.)
GENERALIZED RECURSION THEORY II

RECURSIVE FUNCTIONALS AND QUANTIFIERS OF FINITE TYPES REVISITED I

S. C. Kleene

University of Wisconsin, Madison

In the eighteen years since my last visit to Oslo, I have cherished a very pleasant recollection of an evening spent in the home of Thoralf Skolem. I am glad that the younger Norwegian logicians are keeping up the tradition so brilliantly started by him. Also I am pleased to observe how many able investigators from all parts have been pursuing higher and generalized recursion theory while I have been looking away.

In the theory developed in my original papers (1959 and 1963) on the subject I am revisiting, there is a partial recursive function $\phi(\sigma^2,\underline{a})$ such that $\phi(\lambda\tau^1\ \theta(\underline{a},\tau^1),\underline{a})$ is a partial recursive function of $\underline{a}$ for no completely defined $\theta(\underline{a},\tau^1)$ (1963 LVI p. 110). A suggestion there (p.111) for extending the theory to avoid this anomaly was not then pursued in depth.

Platek in 1966 (to which my attention was drawn by Gandy 1967 p. 239) used my 1963 p. 110 example to show that the first recursion theorem (Kleene IM 1952a p. 348) does not hold in my 1959, 1963 theory (as I also did in 1963 LXVI pp. 120, 125). Platek developed a theory avoiding these anomalies by using what he called "hereditarily consistent" functionals.

The present treatment (which for me is a resurrection of my 1963 p. 111 suggestion) gives the first recursion theorem a central role. I had already been thinking along these lines before encountering Gandy 1967 and Platek 1966. The exact relationship of the present treatment to Platek's and other work since 1963 remains to be clarified.

1. The schemata for partial recursive functionals. 1.1. To smooth the path into the revisited theory, I shall expound it primarily for functions of the lowest four types. (The anomalies in the original theory appeared already at the third type, 2.) But I shall look further, indicating how to deal with higher types. Incidentally, the reader may observe that what is done in the theory with just the lowest four types works with the obvious omissions e.g. for just the lowest two; this will be used in 3.2 (in proving (XV.3.1)).

First, the four types of objects must be specified. I shall begin by taking them as before (but in Part II I shall extend or modify them): the natural numbers 0, 1, 2, ··· (type 0); the total one-place functions from the natural numbers into the natural numbers (type 1); and similarly for $j = 2$ and 3, the total one-place functions from type j-1 into the natural numbers (types 2 and 3). The word "total" emphasizes that the domain of definition of the type-j objects ($j > 0$) is the whole of type j-1. As variables for objects of type 0, I use as before a, b, ... , a_1, a_2, ... , α^0, β^0, ... , α_1^0, α_2^0, ... ; for type 1, α, β, ... , α_1, α_2, ... , α^1, β^1, ... , α_1^1, α_2^1, ... ; for type 2, F, G, ... , α^2, β^2, ... , α_1^2, α_2^2, ... ; for type 3, α^3, β^3, ... , α_1^3, α_2^3, ...

Secondly, we shall deal also with partial functions, such as $\phi(\mathfrak{A})$ (or $\theta(\mathfrak{A})$), where $\mathfrak{A}$ is a list of distinct variables of the four types, and for each tuple of values of $\mathfrak{A}$, $\phi(\mathfrak{A})$ (or $\theta(\mathfrak{A})$) is either undefined or has a natural number as value. "Partial" functions are thus ones whose domain of definition may be a proper subset of the ostensible range of the variable, or for functions of more than one variable of the Cartesian product of those ranges.[1]

[1]Contrary to a statement in a well-known text book, the early workers did not begin "by way of partial functions". Partial functions were introduced into recursion theory by Kleene 1938, and received rather scant attention until after they were publicized in 1943 §§ 6 ff. and IM 1952a Chapter XII. Two earlier occasions on which their introduction was positively avoided were: first, when Church in 1936 p. 352 used "potentially recursive" functions; and again, when Turing in 1936-7 p. 254 and 1937 p. 160 defined the computability of a one-place number-theoretic function f by use of a machine tape on which the value $f(n)$ is represented by the number of 1's printed between the n-th and n+1-st 0 (for $n = 0$, before the 1-st 0), which makes no sense if f is not total.

In writing expressions for (partial) functions, only the order of the variables within a type shall matter; e.g. "$\phi(\underline{a}_1,\underline{a}_2,\alpha^1,\alpha^2)$" means the same as "$\phi(\underline{a}_1,\alpha^1,\alpha^2,\underline{a}_2)$".

Thirdly, we shall have <u>functionals</u> $\phi(\Theta;\mathfrak{a})$ (or $\phi^{\Theta}(\mathfrak{a})$) where Θ is a list of functions and $\mathfrak{a}$ is a list of distinct variables. Given a list Θ of partial functions (as above), ϕ gives a partial function $\lambda\mathfrak{a}\,\phi(\Theta;\mathfrak{a})$ depending on what the partial functions Θ are (so given $\mathfrak{a}$ also, $\phi(\Theta;\mathfrak{a})$ is undefined or a natural number). In the important case that Θ is empty, $\lambda\mathfrak{a}\,\phi(\mathfrak{a})$, or in the familiar "ambiguous value" notation $\phi(\mathfrak{a})$, is simply a fixed partial function. Also, when we have a fixed list Θ in mind throughout a discussion, we may (after so announcing) write simply "ϕ" or "$\phi(\mathfrak{a})$" for the partial function $\lambda\mathfrak{a}\,\phi(\Theta;\mathfrak{a})$.

Our type-2 and 3 "objects" are of course really functionals; and so are our "functions" when they have variables of types ≥ 1. However, it is convenient for us to speak of our three categories as "objects" (of four types), "functions" and "functionals".

<u>1.2.</u> I aim to generate a class of functions $\phi(\mathfrak{a})$ which shall coincide with all the partial functions which are "computable" or "effectively decidable", so that Church's 1936 thesis (IM § 62) will apply with the higher types included (as well as to partial functions, IM p. 332).

In the process of generating these functions, we shall use functionals $\lambda\Theta\lambda\mathfrak{a}\,\phi(\Theta;\mathfrak{a})$ by which such a function $\lambda\mathfrak{a}\,\phi(\Theta;\mathfrak{a})$ is generated from previously generated such functions. And when Θ are not previously generated such functions, but say "assumed" functions, $\lambda\mathfrak{a}\,\phi(\Theta;\mathfrak{a})$ will be "computable from Θ".

Two key ideas will be built into the present formulation. The first concerns how we use the function arguments and assumed functions. We shall use the function arguments by supposing e.g. that the value $\alpha^2(\alpha^1)$ of a type-2 argument α^2 will be made available on demand, as by an "oracle" after Turing 1939, whenever in the course of a computation we have in hand an argument α^1 for α^2.[2] Similarly,

---

[2]The idea, and imagery, of an "oracle" was introduced by Turing 1939 pp. 172-173 in defining the computability of one fixed number-theoretic predicate from another. (He could just as well have talked about their representing functions,

the value $\theta(\mathfrak{b})$, if it exists, of an assumed function θ is to be made available by an "oracle" whenever we have in hand a tuple $\mathfrak{b}$ of values of its variables.

The second idea (and here I am proceeding somewhat differently than in 1959) is that, in the process of defining computable functions, we begin with certain rudimentary, indubitably computable functions, and with certain functionals which indubitably generate computable functions from computable functions. Thence we progressively enlarge the class of functions recognized as computable, by describing how to find the values of a new function by the use of functions already recognized as computable and of functionals already recognized as preserving computability. In doing this, we may operate on a tuple of arguments of the new function, and on values of it already obtained as a subfunction of it, to obtain its value for that tuple of arguments. I.e., we may obtain $\phi(\mathfrak{a})$ for a given $\mathfrak{a}$ by having already computed $\phi(\mathfrak{b})$ for enough tuples of arguments $\mathfrak{b}$ to use as the values of $\eta(\mathfrak{b})$ to satisfy, for the given $\mathfrak{a}$, the requirements of a functional $\lambda\eta\mathfrak{a}\psi(\eta;\mathfrak{a})$ for $\psi(\eta;\mathfrak{a})$ to be defined, the value of which we then take as $\phi(\mathfrak{a})$. Thus

$$\phi(\mathfrak{a}) \simeq \psi(\lambda\mathfrak{a}\,\phi(\mathfrak{a});\mathfrak{a})$$

is a "recursion" for ϕ in the most general sense. (Here "$\ldots \simeq \ldots$" means that if either side is defined, so is the other with the same value.) In the recursion theory of functions of number-variables only, the proposition that the minimal solution for ϕ of the foregoing equation (which there was just what the recursive computation gives) is partial recursive was called the "first recursion theorem" (IM p. 348). Similarly, with Θ on both sides of the equation, so that we are

---

taking as values 0 for truth and 1 for falsity.) Kleene 1952 (at the 1950 International Congress) adapted this idea to define the recursiveness (or computability) of a function $\phi(\underline{a}_1,\ldots,\underline{a}_{\underline{n}_0},\alpha_1,\ldots,\alpha_{\underline{n}_1})$ with type-1 variables by adding the idea of <u>uniformity</u>, whereby the function $\lambda\underline{a}_1\ldots\underline{a}_{\underline{n}_0}\,\phi(\underline{a}_1,\ldots,\underline{a}_{\underline{n}_0},\alpha_1,\ldots,\alpha_{\underline{n}_1})$ is computed from functions $\alpha_1,\ldots,\alpha_{\underline{n}_1}$ by a process which is <u>uniform</u> in $\alpha_1,\ldots,\alpha_{\underline{n}_1}$ as $\alpha_1,\ldots,\alpha_{\underline{n}_1}$ vary. That is, the oracle for each $\alpha_{\underline{i}}$ is questioned, and its answers are used (Turing's pronoun), in the same way whatever the function $\alpha_{\underline{i}}$ is; only the oracle's answers vary with the function $\alpha_{\underline{i}}$. Kleene 1959 extended the idea to arguments $\alpha^{\underline{j}}$ of types $\underline{j} > 1$.

generating computable functionals. Then Θ can be retained as assumed functions for entertaining computability relative to Θ, or some or all of them can ultimately be phased out by serving as the η's in later recursions.

For such a recursion to work as we intend, $\psi(\eta;\mathfrak{a})$ must be monotone in its function argument η; i.e. if $\bar{\eta} \supset \eta$ (regarding functions as sets of ordered n+1-tuples) and $\psi(\eta;\mathfrak{a})$ is defined, then $\psi(\bar{\eta};\mathfrak{a})$ is defined with the same value. We don't want $\psi(\eta;\mathfrak{a})$ to be first defined and made the value of $\phi(\mathfrak{a})$, only to have $\psi(\eta;\mathfrak{a})$ become undefined, or defined with a different value, later when more values of $\phi(\mathfrak{b})$ become available to be used as the values of $\eta(\mathfrak{b})$. But we shall start with only monotone functionals, and generate only others which are monotone.

1.3. I now put these ideas into execution by listing some schemata for defining functionals $\phi(\Theta;\mathfrak{a})$ (functions when Θ is empty). The numbering of the schemata agrees with that in 1959 for ones which correspond or agree, and differs for others.

Say that $\Theta = (\theta_1,\ldots,\theta_l)$ ($l \geq 0$) where θ_t ($t = 1,\ldots,l$) is a function of m_{t0}, m_{t1}, m_{t2}, m_{t3} variables of the types 0, 1, 2, 3, respectively. We use $\langle a_0,\ldots,a_n\rangle = \Pi_{i\leq n}\, p_i^{a_i}$ (= 1 for $n = -1$); so, for $0 \leq i \leq n$, $(\langle a_0,\ldots,a_n\rangle)_i = a_i$ (cf. 1959 p. 7, IM pp. 224, 230). Now write $m_t = \langle m_{t0},m_{t1},m_{t2},m_{t3}\rangle$ for $t = 1,\ldots,l$. The list Θ may vary. Let $m = \langle m_1,\ldots,m_l\rangle$ and $n = \langle n_0,n_1,n_2,n_3\rangle$.

For each schema, the number shown at the right is the index of an application of it, and tells everything about the application. Thus, for the first schema S0, the index $\langle 0,m,n,t\rangle$ describes the schema by 0, the number of functions Θ and the numbers of their variables by m, the numbers of variables in $\mathfrak{a}$ (which is written as $(\mathfrak{b},\mathfrak{c})$) by n, and which of the functions Θ is used on the right by t.

In S1.0, $\mathfrak{a}$ is written as $(a,\mathfrak{b})$, so $\mathfrak{b}$ is n_0-1,n_1,n_2,n_3 distinct variables of the respective types 0,1,2,3; and similarly in some other schemata.

For S4.j (j = 0,1,2,3), the g and h in the index are to be indices of functions ψ and χ introduced by previous applications of the schemata, where ψ has one more type-j variable (at the beginning of its list) than ϕ, and for $j > 0$ χ has one more type-j-1 variable.

In S6.j (j = 0, 1, 2, 3), $\mathfrak{a}$ comes from $\mathfrak{a}_1$ by moving the k+1-st type-j variable to the front of the list; and $\underline{g}$ is an index of ψ.

In S11, ψ is a functional $\psi(\eta,\Theta;\mathfrak{a})$ depending on one more function argument η than ϕ has (for a function of the same numbers of variables as ϕ) at the beginning of its list of function arguments.

S0	$\phi(\Theta;\mathfrak{b},\mathfrak{T}) \simeq \theta_{\underline{t}}(\mathfrak{b})$	$\langle 0,\underline{m},\underline{n},\underline{t}\rangle$
S1.0	$\phi(\Theta;\underline{a},\mathfrak{b}) \simeq \underline{a}'\ [= \underline{a}+1]$	$\langle 1,\underline{m},\underline{n},0\rangle$
S1.1	$\phi(\Theta;\underline{a},\mathfrak{b}) \simeq \underline{a}\dot{-}1\ \left[= \begin{cases} \underline{a}-1 & \text{if } \underline{a} > 0 \\ 0 & \text{if } \underline{a} = 0 \end{cases}\right]$	$\langle 1,\underline{m},\underline{n},1\rangle$
S2.0	$\phi(\Theta;\mathfrak{a}) \simeq 0$	$\langle 2,\underline{m},\underline{n}\rangle$
S3	$\phi(\Theta;\underline{a},\mathfrak{b}) \simeq \underline{a}$	$\langle 3,\underline{m},\underline{n}\rangle$
S4.0	$\phi(\Theta;\mathfrak{a}) \simeq \psi(\Theta;\chi(\Theta;\mathfrak{a}),\mathfrak{a})$	$\langle 4,\underline{m},\underline{n},0,\underline{g},\underline{h}\rangle$
S4.$\underline{j}$ ($\underline{j}$ = 1,2,3)	$\phi(\Theta;\mathfrak{a}) \simeq \psi(\Theta;\lambda\beta^{\underline{j}-1}\ \chi(\Theta;\beta^{\underline{j}-1},\mathfrak{a}),\mathfrak{a})$	$\langle 4,\underline{m},\underline{n},\underline{j},\underline{g},\underline{h}\rangle$
S5.1	$\phi(\Theta;\underline{a},\underline{b},\underline{c},\mathfrak{b}) \simeq \mathrm{cs}(\underline{a},\underline{b},\underline{c})\ \left[= \begin{cases} \underline{b} & \text{if } \underline{a} = 0 \\ \underline{c} & \text{if } \underline{a} > 0 \end{cases}\right]$	$\langle 5,\underline{m},\underline{n},1\rangle$
S6.$\underline{j}$ ($\underline{j}$ = 0,1,2,3)	$\phi(\Theta;\mathfrak{a}) \simeq \psi(\Theta;\mathfrak{a}_1)$	$\langle 6,\underline{m},\underline{n},\underline{j},\underline{k},\underline{g}\rangle$
S7.$\underline{j}$ ($\underline{j}$ = 1,2,3)	$\phi(\Theta;\alpha^{\underline{j}},\alpha^{\underline{j}-1},\mathfrak{b}) \simeq \alpha^{\underline{j}}(\alpha^{\underline{j}-1})$	$\langle 7,\underline{m},\underline{n},\underline{j}\rangle$
S11	$\phi(\Theta;\mathfrak{a}) \simeq \psi(\lambda\mathfrak{a}\,\phi(\Theta;\mathfrak{a}),\Theta;\mathfrak{a})$ $[\simeq \psi(\phi,\Theta;\mathfrak{a})$ briefly]	$\langle 11,\underline{m},\underline{n},\underline{g}\rangle$

Before fixing (in 1.4, 2.2-2.5) how these schemata will be used, let us observe how their selection accords with the heuristics in 1.2.

S11 renders the general form of recursion (or the "first recursion theorem").

Our type-0 objects are the natural numbers. One can surely ask that the natural number 0 be available (thus S2.0); and that for any list of arguments with the number $\underline{a}$ first, there should be available to the computer the number $\underline{a}$ itself, its immediate successor $\underline{a}'$, and its immediate predecessor if $\underline{a} > 0$ or 0 if $\underline{a} = 0$ (thus S3, S1.0, S1.1).

The special role given in S3, S1.0, S1.1 to the number variable $\underline{a}$ at the beginning of the list $\mathfrak{a}$, and also in some other schemata to one or several first variables, is offset by our having S6.0-S6.3, successive applications of which will enable us to effect any reordering of the variables.

In a computation we should be able to make cases according as a number $\underline{a}$ is 0 or > 0, choosing to go one route say to b, or another say to c, accordingly; so we have S5.1 (where "cs" is for "case").

S4.0-S4.3 allow the result of a computation by χ to be fed into another computation by ψ ("composition" of computations). By S4.1, indeed not just a computed number but a computed function $\lambda\beta^0\ \chi(\Theta;\beta^0,\mathfrak{a})$ (one whose value for each β^0 is computed by χ) is made a type-1 function argument for the further computation by ψ; and similarly by S4.2 and S4.3 for type-2 and 3 function arguments. (In 1959, we did not boldly put the schemata S4.j for j > 0 in at the beginning, but only special cases S8.2, S8.3 of S4.1, S4.2; and we depended there on a theorem, XXIII p. 21, rather laboriously proved under certain caveats, in other cases.)

Finally, S7.1-S7.3 give us the Turing oracle-principle by which we can ask for, and receive, the value of a function argument α^j (j = 1,2,3) for an argument α^{j-1}; and S0 similarly gives us the value of an assumed function θ_t (one of Θ) for arguments $\mathfrak{b}$.

What else could one need for computation?

1.4. I postpone for a bit describing exactly how computation with these schemata is to be conducted. It will suffice for the nonce to know that the expressions on the two sides of each schema will receive the same value, if either receives a value, for given values of the variables $\mathfrak{a}$ and of the functions Θ.

First, we must consider how the schemata are used in concert. A functional $\lambda\Theta\lambda\mathfrak{a}\ \phi(\Theta;\mathfrak{a})$ will be called partial recursive, or the function $\lambda\mathfrak{a}\ \phi(\Theta;\mathfrak{a})$ partial recursive in Θ (so if Θ is empty, $\lambda\mathfrak{a}\ \phi(\mathfrak{a})$, or simply ϕ, is partial recursive), iff $\phi(\Theta;\mathfrak{a})$ is introduced by a succession of applications of the schemata. Say the applications introduce successively functionals (or functions) $\phi_1,\ldots,\phi_p$ where ϕ_p is ϕ. Thus, for each i (i = 1,...,p), ϕ_i may be introduced outright by one of S0, S1.0, S1.1, S2.0, S3. S5.1, S7.j. In the cases of the other schemata, the ψ on the right (S6.j, S11), or the ψ and χ on the right (S4.j), must come from among $\phi_1,\ldots,\phi_{i-1}$. The list of indices $z_1,\ldots,z_p$ for the schema applications introducing successively $\phi_1,\ldots,\phi_p$ contains complete in itself all the details of the definition of ϕ from Θ. Indeed, just $z = z_p$ does!

I call $\phi_1,\ldots,\phi_p$ a partial recursive derivation of ϕ from Θ, and for Θ empty a partial recursive description of ϕ (cf. 1959, IM pp. 220, 224).

Now I shall give a specific example $\phi_1,\ldots,\phi_{18}$ of a partial recursive derivatio of a function $\phi(\Theta;\underline{a})$ from Θ, with $\underline{p} = 18$, $\underline{l} = 1$, $\underline{m}_{10} = 2$, $\underline{n}_0 = 1$, $\underline{m}_{11} = \underline{m}_{12} = \underline{m}_{13} = \underline{n}_1 = \underline{n}_2 = \underline{n}_3 = 0$. To save space in showing the indices $\underline{z}_1,\ldots,\underline{z}_{18}$ at the right, we write $\underline{m} = \langle\langle 2,0,0,0\rangle\rangle$, $\overline{\underline{m}} = \langle\langle 2,0,0,0\rangle,\langle 2,0,0,0\rangle\rangle$, and remember that, for any $\underline{k}$, $\langle\underline{k}\rangle = \langle\underline{k},0,0,0\rangle$. The four places shown in each index accord with the indexing in our theory of functions of variables only of types 0, 1, 2, 3; but in this example, where only type-0 variables occur, the same indices apply in the theory with any other $\underline{r} \geqq 0$ as the maximum type, since $\langle\underline{k}\rangle = \langle\underline{k},0,0,0\rangle = \langle\underline{k},0\rangle$, etc.

$\phi_1(\eta,\theta;\underline{c},\underline{d},\underline{e},\underline{y},\underline{b}) \simeq cs(\underline{c},\underline{d},\underline{e})$	$\underline{z}_1 = \langle 5,\overline{\underline{m}},\langle 5\rangle,1\rangle$
$\phi_2(\eta,\theta;\underline{y},\underline{b},\underline{d},\underline{e}) \simeq \theta(\underline{y},\underline{b})$	$\underline{z}_2 = \langle 0,\overline{\underline{m}},\langle 4\rangle,2\rangle$
$\phi_3(\eta,\theta;\underline{e},\underline{y},\underline{b},\underline{d}) \simeq \phi_2(\eta,\theta;\underline{y},\underline{b},\underline{d},\underline{e})$	$\underline{z}_3 = \langle 6,\overline{\underline{m}},\langle 4\rangle,0,3,\underline{z}_2\rangle$
$\phi_4(\eta,\theta;\underline{d},\underline{e},\underline{y},\underline{b}) \simeq \phi_3(\eta,\theta;\underline{e},\underline{y},\underline{b},\underline{d})$	$\underline{z}_4 = \langle 6,\overline{\underline{m}},\langle 4\rangle,0,3,\underline{z}_3\rangle$
$\phi_5(\eta,\theta;\underline{d},\underline{e},\underline{y},\underline{b}) \simeq \phi_1(\eta,\theta;\phi_4(\eta,\theta;\underline{d},\underline{e},\underline{y},\underline{b}),\underline{d},\underline{e},\underline{y},\underline{b})$	$\langle 4,\overline{\underline{m}},\langle 4\rangle,0,\underline{z}_1,\underline{z}_4\rangle$
$\phi_6(\eta,\theta;\underline{e},\underline{y},\underline{b}) \simeq 0$	$\langle 2,\overline{\underline{m}},\langle 3\rangle\rangle$
$\phi_7(\eta,\theta;\underline{e},\underline{y},\underline{b}) \simeq \phi_5(\eta,\theta;\phi_6(\eta,\theta;\underline{e},\underline{y},\underline{b}),\underline{e},\underline{y},\underline{b})$	$\langle 4,\overline{\underline{m}},\langle 3\rangle,0,\underline{z}_5,\underline{z}_6\rangle$
$\phi_8(\eta,\theta;\underline{a},\underline{y},\underline{b}) \simeq \underline{a}'$	$\langle 1,\overline{\underline{m}},\langle 3\rangle,0\rangle$
$\phi_9(\eta,\theta;\underline{z},\underline{b},\underline{y}) \simeq \eta(\underline{z},\underline{b})$	$\langle 0,\overline{\underline{m}},\langle 3\rangle,1\rangle$
$\phi_{10}(\eta,\theta;\underline{y},\underline{z},\underline{b}) \simeq \phi_9(\eta,\theta;\underline{z},\underline{b},\underline{y})$	$\langle 6,\overline{\underline{m}},\langle 3\rangle,0,2,\underline{z}_9\rangle$
$\phi_{11}(\eta,\theta;\underline{z},\underline{y},\underline{b}) \simeq \phi_{10}(\eta,\theta;\underline{y},\underline{z},\underline{b})$	$\langle 6,\overline{\underline{m}},\langle 3\rangle,0,1,\underline{z}_{10}\rangle$
$\phi_{12}(\eta,\theta;\underline{y},\underline{b}) \simeq \underline{y}'$	$\langle 1,\overline{\underline{m}},\langle 2\rangle,0\rangle$
$\phi_{13}(\eta,\theta;\underline{y},\underline{b}) \simeq \phi_{11}(\eta,\theta;\phi_{12}(\eta,\theta;\underline{y},\underline{b}),\underline{y},\underline{b})$	$\langle 4,\overline{\underline{m}},\langle 2\rangle,0,\underline{z}_{11},\underline{z}_{12}\rangle$
$\phi_{14}(\eta,\theta;\underline{y},\underline{b}) \simeq \phi_8(\eta,\theta;\phi_{13}(\eta,\theta;\underline{y},\underline{b}),\underline{y},\underline{b})$	$\langle 4,\overline{\underline{m}},\langle 2\rangle,0,\underline{z}_8,\underline{z}_{13}\rangle$
$\phi_{15}(\eta,\theta;\underline{y},\underline{b}) \simeq \phi_7(\eta,\theta;\phi_{14}(\eta,\theta;\underline{y},\underline{b}),\underline{y},\underline{b})$	$\langle 4,\overline{\underline{m}},\langle 2\rangle,0,\underline{z}_7,\underline{z}_{14}\rangle$
$\phi_{16}(\theta;\underline{y},\underline{b}) \simeq \phi_{15}(\phi_{16},\theta;\underline{y},\underline{b})$	$\langle 11,\underline{m},\langle 2\rangle,\underline{z}_{15}\rangle$
$\phi_{17}(\theta;\underline{b}) \simeq 0$	$\langle 2,\underline{m},\langle 1\rangle\rangle$
$\phi_{18}(\theta;\underline{b}) \simeq \phi_{16}(\theta;\phi_{17}(\theta;\underline{b}),\underline{b})$	$\langle 4,\underline{m},\langle 1\rangle,0,\underline{z}_{16},\underline{z}_{17}\rangle$

In fact, $\phi_{18}(\theta;\underline{b}) \simeq \mu\underline{y}[\theta(\underline{y},\underline{b})=0]$ (IM pp. 279, 329, or (XIII) in 3.1 below). To see this, one can first verify quite routinely, working backward (upward) from ϕ_{15}, that the subderivation $\phi_1,\ldots,\phi_{15}$ amounts simply to

$$\phi_{15}(\eta,\theta;\underline{y},\underline{b}) \simeq \begin{cases} 0 & \text{if } \theta(\underline{y},\underline{b}) = 0, \\ \eta(\underline{y}',\underline{b})' & \text{if } \theta(\underline{y},\underline{b}) > 0. \end{cases}$$

So

$$\phi_{16}(\theta;\underline{y},\underline{b}) \simeq \left\{\begin{matrix} 0 & \text{if } \theta(\underline{y},\underline{b}) = 0 \\ (\{\lambda \underline{yb}\ \phi_{16}(\theta;\underline{y},\underline{b})\}(\underline{y}',\underline{b}))' & \text{if } \theta(\underline{y},\underline{b}) > 0 \end{matrix}\right\}$$
$$\simeq \begin{cases} 0 & \text{if } \theta(\underline{y},\underline{b}) = 0, \\ \phi_{16}(\theta;\underline{y}',\underline{b})' & \text{if } \theta(\underline{y},\underline{b}) > 0. \end{cases}$$

Thence, with

$$\phi_{18}(\theta;\underline{b}) \simeq \phi_{16}(\theta;0,\underline{b}),$$

the result is easily seen.

In this example, $\phi_1,\ldots,\phi_{15}$ are introduced as functions of number variables partial recursive in η,θ, where η,θ can each be any assumed partial function of two number variables. But by application of S11, ϕ_{16} is partial recursive just in θ; and then ϕ_{17} and ϕ_{18} are also. Thus, if we had before us just the derivation $\phi_1,\ldots,\phi_{15}$ and didn't know of the coming application of S11, we would be asking for values of two assumed functions η,θ in our computations. But when $\phi_1,\ldots,\phi_{15}$ is built into the derivation of ϕ_{18}, the η comes to be identified with ϕ_{16} (or more explicitly with $\lambda\underline{yb}\ \phi_{16}(\theta;\underline{y},\underline{b})$). It is ϕ_{18} we are really interested in (and not $\phi_1,\ldots,\phi_{15}$ per se). At the step introducing ϕ_9, which is where η enters, we shall feed ϕ_{16} back in rather than using a value of η as an independent assumed function. We could have emphasized this by writing ϕ_{16} in place of η in $\phi_1,\ldots,\phi_{15}$; but that could be puzzling to a person reading the derivation forward for the first time.

We shall usually be interested, ultimately, in what functions ϕ are partial recursive in a fixed list Θ, either of fixed assumed functions or possibly of variable assumed functions (function variables). So we shall want to define computation relative to completed derivations $\phi_1,\ldots,\phi_{\underline{p}}$ of functions $\phi = \phi_{\underline{p}}$ from Θ. This can require, as we have just seen, looking forward as well as backward in handling a $\phi_{\underline{i}}$ with $\underline{i} < \underline{p}$.

To make the presentation of our theory as straightforward as possible, we shall engage in a bit of canonization, even though this may result in some redundancy in the practice.

So we shall say a partial recursive derivation $\phi_1,\ldots,\phi_{\underline{p}}$ from Θ (or a partial recursive description, when Θ is empty) is canonical iff it has the following property, defined by recursion on $\underline{p}$. If $\phi_{\underline{p}}$ is given outright (by one of S0, S1.0, S1.1, S2.0, S3, S5.1, S7$\underline{j}$), then $\underline{p} = 1$ and the unit sequence $\phi_{\underline{p}}$ is the whole derivation. If $\phi_{\underline{p}}$ is given from one previously derived function ψ (by S6.$\underline{j}$ or S11), the derivation $\phi_1,\ldots,\phi_{\underline{p}}$ is $\psi_1,\ldots,\psi_{\underline{p}-1},\phi_{\underline{p}}$ where $\psi_1,\ldots,\psi_{\underline{p}-1}$ is a canonical derivation of ψ. If $\phi_{\underline{p}}$ is given from two previously derived functions ψ and χ (by S4.$\underline{j}$), the derivation $\phi_1,\ldots,\phi_{\underline{p}}$ is $\psi_1,\ldots,\psi_{\underline{q}},\chi_1,\ldots,\chi_{\underline{r}},\phi_{\underline{p}}$ where $\psi_1,\ldots,\psi_{\underline{q}}$ is a canonical derivation of ψ and $\chi_1,\ldots,\chi_{\underline{r}}$ is a canonical derivation of χ ($\underline{p} = \underline{q}+\underline{r}+1$). The above derivation $\phi_1,\ldots,\phi_{18}$ is canonical.

There is no difficulty in defining a primitive recursive predicate $\mathrm{Ix}(\underline{z})$ of ordinary recursion theory (ORT; e.g. IM Chapter IX) which says that $\underline{z}$ is an index of a function partial recursive in Θ, where $\underline{m} = (\underline{z})_1$ tells the number $\underline{l}$, and respective numbers of variables, of the functions $\Theta = (\theta_1,\ldots,\theta_{\underline{l}})$ as in 1.3. It is done similarly to 1959 XIX pp. 17-18. For definiteness, we suppose it done here for our theory with just the first four types 0, 1, 2, 3 of objects.

An index $\underline{z}$ of ϕ from Θ determines uniquely a canonical derivation $\phi_1,\ldots,\phi_{\underline{p}}$ ($\phi = \phi_{\underline{p}}$) and thence the quantity $\langle \underline{z}_1,\ldots,\underline{z}_{\underline{p}}\rangle$ ($\underline{z} = \underline{z}_{\underline{p}}$) where $\underline{z}_1,\ldots,\underline{z}_{\underline{p}}$ are the indices of $\phi_1,\ldots,\phi_{\underline{p}}$ in that derivation. There is a primitive recursive function $\mathrm{cd}(\underline{z})$ ("canonical derivation") such that, if $\mathrm{Ix}(\underline{z})$, then $\mathrm{cd}(\underline{z}) = \langle \underline{z}_1,\ldots,\underline{z}_{\underline{p}}\rangle$; indeed (cf. IM pp. 228-231)

$$\mathrm{cd}(\underline{z}) = \begin{cases} \langle \underline{z}\rangle \text{ if } \mathrm{Ix}(\underline{z})\ \&\ (\underline{z})_0=0,1,2,3,5 \text{ or } 7, \\ \mathrm{cd}((\underline{z})_5)*\langle \underline{z}\rangle \text{ if } \mathrm{Ix}(\underline{z})\ \&\ (\underline{z})_0=6, \\ \mathrm{cd}((\underline{z})_3)*\langle \underline{z}\rangle \text{ if } \mathrm{Ix}(\underline{z})\ \&\ (\underline{z})_0=11, \\ \mathrm{cd}((\underline{z})_4)*\mathrm{cd}((\underline{z})_5)*\langle \underline{z}\rangle \text{ if } \mathrm{Ix}(\underline{z})\ \&\ (\underline{z})_0=4, \\ 0 \text{ otherwise (indeed, if } \overline{\mathrm{Ix}}(\underline{z})). \end{cases}$$

As we have observed, in reading a derivation $\phi_1,\ldots,\phi_{\underline{p}}$ (which we now always take to be canonical) of $\phi_{\underline{p}} = \phi_{\underline{p}}(\Theta;\mathfrak{A})$ forward, we have various lists $\Theta_1,\Theta_2,\ldots,\Theta_{\underline{p}}$ of assumed functions for $\phi_1,\phi_2,\ldots,\phi_{\underline{p}}$, where $\Theta = \Theta_{\underline{p}}$ is the final list. But when the whole derivation has been read, all of the functions in each $\Theta_{\underline{i}}$ except Θ will have been identified by S11 with functions from among $\phi_2,\ldots,\phi_{\underline{p}}$. So, given $\phi_1,\ldots,\phi_{\underline{p}}$ with an index $\underline{z}$ of $\phi_{\underline{p}}$, we can write each $\Theta_{\underline{i}} = (\Psi_{\underline{i}},\Theta)$ where $\Psi_{\underline{i}}$ is a list (possibly with repetitions) from among $\phi_2,\ldots,\phi_{\underline{p}}$; say $\Psi_{\underline{i}} = (\phi_{\underline{k}_{\underline{i}1}},\ldots,\phi_{\underline{k}_{\underline{iq}_{\underline{i}}}})$ $(\underline{q}_{\underline{i}} \geq 0)$. Let $\underline{s}_{\underline{i}} = \langle \underline{k}_{\underline{i}1},\ldots,\underline{k}_{\underline{iq}_{\underline{i}}}\rangle$ and $\underline{s} = \langle \underline{s}_1,\ldots,\underline{s}_{\underline{p}}\rangle$. It is an easy exercise to get a primitive recursive function ps($\underline{z}$) ("$\underline{\text{psi}}$") such that, if Ix($\underline{z}$), then ps($\underline{z}$) = $\underline{s}$ = $\langle \underline{s}_1,\ldots,\underline{s}_{\underline{p}}\rangle$. For, from $\underline{z}$ we can get cd($\underline{z}$) = $\langle \underline{z}_1,\ldots,\underline{z}_{\underline{p}}\rangle$, and $\underline{z}_1,\ldots,\underline{z}_{\underline{p}}$ determine how functions from $\phi_2,\ldots,\phi_{\underline{p}}$ are added at the front of the list Θ via the applications of S11 read backwards from the end of the derivation $\phi_1,\ldots,\phi_{\underline{p}}$ toward the beginning. In any case, the whole structure is given by the index $\underline{z}$, and its details can be unravelled primitive recursively.

2. The rules for computation by the schemata. 2.1. Thus far we have been rather uncritically assuming that the schema applications introducing successively $\phi_1,\ldots,\phi_{\underline{p}}$ define these functions from Θ.

In this section, we shall formalize a specific computation process to fulfil this assumption. There are some options requiring careful handling.

The problem is this. Suppose $\phi = \phi_{\underline{p}}$ has been described from Θ by $\phi_1,\ldots,\phi_{\underline{p}}$, which we take as canonical. Indeed, we need only know the index $\underline{z}$ of $\phi = \phi_{\underline{p}}$, and all the rest falls out. What, then, is the value of $\phi(\Theta;\mathfrak{A})$ for a given choice of the assumed functions Θ and a given assignment of values to the variables $\mathfrak{A}$? Until we change in Part II, $\mathfrak{A}$ are to be variables each ranging over one of our four types of 1.1.

The use of S11 entails that at some steps some assumed functions, specifically those identified with one of $\phi_2,\ldots,\phi_{\underline{p}}$ via S11, are non-total partial functions, or at least are used in the computation by the ψ of S11 in a stage of being only incompletely determined. This is the case (if S11 is used non-trivially) even if the list Θ for $\phi = \phi_{\underline{p}}$ consists of only total functions and even if $\phi_{\underline{p}}$ itself is total.

With Alfred E. Neuman ("What - Me Worry?"), I shall not let this worry me (much, until Part II). I worked like MAD to get the address ready for June 13, 1977 and the typescript by January.

Here is a simple example. Suppose that, for a given value of the variable $\underline{a}$, we are to compute the value of $\phi(\underline{a})$ where $\phi(\underline{a}) \simeq \psi(\chi(\underline{a}),\underline{a})$ by S4.0. In 1959, we did not take any step toward "reducing" the whole, $\psi(\chi(\underline{a}),\underline{a})$, on the basis of the definition of ψ without first calculating $\chi(\underline{a})$. That is, we demanded a subordinate calculation, as shown below the (first) horizontal line

$$\begin{array}{lllll} \phi(\underline{a}) & \text{——} & \psi(\chi(\underline{a}),\underline{a}) & \text{——} & \psi(\underline{u},\underline{a}) \quad \ldots \quad , \\ & & & \searrow & \chi(\underline{a}) \quad \ldots \quad \underline{u} \end{array}$$

to determine (in general, via a whole "computation subtree") a value $\underline{u}$ for $\chi(\underline{a})$; and only if we obtained such a $\underline{u}$ did we continue in the "upper branch" by passing from $\psi(\chi(\underline{a}),\underline{a})$ to $\psi(\underline{u},\underline{a})$ and then thinking about the definition of ψ. Now (1977) we shall proceed differently. We won't worry about what $\chi(\underline{a})$ is unless and until we are forced to do so. Instead, we shall plow right ahead to evaluate ψ (if we can) carrying along $\chi(\underline{a})$ as its first argument into whatever the schema application for ψ gives. Say e.g. $\psi(\underline{b},\underline{a}) \simeq \rho(\underline{a},\underline{b})$ by S6.0, and $\rho(\underline{a},\underline{b}) \simeq \underline{a}$ by S3. Our computation will go now

$$\phi(\underline{a}) \text{ —— } \psi(\chi(\underline{a}),\underline{a}) \text{ —— } \rho(\underline{a},\chi(\underline{a})) \text{ —— } \underline{a},$$

to give, as the value of $\phi(\underline{a})$, whatever was the value assigned to $\underline{a}$. We get this even if $\chi(\underline{a})$ is undefined, which in 1959 would have made $\phi(\underline{a})$ undefined.

If one really wants a function $\phi(\underline{a})$ whose value is that of $\underline{a}$ if $\chi(\underline{a})$ is defined for that value, and is undefined otherwise, he can take $\phi(\underline{a}) \simeq \mathrm{cs}(\chi(\underline{a}),\underline{a},\underline{a})$, as can easily be done with our schemata.

We do the like, for example, with $\phi(\underline{a}) \simeq \psi(\lambda\underline{b}\ \chi(\underline{b},\underline{a}),\underline{a})$, which as we remarked in 1.3 we did not allow directly by our schemata in 1959. If, as is easily done, $\psi(\alpha,\underline{a}) \simeq \alpha(0)$, we would get a value of $\phi(\underline{a})$ for a given value of $\underline{a}$, even with $\chi(\underline{b},\underline{a})$ defined only for $\underline{b} = 0$, so that $\lambda\underline{b}\ \chi(\underline{b},\underline{a})$ would not be a type-1 object.

The procedure in 1959 had the elegant feature that at every step in the computation we were dealing with a function (or its index) applied simply to a tuple of arguments of our several types. (Thus, at the bottom of 1959 p. 22, we could "avoid syntactical considerations".)

Here instead, we may get up to our necks in complicated syntax. After several applications of S4.$\underline{j}$ in a row, we will be far from having simply one of $\phi_1,\ldots,\phi_{\underline{p}}$

applied to variables having assigned values (maybe values previously computed). But we will get things more often defined. And after all, since the invention of Gödel numbering (or of "coding" or "indexing"),[3] complicated syntax should not frighten a logician. Our computations, I am sure you will agree, will be as legitimate as computations as were the more elegant ones of 1959.

There is herewith a new problem of interpretation, which we will address in Part II. As a crude analogy, we might be interested in a function $\phi(\underline{a})$ on even values of $\underline{a}$, but use also odd values of variables of auxiliary functions in its computation.

2.2. As proposed, in the computation of the value of a function $\phi(\Theta;\mathfrak{a})$, we will work with formal expressions beginning with "$\phi(\Theta;\mathfrak{a})$" itself. We will operate with them syntactically, and with assignments Ω of values of our four types to their free variables as well as of partial functions of lists of variables of those types to the function symbols "θ_1",...,"$\theta_{\underline{l}}$" in "Θ". This falls within model theory, where one talks about formal expressions under assignments to their free variables and other symbols.

At this moment, we are addressing immediately only the question of how to compute the value of "$\phi(\Theta;\mathfrak{a})$" (under any given assignment Ω to $\mathfrak{a}$ and Θ) for the particular function $\phi = \phi_{\underline{p}}$ introduced by one given canonical derivation $\phi_1,\ldots,\phi_{\underline{p}}$ from Θ determined by an index $\underline{z}$ of ϕ. (In 2.5 we shall liberalize.)

The expressions which our proposed strategy then obliges us to consider

---

[3] Gödel 1931; also introduced independently by Tarski in what became his monograph 1933 on the truth concept in formalized languages. Kleene in 1936 (inspired by Gödel 1931) seems to have been the third to apply it.

It was pointed out to me in Oslo on June 13, 1977 by David MacQueen that computer scientists have developed the theory of working with the expressions of a given formal language as a generalized arithmetic to the extent that it is not necessary to reduce the language to the simple arithmetic of the natural numbers by a Gödel numbering or indexing. Cf. Levin 1974, Scott 1976. (An early example of generalized arithmetic is in IM § 50.) As I am making scarcely any work out of the Gödelization (just a bit in 3.2), I have chosen not to redo my treatment in the light of this suggestion, or seriously to evaluate which format is the simpler over-all.

will evidently contain no function symbols other than for the functions $\theta_1,\ldots,\theta_{\underline{l}}$, $\phi_1,\ldots,\phi_{\underline{p}}$ and the three particular functions used in the schemata. So we shall define the class of all (significant) expressions constructible using only those function symbols, "0", "λ", variables of our four types, commas and parentheses. We shall assume available an ω-list of each of our four types of variables.

The context should make it clear when we are using "$\phi_{\underline{i}}$" to name a function and when to name a function symbol. Let us not make syntax hard, as all of us (speaker-writer, listeners and readers) have surely sometime learned what is involved, either by writing or reading an explicit treatment.

Since, for each $\underline{i}$, the $\Theta_{\underline{i}}$ of $\phi_{\underline{i}}(\Theta_{\underline{i}};\mathcal{O}_{\underline{i}})$ in the derivation $\phi_1,\ldots,\phi_{\underline{p}}$ of $\phi(\Theta;\mathcal{O})$ $= \phi_{\underline{p}}(\Theta;\mathcal{O})$ is determined, including the identifications by S11 of all of $\Theta_{\underline{i}}$ but Θ with a list $\Psi_{\underline{i}}$ from $\phi_2,\ldots,\phi_{\underline{p}}$ (end 1.4), we can without confusion write simply "$\phi_{\underline{i}}(\mathcal{O}_{\underline{i}})$" instead of "$\phi_{\underline{i}}(\Theta_{\underline{i}};\mathcal{O}_{\underline{i}})$" or "$\phi_{\underline{i}}^{\Theta_{\underline{i}}}(\mathcal{O}_{\underline{i}})$".

So, relative to a number $\underline{z}$ such that $Ix(\underline{z})$, which $\underline{z}$ determines a canonical derivation $\phi_1,\ldots,\phi_{\underline{p}}$ with indices $\underline{z}_1,\ldots,\underline{z}_{\underline{p}}$, and relative to a list $\Theta = (\theta_1,\ldots,\theta_{\underline{l}}$ (described as to the number $\underline{l}$ and types by $\underline{m} = (\underline{z})_1$), here is our syntactical definition of $\underline{j}$-expression ($\underline{j} = 0,1,2,3$).

E1. A type-$\underline{j}$ variable $\alpha^{\underline{j}}$ is a $\underline{j}$-expression ($\underline{j} = 0,1,2,3$).

E2. If $A_1,\ldots,A_{\underline{n}_{\underline{i}0}}$ are 0-expressions, $B_1,\ldots,B_{\underline{n}_{\underline{i}1}}$ are 1-expressions, $C_1,\ldots,C_{\underline{n}_{\underline{i}2}}$ are 2-expressions, and $D_1,\ldots,D_{\underline{n}_{\underline{i}3}}$ are 3-expressions (where $(\underline{z}_{\underline{i}})_2 = \langle\underline{n}_{\underline{i}0},\underline{n}_{\underline{i}1},\underline{n}_{\underline{i}2},\underline{n}_{\underline{i}3}\rangle$), then $\phi_{\underline{i}}(A_1,\ldots,A_{\underline{n}_{\underline{i}0}},B_1,\ldots,B_{\underline{n}_{\underline{i}1}},C_1,\ldots,C_{\underline{n}_{\underline{i}2}},D_1,\ldots,D_{\underline{n}_{\underline{i}3}})$ is a 0-expression ($\underline{i} = 1,\ldots,\underline{p}$).

E3. If A is a 0-expression, and $\alpha^{\underline{j}}$ is a type-$\underline{j}$ variable, then $\lambda\alpha^{\underline{j}}$ A is a $\underline{j}$+1-expression ($\underline{j} = 0,1,2$), called a type-$\underline{j}$+1 λ-functor.

E4. If A is a $\underline{j}$+1-expression, and B is a $\underline{j}$-expression, then $\{A\}(B)$ (often abbreviated A(B)) is a 0-expression ($\underline{j} = 0,1,2$).

E5. 0 is a 0-expression.

E6. If A is a 0-expression, so are (A)' and $(A)^{\pm}1$ (often abbreviated A' and $A^{\pm}1$). If A, B and C are 0-expressions, so is cs(A,B,C).

E7. If $\underline{l} > 0$, we have a clause like E2, with $\theta_{\underline{t}},\underline{m}_{\underline{t}0},\underline{m}_{\underline{t}1},\underline{m}_{\underline{t}2},\underline{m}_{\underline{t}3}$ for $\underline{t} = 1,\ldots,\underline{l}$ (where $(\underline{z})_{1,\underline{t}-1} = \langle\underline{m}_{\underline{t}0},\underline{m}_{\underline{t}1},\underline{m}_{\underline{t}2},\underline{m}_{\underline{t}3}\rangle$) instead of $\phi_{\underline{i}},\underline{n}_{\underline{i}0},\underline{n}_{\underline{i}1},\underline{n}_{\underline{i}2},\underline{n}_{\underline{i}3}$ for $\underline{i} = 1,\ldots,\underline{p}$.

2.3. In computing, after fixing $\underline{z}$ (and thence $\phi_1, \ldots, \phi_p$, etc.), we take any 0-expression E (under the resulting definition in 2.2) and an assignment Ω (as above), and seek the numeral W (i.e. $0' \cdots '$ with $\underline{w} \geq 0$ accents) as expression for the answer "$\underline{w}$" (if it exists) to the question "What is the value of E under Ω?". (Incidentally, we may abbreviate $0', 0'', \ldots$ as $1, 2, \ldots$.)

We shall represent a computation as in the form of a tree with E at the initial (leftmost) vertex. We draw our trees lying on their sides (as in 1959 and as some in Madison after the ice storm of March 4, 1976). The trees branch to the right.

The principal branch, ending (if the computation can be completed) in the numeral W expressing the value $\underline{w}$ of E under Ω, runs horizontally rightward from the initial vertex. (We didn't show it quite so in 1959 p. 22.) When a numeral is recognized as the value of an expression, whose computation is therewith completed, we shall celebrate by flagging it there with a "$\dagger$". A flagged numeral gives the value of the expression at each vertex along the branch running leftward from it as far as the branch runs horizontally, under the assignment in force at the vertex.

We shall define completed computation trees inductively. If E is 0, we flag E immediately. Otherwise, we graft onto the vertex carrying E subtrees issuing from next vertices, of which there are either one along the horizontal (principal) branch, or also others below it.

Yes, I have been a successful grafter, with apples and plums. The ripe fruit at the ends of all the branches of a completed computation tree will be flagged numerals.

In the simplest case, when we need a new variable, we will choose the next unused one in the list of variables of its type (assumed in 2.2). The computation of an expression E will be based on having an assignment Ω of values to the free variables of E and of appropriate functions to all the function symbols in the list Θ. As we define computation trees moving leftward (the computations proceeding rightward), the list of the free variables of the expression E at the vertex we have before us will vary. It will be simplest now to think of Ω as an assignment to exactly the free variables of the E of the moment, but to all the function symbols in Θ (which list remains fixed, after fixing $\underline{z}$ and thus fixing the class of $\underline{j}$-expressions for $\underline{j} = 0,1,2,3$).

In special circumstances, we may wish to reserve a finite list $\underline{V}$ of variables that will not be introduced as new free variables in the computation and which may have values assigned to them throughout a discussion.

2.4. Now we are ready to give the inductive definition of being the computation tree for E under Ω, by cases corresponding to the cases in the definition of 0-expression in 2.2.

E1: E is a type-0 variable $\underline{a}$. The tree is completed by adding one vertex horizontally rightward bearing the numeral R for the value $\underline{r}$ of $\underline{a}$ in the assignment Ω, flagged:

$$\underline{a} \text{ —— } R\dagger$$

E2: E is $\phi_{\underline{i}}(A_1,\ldots,D_{\underline{n}_{\underline{i}3}})$ as under E2. We pass from E horizontally rightward to a next vertex bearing the result F of the same substitution for the variables on the right of the schema (determined by $\underline{z}_{\underline{i}}$) that introduced $\phi_{\underline{i}}$ as gives E on the left. There is indeed one detail in the schemata (after having fixed the symbols "ϕ_1",...,"$\phi_{\underline{p}}$","θ_1",...,"$\theta_{\underline{l}}$") which isn't determined by the index $\underline{z}$ of $\phi = \phi_{\underline{p}}$, namely what variable is the $\beta^{\underline{j}-1}$ of an application of S4.$\underline{j}$ ($\underline{j} = 1,2,3$). We pick it here to be the first variable in the list of its type (other than a variable in a reserved list $\underline{V}$) which does not occur free in E. Thereby, this step will not result in a "collision". Should $\phi_{\underline{i}}$ be introduced by S0, then if $\underline{t} \leq \underline{q}_{\underline{i}} = lh(\underline{s}_{\underline{i}}) = lh((\underline{s})_{\underline{i}-1}) = lh((ps(\underline{z}))_{\underline{i}-1})$ (end 1.4), then the "$\theta_{\underline{t}}$" we use here for the right side of the schema application is to be the symbol for the function $\phi_{\underline{k}_{\underline{i}\underline{t}}}$ from $\phi_2,\ldots,\phi_{\underline{p}}$ fed back in by S11; if $\underline{t} > \underline{q}_{\underline{i}}$, the "$\theta_{\underline{t}}$" is $\theta_{\underline{t}-\underline{q}_{\underline{i}}}$ from the list $\theta_1,\ldots,\theta_{\underline{l}}$. Clearly, F will again be a 0-expression. (Likewise, in all the other steps.) Anticipating ultimate "success" if we are describing computation (proceeding rightward), or having "success" by the hyp. ind. if we are giving the inductive definition of a completed computation tree (proceeding leftward), the subtree we graft onto the vertex bearing E is say "F ... W†", and the whole result of the step is:

$$\phi_{\underline{i}}(A_1,\ldots,D_{\underline{n}_{\underline{i}3}}) \text{ —— } F \;\ldots\; W\dagger.$$

E4.λ: E is a 0-expression under E4 of the form $\{\lambda\alpha^{\underline{j}}\ A\}(B)$ ($\underline{j} = 0,1,2$). We pass rightward to $\underline{S}_B^{\alpha^{\underline{j}}}\ A|$, the result of substituting B for the free occurrences of $\alpha^{\underline{j}}$ in A. If this would result in a "collision", i.e. if free variables of some substituted parts B would become bound by λ-prefixes $\lambda\beta^{\underline{k}}$ in A, we first change such bound variables $\beta^{\underline{k}}$ in respective parts $\lambda\beta^{\underline{k}}$ C of A to avoid this happening, using the first eligible variables not occurring free in E from the appropriate lists, doing this in order on the offending $\lambda\beta^{\underline{k}}$'s from left to right in A.

$$\{\lambda\alpha^{\underline{j}}\ A\}(B) \;\text{———}\; \underline{S}_B^{\alpha^{\underline{j}}}\ A| \quad \ldots \quad W\dagger.$$

E4.1: E is a 0-expression under E4 of the form $\alpha^1(B)$. There will be two next vertices. For the tree to be completed, the lower next vertex must begin a subtree leading to a value $\underline{r}$ of B represented by the flagged numeral R†; and then the upper next vertex (horizontally rightward) receives the flagged numeral N† for the value $\underline{n}$ of $\alpha^1(\underline{r})$ under Ω, i.e. for the value $\underline{n}$, for $\underline{r}$ as argument, of the function assigned by Ω to the function variable α^1.

$\alpha^1(B)$ ——— N†.
 \
  B ... R†

E4.2: E is $\alpha^2(B)$.

$\alpha^2(B)$ ——— M†.

. . .

B($\underline{c}$) ... $\underline{c}$ evaluated by $\underline{r}$ ... $N_{\underline{r}}\dagger$

. . .

B($\underline{c}$) ... $\underline{c}$ evaluated by 1 ... $N_1\dagger$

B($\underline{c}$) ... $\underline{c}$ evaluated by 0 ... $N_0\dagger$

The assignments $\Omega_{\underline{r}}$ for the lower next vertices include respectively (from the lowest upward) $0,1,\ldots,\underline{r},\ldots$ as the values of the new variable $\underline{c}$ (to be picked as the first in the list of type-0 variables not occurring free in B and not in a reserved list $\underline{V}$). B may or may not contain α^2 free; if not, the assignments $\Omega_{\underline{r}}$ lack a value for α^2. If all the lower subtrees can be completed, a type-1 function α^1 is determined as $\{<0,\underline{n}_0>,<1,\underline{n}_1>,\ldots,<\underline{r},\underline{n}_{\underline{r}}>,\ldots\}$ where $\underline{n}_0,\underline{n}_1,\ldots,\underline{n}_{\underline{r}},\ldots$ are the numbers expressed by the numerals $N_0,N_1,\ldots,N_{\underline{r}},\ldots$ (in effect, α^1 is the value of B). Then we can complete the computation as shown, where M is the numeral for the value of $\alpha^2(\alpha^1)$ under the assignment in Ω to α^2 .

E4.3: E is $\alpha^3(B)$. Similarly, using some well-ordering $\alpha_0, \alpha_1, \ldots, \alpha_\zeta, \ldots$

($\zeta < \xi$) of all the type-1 objects.

$$\alpha^3(B) \rule{2em}{0.4pt} S\dagger.$$

$\cdot \cdot \cdot$

$\quad\quad B(\gamma) \quad \ldots \quad \gamma \text{ evaluated by } \alpha_\zeta \ldots M_\zeta\dagger$

$\cdot \cdot \cdot$

$\quad\quad B(\gamma) \quad \ldots \quad \gamma \text{ evaluated by } \alpha_1 \ldots M_1\dagger$

$\quad\quad B(\gamma) \quad \ldots \quad \gamma \text{ evaluated by } \alpha_0 \ldots M_0\dagger$

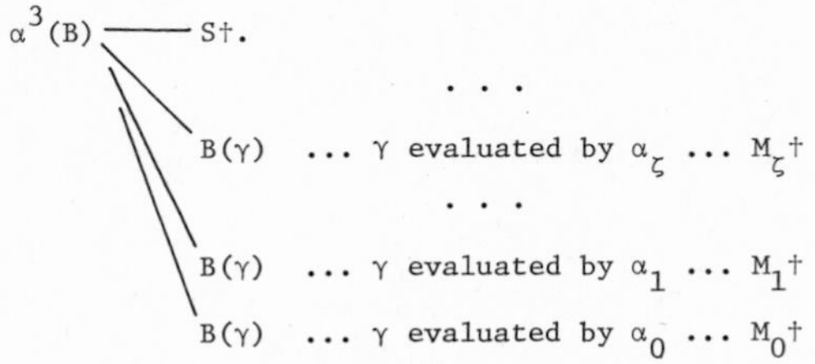

If all the lower subtrees can be completed, S is the numeral for $\alpha^3(\alpha^2)$, where $\alpha^2 = \{<\alpha_\zeta, \underline{m}_\zeta> \mid \zeta < \xi\}$, for the value of α^3 in the assignment Ω.

E5: E is 0. We simply flag the 0.

$$0\dagger.$$

E6.': E is A'.

$$A' \rule{2em}{0.4pt} N'\dagger.$$
$$A \quad \ldots \quad N\dagger$$

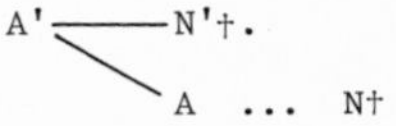

We start a subordinate calculation for A. If it terminates in a flagged numeral N†, the upper next vertex receives the flagged numeral N'†.

You may wonder why, in the special case that A is a numeral, we don't flag A' at once. (Our choice not to do so meets the needs of Part II.) Practically, confronted with the numeral A' for 10^9 say, we would have to dissect it by an effective process, e.g. by a Turing machine, accent by accent, to confirm that A is a numeral. The flags would then march up on the right through 10^9 steps by E6.'.

E6.$\doteq$: E is A$\doteq$1.

$$A\doteq1 \rule{2em}{0.4pt} M\dagger \quad (\underline{m} = \underline{n}\doteq1).$$
$$A \quad \ldots \quad N\dagger$$

E6.cs: E is cs(A,B,C).

$$cs(A,B,C) \rule{2em}{0.4pt} \begin{cases} B \quad \ldots \quad W\dagger \text{ if } \underline{n} = 0, \\ C \quad \ldots \quad W\dagger \text{ if } \underline{n} > 0, \end{cases}$$
$$A \quad \ldots \quad N\dagger$$

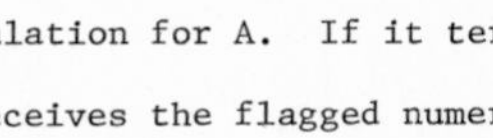

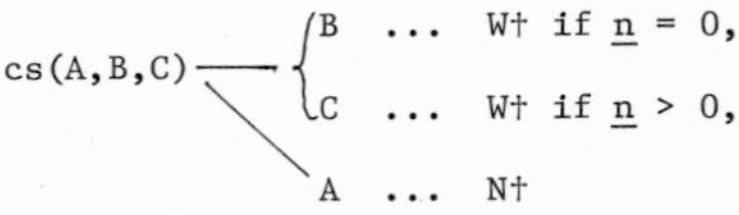

The assignment for A is Ω less values assigned to variables free only in B,C; and similarly for B or C.

E7: E is $\theta_{\underline{t}}(A_1,\ldots,D_{\underline{m}_{\underline{t}3}})$. For example, say E is $\theta_{\underline{t}}(A,B_1,B_2,C,D)$ ($\underline{m}_{\underline{t}0} = \underline{m}_{\underline{t}2} = \underline{m}_{\underline{t}3} = 1$, $\underline{m}_{\underline{t}2} = 2$). A potpourri of E4.1-E4.3. There must be subordinate calculations for A, for $B_1(\underline{c}_1)$ and $B_2(\underline{c}_2)$ for each type-0 value of $\underline{c}_1$ or $\underline{c}_2$, for $C(\gamma)$ for each type-1 value of γ, and for $D(\gamma^2)$ for each type-2 value of γ^2, which together we suppose well-ordered upward. If these can all be completed, they determine objects $\underline{a}$, α_1, α_2, α^2, α^3 of types 0,1,1,2,3 as values of A,B_1,B_2,C,D; and the oracle for $\theta_{\underline{t}}$ (supposed given by Ω) will then produce a value $\underline{w}$ for E if $\theta_{\underline{t}}$ is defined for the arguments $\underline{a},\alpha_1,\alpha_2,\alpha^2,\alpha^3$, and then horizontally: E —— $\underline{w}\dagger$.

This completes the list of cases for the definition of the class of completed computation trees, as generated inductively proceeding from right to left. Such trees are given outright by E1 or E5 (or by E7 for $\underline{m}_{\underline{t}0}+\underline{m}_{\underline{t}1}+\underline{m}_{\underline{t}2}+\underline{m}_{\underline{t}3} = 0$),[7] and in the other cases from the existence of other such trees. The subtree issuing from the upper next vertex in the cases of E1, E4.1, E4.2, E4.3, E6.', E6.∸ and E7 is only that vertex itself, and is not a "computation subtree", i.e. one which itself is a completed computation tree for a 0-expression under an assignment, unless "accidentally" by E5 used "out of context". Otherwise, the subtrees issuing from the next vertices are computation subtrees given by the hyp. ind.

Looking only at the left and right ends of the principal branch of any computation tree, we get the inductive definition of the predicate $E \simeq_{\Omega} \underline{w}$ (or $E \simeq \underline{w}$ under Ω) saying that $\underline{w}$ is the value of E under Ω (false, if the value does not exist).[4]

Looking from left (starting with a given E and Ω) toward right, we have defined computation as a (in general, infinite) reduction procedure, channeled through the schemata, with some steps ("pure reduction steps") determined by just the expression E of the moment, and with others ("evaluation steps") depending also

---

[4]Am I guilty of mixing languages in writing "$E \simeq_{\Omega} \underline{w}$"? For, "E" denotes (E is) a 0-expression of our formal object language, while "$\underline{w}$" denotes ($\underline{w}$ is) a natural number in our own language (the observer's language). But I am thinking of "$\simeq$" here as a connective (asserting equality) not between E and $\underline{w}$, but between <u>the value of</u> E <u>under</u> Ω (which must then exist) and $\underline{w}$. And "$E_1 \simeq E_2$ under Ω" asserts that E_1 and E_2 have the same value (or both no value) under Ω.

on the assignment Ω of the moment. Proceeding from left to right, the possibility of making a step horizontally may be contingent on the completion and results of subcomputations rightward from lower next vertices. Our claim is that this is well-defined as a (usually transfinite) computation process (such as was previously exemplified in 1959), even though now we do not have an interpretation from among our types 0,1,2,3 for each subexpression (e.g. every argument of a $\phi_{\underline{i}}$) entertained enroute in a completed computation.

2.5. In 2.2-2.4, in giving the definition of $\underline{j}$-expression and the rules for computing the value of a 0-expression E under an assignment Ω, our treatment related to some one specific canonical derivation $\phi_1,\ldots,\phi_{\underline{p}}$ from Θ as determined by an index $\underline{z}$. We made it so because the handling of $\phi_{\underline{i}}(A_1,\ldots,D_{\underline{n}_{\underline{i}3}})$ under E2 can depend on the remainder $\phi_{\underline{i}+1},\ldots,\phi_{\underline{p}}$ of the derivation, namely when S11 is used in that remainder. We had set ourselves the specific problem of defining how to compute $\phi(\Theta;\mathfrak{A})$ under Ω given an index $\underline{z}$ of ϕ.

When we are interested in "collecting" functions partial recursive in one and the same Θ (i.e. in recognizing some functions as such, adding others, etc.), it may be inconvenient to have $\underline{j}$-expression and the computation rules defined only for the $\underline{p}$ function symbols $\phi_1,\ldots,\phi_{\underline{p}}$ in the derivation fixed by one index $\underline{z}$.

We can liberalize the "1-fold" definitions in 2.2-2.4 to get "$\underline{q}$-fold" definitions, using simultaneously any finite number $\underline{q}$ of canonical derivations, say $\phi_1^{(1)},\ldots,\phi^{(1)}_{\underline{p}^{(1)}}$ and ... and $\phi_1^{(\underline{q})},\ldots,\phi^{(\underline{q})}_{\underline{p}^{(\underline{q})}}$, determined by respective indices $\underline{z}^{(1)},\ldots,\underline{z}^{(\underline{q})}$. We simply extend E2 in 2.2 so that, for each $\underline{k}=1,\ldots,\underline{q}$, each $\phi_{\underline{i}}^{(\underline{k})}(A_1,\ldots,D_{\underline{n}^{(\underline{k})}_{\underline{i}3}})$ counts as a 0-expression for each $\underline{i}=1,\ldots,\underline{p}^{(\underline{k})}$. Then the computation rule E2 in 2.4 will reduce $\phi_{\underline{i}}^{(\underline{k})}(A_1,\ldots,D_{\underline{n}^{(\underline{k})}_{\underline{i}3}})$ in accordance with the schema application introducing $\phi_{\underline{i}}^{(\underline{k})}$ in the respective derivation $\phi_1^{(\underline{k})},\ldots,\phi^{(\underline{k})}_{\underline{p}^{(\underline{k})}}$ (whether or not $A_1,\ldots,D_{\underline{n}^{(\underline{k})}_{\underline{i}3}}$ contain function symbols coming from others of the $\underline{q}$ derivations).

It should be clear that in this $\underline{q}$-fold treatment, the computation of a 0-expression E which has in it only $\phi_{\underline{i}}^{(\underline{k})}$'s for one $\underline{k}$ will be exactly the same as under the 1-fold treatment of 2.2-2.4 when applied to the present $\phi_1^{(\underline{k})},\ldots,\phi^{(\underline{k})}_{\underline{p}^{(\underline{k})}}$, $\underline{z}^{(\underline{k})}$ as its $\phi_1,\ldots,\phi_{\underline{p}},\underline{z}$. The other $\underline{q}$-1 derivations, and the symbols for the functions in them, don't get into the act for such an E. In particular, the computation of each $\phi^{(\underline{k})}(\mathfrak{A}^{(\underline{k})})$ (where $\phi^{(\underline{k})} = \phi^{(\underline{k})}_{\underline{p}^{(\underline{k})}}$) for a given assignment $\Omega^{(\underline{k})}$ to $\mathfrak{A}^{(\underline{k})}$ and to the symbols Θ is unaltered by the widening of the language and rules.

By taking $\underline{q}$ large enough, we thus have a definition of computation encompassing simultaneously as many particular functions partial recursive in one and the same Θ as we may wish to use in a given discussion. But, of course, at the cost of some redundancy.[5]

3. Some functions which are partial recursive; the enumeration theorem and consequences. 3.1. It seems appropriate to give the following propositions now, even though the final justification of some steps in their proofs must await the results of Part II, as we shall point out at several places.

In (I), (II), (III) and (V), we list properties of relative partial recursiveness analogous to the general properties of deducibility "⊢" IM p. 89 (i)-(iv) (cf. IM p. 224 ¶ 2).

(I) (Reflexivity.) If θ is one of Θ, then θ is partial recursive in Θ.

Proof. Simply use a derivation with $\underline{p} = 1$ to introduce θ by S0.

(II) (Addition of assumed functions.) If ϕ is partial recursive in Θ, then ϕ is partial recursive in Θ,Σ.

Proof. All the schema applications in a partial recursive derivation of ϕ from Θ can be rewritten with Θ replaced by Θ,Σ. In the indices, each second member $\underline{m}_{\underline{i}}$ is replaced by $\overline{\underline{m}}_{\underline{i}}$ describing the functions $\overline{\Theta}_{\underline{i}} = (\Psi_{\underline{i}},\Theta,\Sigma)$ as $\underline{m}_{\underline{i}}$ did $\Theta_{\underline{i}} = (\Psi_{\underline{i}},\Theta)$.

(III) (Permutation of assumed functions, and suppression of repetitions.) If ϕ is partial recursive in Θ, then ϕ is partial recursive in Σ where Σ comes from Θ by permuting the functions Θ and (optionally) omitting functions which are duplicates of those remaining.

Proof. Similarly, replacing $\underline{m}_{\underline{i}}$'s referring to $\Theta_{\underline{i}} = (\Psi_{\underline{i}},\Theta)$ by $\overline{\underline{m}}_{\underline{i}}$'s referring to

---

[5] It is fairly obvious how one can then reduce such redundancy, as by coalescing groups of function symbols which are in fact handled identically by the computation rules. We shall not take the space now to develop this theme.

$\bar{\Theta}_{\underline{i}} = (\Psi_{\underline{i}},\Sigma)$, and changing the $\underline{t}$'s correspondingly in indices for S0.

(IV) (Addition of variables.) If $\psi(\mathfrak{b})$ is partial recursive in Θ, so is $\phi(\mathfrak{b},\mathfrak{c}) \simeq \psi(\mathfrak{b})$ where $\mathfrak{b},\mathfrak{c}$ are distinct variables. (Cf. 1959 III p. 4.)

Proof. We rewrite the schema applications in the derivation $\psi_1,\ldots,\psi_{\underline{p}}$ of ψ adding $\mathfrak{c}$ at the right as further arguments of $\psi_1,\ldots,\psi_{\underline{p}}$ and of assumed functions to be identified by S11 later in the derivation with one of $\psi_2,\ldots,\psi_{\underline{p}}$. The result is a derivation $\phi_1,\ldots,\phi_{\underline{p}}$ of ϕ. Of course, the variables in $\psi_1,\ldots,\psi_{\underline{p}}$ must first be chosen to be distinct from all of $\mathfrak{c}$.

Thus, if $\psi_1,\ldots,\psi_{\underline{p}}$ be the $\phi_1,\ldots,\phi_{18}$ of 1.4, and none of $\underline{a},\underline{c},\underline{d},\underline{e},\underline{y},\underline{z}$ occur in $\mathfrak{c}$ ($\underline{b}$ doesn't, if it is the $\mathfrak{b}$ of (IV)), each occurrence of $\phi_{\underline{i}}(\ldots)$ ($\underline{i} = 1,\ldots,18$) in the schema applications is rewritten as $\phi_{\underline{i}}(\ldots,\mathfrak{c})$, and in the application for ϕ_9 the right side becomes $\eta(\underline{z},\underline{b},\mathfrak{c})$ (but for ϕ_2 the right side is not changed).

Clearly, a computation via $\psi_1,\ldots,\psi_{\underline{p}}$ of $\psi(\mathfrak{b})$ under a given assignment Ω will be isomorphic to a computation via $\phi_1,\ldots,\phi_{\underline{p}}$ of $\phi(\mathfrak{b},\mathfrak{c})$ under Ω-plus-any-assignment-to-$\mathfrak{c}$, with the variables $\mathfrak{c}$ riding through the steps unaltered as the final arguments of each $\phi_{\underline{i}}$. ∇

A function $\bar{\phi}(\mathfrak{a})$ is an extension of $\phi(\mathfrak{a})$, iff, for each tuple of values for which $\phi(\mathfrak{a})$ is defined, $\bar{\phi}(\mathfrak{a})$ is defined with the same value (but $\bar{\phi}(\mathfrak{a})$ may be defined for some more tuples); equivalently, considering $\underline{n}$-place functions ϕ and $\bar{\phi}$ as sets of $\underline{n}$+1-tuples, iff $\bar{\phi} \supset \phi$.

(V) (Extended transitivity.) If ϕ is partial recursive in Θ, and each of Θ is partial recursive in Σ, then an extension $\bar{\phi}$ of ϕ is partial recursive in Σ.

Proof. Toward obtaining a partial recursive derivation $\phi_1,\ldots,\phi_{\underline{p}}$ from Σ of an extension $\bar{\phi}$ of ϕ, the natural approach is this. Take a partial recursive derivation $\phi_1,\ldots,\phi_{\underline{p}}$ of ϕ from Θ, and in it replace each application of S0 to introduce an assumed function $\theta_{\underline{t}}$ (one of Θ) by a partial recursive derivation of that $\theta_{\underline{t}}$ from Σ, after using (IV) if necessary to add to the variables of $\theta_{\underline{t}}$ the $\mathfrak{c}$ of the application of S0. This may involve rewriting a given $\theta_{\underline{t}}$ with various additional variables $\mathfrak{c}$ to make it "compatible" with the $\phi_{\underline{i}}$'s of various applications of S0 for that $\underline{t}$. Each "expanded" $\theta_{\underline{t}}$ then has its own derivation from Σ, which can be sandwiched into $\phi_1,\ldots,\phi_{\underline{p}}$ at the proper place, identifying $\theta_{\underline{t}}$

with the respective $\phi_{\underline{i}}$, which becomes say $\overline{\phi_{\underline{i}}}$ in the proposed derivation $\overline{\phi_1},\ldots,\overline{\phi_{\underline{p}}}$ from Σ.

The theorem hypothesis means that we are dealing with a given choice of Θ,Σ, under which each $\theta_{\underline{t}}$ is partial recursive in Σ. The choice being fixed, suppose that, via the derivation $\phi_1,\ldots,\phi_{\underline{p}}$ of ϕ from Θ, $\phi(\mathfrak{A})$ receives the value $\underline{w}$ for a given set of values of $\mathfrak{A}$.

Consider any place in the computation of this value $\underline{w}$ where E7 is used to obtain a value $\underline{v}$ of a $\theta_{\underline{t}}$-expression. Using the illustration for E7 in 2.4, suppose the application of S0 for the immediately preceding step by E2 reads $\phi_{\underline{i}}(\mathcal{L},\zeta) \simeq \theta_{\underline{t}}(\mathcal{L})$ (with ζ as the $\mathcal{L}$), so the E2-step is say

$$\phi_{\underline{i}}(A,B_1,B_2,C,D,Z) \text{ ——— } \theta_{\underline{t}}(A,B_1,B_2,C,D).$$

In using E7, we only get $\underline{v}$ from our $\theta_{\underline{t}}$-oracle after subordinate calculations have been completed determining a quintuple of values $\underline{a},\alpha_1,\alpha_2,\alpha^2,\alpha^3$ for A,B_1,B_2,C,D, this quintuple being one for which $\theta_{\underline{t}}$ is defined. Now consider what happens at the corresponding place in attempting to compute $\overline{\phi}(\mathfrak{A})$ via $\overline{\phi_1},\ldots,\overline{\phi_{\underline{p}}}$. Our hypothesis that $\theta_{\underline{t}}$ is partial recursive in Σ tells us that there is a computation of $\theta_{\underline{t}}(\varepsilon^0,\varepsilon_1,\varepsilon_2,\varepsilon^2,\varepsilon^3)$ from Σ, via the partial recursive derivation of $\theta_{\underline{t}}$ from Σ, under the assignment of $\underline{a},\alpha_1,\alpha_2,\alpha^2,\alpha^3$ to the variables $\varepsilon^0,\varepsilon_1,\varepsilon_2,\varepsilon^2,\varepsilon^3$, producing the value $\underline{v}$; and hence one of $\theta_{\underline{t}}(\varepsilon^0,\varepsilon_1,\varepsilon_2,\varepsilon^2,\varepsilon^3,\zeta)$ under the same assignment plus any assignment to ζ. Upon identifying $\theta_{\underline{t}}$ with $\overline{\phi_{\underline{i}}}$, this becomes a computation of $\overline{\phi_{\underline{i}}}(\varepsilon^0,\varepsilon_1,\varepsilon_2,\varepsilon^2,\varepsilon^3,\zeta)$ from Σ via $\overline{\phi_1},\ldots,\overline{\phi_{\underline{p}}}$. It is surely plausible that there is also a computation of $\overline{\phi_{\underline{i}}}(A,B_1,B_2,C,D,Z)$ via $\overline{\phi_1},\ldots,\overline{\phi_{\underline{p}}}$ with the same result $\underline{v}$. That this is so will be established in Part II. In computing $\overline{\phi}(\mathfrak{A})$ from Σ via $\overline{\phi_1},\ldots,\overline{\phi_{\underline{p}}}$, this computation replaces the aforesaid E2- and E7-steps (including the subtrees for the E7-step) used in computing $\phi(\mathfrak{A})$ from Θ via $\phi_1,\ldots,\phi_{\underline{p}}$.

However, while our $\theta_{\underline{t}}$-oracle, before giving us a value $\underline{v}$ of $\theta_{\underline{t}}(A,B_1,B_2,C,D)$ by E7, demands type-0,1,1,2,3 values $\underline{a},\alpha_1,\alpha_2,\alpha^2,\alpha^3$ for A,B_1,B_2,C,D, our computation via the derivation of $\theta_{\underline{t}}$ from Σ (as built into $\overline{\phi_1},\ldots,\overline{\phi_{\underline{p}}}$) might produce a value of $\overline{\phi_{\underline{i}}}(A,B_1,B_2,C,D,Z)$ without having a full quintuple of such values. So there may be tuples of values of $\mathfrak{A}$ for which, in trying to compute $\phi(\mathfrak{A})$ via $\phi_1,\ldots,\phi_{\underline{p}}$, we are stopped at some E7 for want of a full quintuple of scuh values, while in computing

$\bar{\phi}(\mathfrak{A})$ via $\bar{\phi}_1,\ldots,\bar{\phi}_{\bar{\underline{p}}}$ we succeed. Thus $\bar{\phi}$ (= $\bar{\phi}_{\bar{\underline{p}}}$) may be a proper extension of ϕ (= $\phi_{\underline{p}}$). Example 1 (and Remark 1) here, and Example 2 following (XIII), should make the matter clear.

Example 1. In this example, Θ is the single function $\theta(\underline{a},\underline{b}) \simeq \underline{a}$, and Σ is empty. Let $\phi = \phi_{\underline{p}}$ be defined from θ by the partial recursive derivation

$$\begin{array}{ll} \phi_1(\underline{a},\underline{b}) \simeq \theta(\underline{a},\underline{b}) & \text{by S0,} \\ \phi_2(\underline{b},\underline{a}) \simeq \phi_1(\underline{a},\underline{b}) & \text{by S6.0,} \\ \quad \cdots & \\ \phi_{\underline{p}-1}(\underline{a}) & \\ \quad \phi_{\underline{p}}(\underline{a}) \simeq \phi_2(\phi_{\underline{p}-1}(\underline{a}),\underline{a}) & \text{by S4.0,} \end{array}$$

where $\phi_3,\ldots,\phi_{\underline{p}-1}$ is a partial recursive derivation from θ, not using S0 (so it is also a partial recursive description), of a function $\phi_{\underline{p}-1}(\underline{a})$ which is undefined for some values $\underline{r}$ of $\underline{a}$. For any such value (and the given interpretation of θ), the attempted computation of $\phi(\underline{a})$ from θ runs so (using E2, E2, E7 partially, and in the upper subordinate computation E1):

$$\phi_{\underline{p}}(\underline{a}) \text{ —— } \phi_2(\phi_{\underline{p}-1}(\underline{a}),\underline{a}) \text{ —— } \phi_1(\underline{a},\phi_{\underline{p}-1}(\underline{a})) \text{ —— } \theta(\underline{a},\phi_{\underline{p}-1}(\underline{a})) \begin{cases} \underline{a} \text{ —— R}\dagger \\ \phi_{\underline{p}-1}(\underline{a}) \quad \cdots \end{cases}$$

The application of E7 can't be completed because $\phi_{\underline{p}-1}(\underline{a})$ is undefined for the value $\underline{r}$ of $\underline{a}$ in question. The function $\phi = \phi_{\underline{p}}$ partial recursive in θ is indeed

$$\phi(\underline{a}) \simeq \begin{cases} \underline{a} \text{ if } \phi_{\underline{p}-1}(\underline{a}) \text{ is defined,} \\ \text{undefined otherwise.} \end{cases}$$

Now replace the application of S0 by

$$\bar{\phi}_1(\underline{a},\underline{b}) \simeq \underline{a} \qquad \text{by S3.}$$

This gives us a partial recursive description (with $\bar{\underline{p}} = \underline{p}$) of the function $\bar{\phi}(\underline{a}) \simeq \underline{a}$, since the computation (for any value $\underline{r}$ of $\underline{a}$) becomes (by E2, E2, E2, E1)

$$\bar{\phi}_{\underline{p}}(\underline{a}) \text{ —— } \bar{\phi}_2(\bar{\phi}_{\underline{p}-1}(\underline{a}),\underline{a}) \text{ —— } \bar{\phi}_1(\underline{a},\bar{\phi}_{\underline{p}-1}(\underline{a})) \text{ —— } \underline{a} \text{ —— R}\dagger$$

(as in 2.1 ¶ 6, where $\bar{\phi}_{\underline{p}}$, $\bar{\phi}_2$, $\bar{\phi}_{\underline{p}-1}$, $\bar{\phi}_1$ are written ϕ, ψ, χ, ρ).

Remark 1. It is somewhat anomalous to consider transitivity (V) in the present context, where each assumed function θ is interpreted as a function just on our types 0, 1, 2, 3. The $\theta(\underline{a},\underline{b}) \simeq \underline{a}$ in Example 1 is partial recursive (immediately by S3); but as such it is not naturally restricted to $b \in$ type 0, and our rather free-wheeling computation procedure does not so restrict θ when it is a computed, rather than assumed, function.

(VI) (Permutation of variables.) *If* $\psi(\mathfrak{b})$ *is partial recursive in* Θ, *so is* $\phi(\mathfrak{a}) \simeq \psi(\mathfrak{b})$ *where* $\mathfrak{a}$ *is any permutation of* $\mathfrak{b}$ *without repetitions or omissions*. (Cf. 1959 II pp. 3-4.)

Proof. We simply extend the given derivation by some applications of S6.j.

(VII) (Identifications of variables.) *If* $\psi(\mathfrak{b})$ *is partial recursive in* Θ, *so is* $\phi(\mathfrak{a}) \simeq \psi(\mathfrak{b}^*)$ *where* $\mathfrak{b}^*$ *is a list of as many variables of each type as in* $\mathfrak{b}$ *but with repetitions allowed, and* $\mathfrak{a}$ *is the distinct variables in* $\mathfrak{b}^*$ *in any chosen order*.

Proof, by induction on the number of repetitions in $\mathfrak{b}^*$, i.e. on the sum of the number of variables in $\mathfrak{b}^*$ minus the number in $\mathfrak{a}$. The basis (no repetitions) is (VI).

Case 0: a number variable a is repeated in $\mathfrak{b}^*$. Let $\mathfrak{b}'$ come from $\mathfrak{b}^*$ by changing one occurrence of a to a new number variable b. By hyp. ind., we get $\rho(b,\mathfrak{a}) \simeq \psi(\mathfrak{b}')$; by S3 with (VI), $\chi(\mathfrak{a}) \simeq a$; and by S4.0, $\phi(\mathfrak{a}) \simeq \rho(\chi(\mathfrak{a}),\mathfrak{a}) \simeq \rho(a,\mathfrak{a}) \simeq \psi(\mathfrak{b}^*)$. This assumes that in the computation of $\rho(A,\mathfrak{a})$ we get the same results with $\chi(\mathfrak{a})$ and a as the A (where $\chi(\mathfrak{a}) \simeq a$ via S3 with (VI)), which is surely plausible. This assumption will be justified by the theory to be developed in Part II.

Case 1: otherwise, but a type-1 variable α is repeated. By hyp. ind., we get $\rho(\beta,\mathfrak{a}) \simeq \psi(\mathfrak{b}')$ where β is a new type-1 variable replacing one occurrence of α in $\mathfrak{b}^*$; by S7.1 with (VI), $\chi(b,\mathfrak{a}) \simeq \alpha(b)$ (b a variable not in $\mathfrak{a}$); and by S4.1, $\phi(\mathfrak{a}) \simeq \rho(\lambda b\ \chi(b,\mathfrak{a}),\mathfrak{a}) \simeq \rho(\lambda b\ \alpha(b),\mathfrak{a}) \simeq \rho(\alpha,\mathfrak{a}) \simeq \psi(\mathfrak{b}^*)$. This assumes that in the computation of $\rho(B,\mathfrak{a})$ we get the same results with $\lambda b\ \chi(b,\mathfrak{a})$, $\lambda b\ \alpha(b)$ and α as the B, which Part II will justify.

Cases 2, 3. Similarly. ∇

We define a *standard substitution* to be a substitution, for each of the variables β^j of a function ψ except a final list $\mathfrak{a}$ of variables, of a function of $\mathfrak{a}$ of the respective type (formed, when $j > 0$, using the λ-operator); i.e. we have

$\mathfrak{a}$ left over. For example, into $\psi(\underline{a},\alpha,F,\mathfrak{a})$ substitute $\chi_1(\mathfrak{a})$ for $\underline{a}$, $\lambda\underline{b}\ \chi_2(\underline{b},\mathfrak{a})$ for α, $\lambda\beta\ \chi_3(\beta,\mathfrak{a})$ for F to get

$$\phi(\mathfrak{a}) \simeq \psi(\chi_1(\mathfrak{a}),\lambda\underline{b}\ \chi_2(\underline{b},\mathfrak{a}),\lambda\beta\ \chi_3(\beta,\mathfrak{a}),\mathfrak{a}).$$

In a full substitution 1959 p. 6, the final $\mathfrak{a}$ is not present. (Here, unlike 1959, we are putting the bound variables first.)

(VIII) (Standard substitution.) The class of the functions partial recursive in Θ is closed under standard substitution.

Proof, like 1959 V p. 6 except we have $\mathfrak{a}$ extra. By (IV) and (VI), we first express $\chi_1(\mathfrak{a})$, $\chi_2(\underline{b},\mathfrak{a})$ as $\chi_1(\alpha,F,\mathfrak{a})$, $\chi_2(\underline{b},F,\mathfrak{a})$. Then by applications successively of S4.0, S4.1 and S4.2,

$$\psi_1(\alpha,F,\mathfrak{a}) \simeq \psi(\chi_1(\alpha,F,\mathfrak{a}),\alpha,F,\mathfrak{a}) \simeq \psi(\chi_1(\mathfrak{a}),\alpha,F,\mathfrak{a}),$$

$$\psi_2(F,\mathfrak{a}) \simeq \psi_1(\lambda\underline{b}\ \chi_2(\underline{b},F,\mathfrak{a}),F,\mathfrak{a}) \simeq \psi(\chi_1(\mathfrak{a}),\lambda\underline{b}\ \chi_2(\underline{b},\mathfrak{a}),F,\mathfrak{a}),$$

$$\phi(\mathfrak{a}) \simeq \psi_2(\lambda\beta\ \chi_3(\beta,\mathfrak{a}),\mathfrak{a}) \simeq \psi(\chi_1(\mathfrak{a}),\lambda\underline{b}\ \chi_2(\underline{b},\mathfrak{a}),\lambda\beta\ \chi_3(\beta,\mathfrak{a}),\mathfrak{a}).$$

Of course, we have made similar assumptions to those for (VII), as we shall do again below.

(IX) (Full substitution.) The class of the functions partial recursive in Θ is closed under full substitution. (Cf. 1959 V p. 6.)

Proof, similar to (VIII) and illustrated by 1959.

(X) (Explicit definition.) Let E be any 0-expression containing no function symbols except $\theta_1,\ldots,\theta_{\underline{l}},\psi_1,\ldots,\psi_{\underline{q}}$, $'$, $\dot{-}1$, cs, where $\psi_1,\ldots,\psi_{\underline{q}}$ are partial recursive in $\Theta = (\theta_1,\ldots,\theta_{\underline{l}})$, and containing free at most the (distinct) variables $\mathfrak{a}$. Then there is a function ϕ partial recursive in Θ such that $\phi(\mathfrak{a}) \simeq$ E. (Cf. 1959 VII p. 6.)

Proof. For clarity, let us understand that each $\psi_{\underline{k}}$ ($\underline{k} = 1,\ldots,\underline{q}$) is introduced by a different partial recursive derivation $\phi_1^{(\underline{k})},\ldots,\phi_{\underline{p}^{(\underline{k})}}^{(\underline{k})}$ ($\psi_{\underline{k}} = \phi_{\underline{p}^{(\underline{k})}}^{(\underline{k})}$) from Θ, and the formation and computation rules are those of the $\underline{q}$-fold treatment in 2.5. In building up the ϕ to express E, we add other derivations, extending the formation and computation rules accordingly.[5]

We use induction on the number of the symbols in E.

Case E1: E is a number variable. Say the number variables in $\mathfrak{a}$ are, in order, $\underline{a}_1,\ldots,\underline{a}_{\underline{n}_0}$. If E is $\underline{a}_1$, take $\phi(\mathfrak{a}) \simeq \underline{a}_1$ by S3. If E is $\underline{a}_{\underline{i}}$ with $\underline{i} > 1$, take $\mathcal{L}$ to e $\mathfrak{a}$ with $\underline{a}_{\underline{i}}$ moved to the front, let $\phi_1(\mathfrak{b}) \simeq \underline{a}_{\underline{i}}$, and use (VI) to get $(\mathfrak{a}) \simeq \phi_1(\mathfrak{b})$.

Case E2: E is $\psi_{\underline{i}}(A_1,\ldots,A_{\underline{n}_0},B_1,\ldots,B_{\underline{n}_1},C_1,\ldots,C_{\underline{n}_2},D_1,\ldots,D_{\underline{n}_3})$. If $A_1,\ldots,D_{\underline{n}_3}$ are all of them variables but maybe with repetitions, then we get $(\mathfrak{a}) \simeq \psi_{\underline{i}}(A_1,\ldots,D_{\underline{n}_3})$ by (VII) with (IV). Now suppose some of $A_1,\ldots,D_{\underline{n}_3}$ are not variables. Replace each which is not a variable by a distinct new variable to get $\psi_{\underline{i}}(\mathfrak{a}')$. By (VII) with (IV) as already remarked, we get $\phi_1(\mathfrak{b},\mathfrak{a}) \simeq \psi_{\underline{i}}(\mathfrak{a}')$ where $\mathfrak{b}$ are the new variables. By the hyp. ind., each of $A_1,\ldots,A_{\underline{n}_1}$ which is not a variable can be expressed as a $\chi(\mathfrak{a})$; each of $B_1,\ldots,B_{\underline{n}_1}$ which is not a variable, say it is $\lambda\underline{c}$ B, can be expressed as a $\lambda\underline{c}\ \chi(\underline{c},\mathfrak{a})$, by the hyp. ind. applied to B; and similarly with $C_1,\ldots,C_{\underline{n}_2}$ and $D_1,\ldots,D_{\underline{n}_3}$. Now a standard substitution (VIII) for $\mathfrak{b}$ gives $\phi(\mathfrak{a}) \simeq \phi_1(\ldots,\mathfrak{a}) \simeq \psi_{\underline{i}}(A_1,\ldots,D_{\underline{n}_3})$.

Case E4: E is A(B). We can write this as $\phi(\mathfrak{a}) \simeq \phi_0(A,B,\mathfrak{a})$ where $\phi_0(\alpha^{\underline{j}},\alpha^{\underline{j}-1},\mathfrak{a}) \simeq \alpha^{\underline{j}}(\alpha^{\underline{j}-1})$ by S7.$\underline{j}$. Then we can operate with $\phi_0(A,B,\mathfrak{a})$ as under Case E2, noting that for the treatment there the hyp. ind. will be needed now only for C if A (not a variable) is $\lambda\alpha^{\underline{j}-1}$ C, and if B is not a variable, for B if $\underline{j} = 1$, and if $\underline{j} > 1$ for D where B is $\lambda\alpha^{\underline{j}-2}$ D. Each of these has fewer symbols than A(B).

Case E5 is immediate by S2.0. Case E6 is reduced to Case E2 in the same manner as Case E4, and Case E7 likewise after introducing $\phi_0(\mathfrak{b},\mathfrak{a}) \simeq \theta_{\underline{t}}(\mathfrak{b})$ by S0.

(XI) (Definition by cases.) If $\chi_0(\mathfrak{a}),\chi_1(\mathfrak{a}),\ldots,\chi_{\underline{n}}(\mathfrak{a})$ are partial recursive in Θ, so is

$$\phi(\mathfrak{a}) \simeq \begin{cases} \chi_1(\mathfrak{a}) & \text{if } \chi_0(\mathfrak{a}) = 0, \\ \quad \ldots & \\ \chi_{\underline{n}-1}(\mathfrak{a}) & \text{if } \chi_0(\mathfrak{a}) = \underline{n}-2, \\ \chi_{\underline{n}}(\mathfrak{a}) & \text{if } \chi_0(\mathfrak{a}) \geq \underline{n}-1, \end{cases}$$

where $\phi(\mathfrak{a})$ is defined exactly if $\chi_0(\mathfrak{a})$ is defined, say it fits the i-th case, and $\chi_{\underline{i}}(\mathfrak{a})$ is defined (irrespective of whether $\chi_{\underline{j}}(\mathfrak{a})$ is defined for $\underline{j} \neq \underline{i}$).

Proof, e.g. for $\underline{n} = 4$. Using (X), let

$$\phi(\mathfrak{a}) \simeq cs(\chi_0(\mathfrak{a}),\chi_1(\mathfrak{a}),(cs(\chi_0(\mathfrak{a})\dot{-}1,\chi_2(\mathfrak{a}),cs(\chi_0(\mathfrak{a})\dot{-}2,\chi_3(\mathfrak{a}),\chi_4(\mathfrak{a}))))).$$

(XII) (Primitive recursion.) If $\psi(\mathfrak{b})$ and $\chi(\underline{a},\underline{c},\mathfrak{b})$ are partial recursive in Θ, so is the function $\phi(\underline{a},\mathfrak{b})$ defined by

$$\begin{cases} \phi(0,\mathcal{L}) \simeq \psi(\mathcal{L}), \\ \phi(\underline{a}',\mathcal{L}) \simeq \chi(\underline{a},\phi(\underline{a},\mathcal{L}),\mathcal{L}). \end{cases}$$

Proof. (Cf. IM Example 3 p. 350.) The usual proof by induction on $\underline{a}$ that the pair of recursion equations has a unique solution ϕ works here. We get a partial recursive derivation $\phi_1,\ldots,\phi_{\underline{p}}$ from Θ of that function $\phi = \phi_{\underline{p}}$ as follows. First, use (II) and (III) to reconstrue ψ,χ as functions partial recursive in η,Θ (but, as in 2.2 we won't show the Θ explicitly, nor here the η since ψ,χ don't actually depend on η). Next, using (X) for η,Θ as its "Θ", let

$$\rho^{\eta}(\underline{a},\mathcal{L}) \simeq \mathrm{cs}(\underline{a},\psi(\mathcal{L}),\chi(\underline{a}\dot{-}1,\eta(\underline{a}\dot{-}1,\mathcal{L}),\mathcal{L})).$$

Finally, using S11, let

$$\phi(\underline{a},\mathcal{L}) \simeq \rho^{\phi}(\underline{a},\mathcal{L}).$$

To elaborate a bit, by saying that the recursion equations have a unique solution ϕ we mean that, when we treat "ϕ" in them as a symbol for a variable assumed function, call it η, and use the computation rules as from η,Θ, then (for a given interpretation of the function symbols Θ) there is exactly one interpretation ϕ of η that makes both equations hold for all choices of values of $a,\mathcal{L}$. It is a fundamental property of our computation rules 2.4 (to be explored in Part II) that, in computing $\chi(\underline{a},\eta(\underline{a},\mathcal{L}),\mathcal{L})$, $\eta(\underline{a},\mathcal{L})$ will be carried along intact (perhaps being substituted, and the results being again substituted, etc., for variables in other expressions, but never having anything substituted for its variables $\underline{a},\mathcal{L}$), until and unless (as may happen more than once, on different branches) the evaluation by E7 of $\eta(\underline{a},\mathcal{L})$ is called for. In brief, $\chi(\underline{a},\eta(\underline{a},\mathcal{L}),\mathcal{L})$ depends on η only via $\eta(\underline{a},\mathcal{L})$, as one could reasonably expect. The like applies to $\chi(\underline{a}\dot{-}1,\eta(\underline{a}\dot{-}1,\mathcal{L}),\mathcal{L}))$; that is, the value of this will depend on η only via E7 applied to evaluate $\eta(\underline{a}\dot{-}1,\mathcal{L})$. For a given type-0 value of $\underline{a}$, the expression $\underline{a}\dot{-}1$ is defined (and $\mathcal{L}$ will have given values of the appropriate types, the same as the subordinate computations for them under E7 will give). Consequently, the same result will be obtained for $\eta(\underline{a}\dot{-}1,\mathcal{L})$ by E7 with η interpreted by our ϕ as by using S11 to feed in a computation of $\phi(\underline{a}\dot{-}1,\mathcal{L})$ via the partial recursive derivation $\phi_1,\ldots,\phi_{\underline{p}}$ of ϕ. Thus, the phenomenon exhibited in Example 1 (following (V)) will not arise here. ∇

A function is primitive recursive in total functions Θ, iff it is definable from Θ using the schemata we listed in 1.3 except that (XII) (called S5 in 1959 p. 3) used as a "postulated" schema replaces S11, whereupon S1.1 and S5.1 become redundant; primitive recursive then, if Θ is empty.[6] Primitive recursive functions are total, as are functions primitive recursive in total functions Θ.

The present definition of "primitive recursive" functions is equivalent for Θ empty to that in 1959 p. 3 (for types 0,1,2,3) using 1959 Remark 1 p. 6, and for Θ non-empty by 1959 1.8 p. 7. For variables of types 0 and 1 only, it agrees with ORT, by 1959 1.7 pp. 6-7.

Since we didn't use S11 in establishing (I)-(XI), they hold for primitive recursiveness (with $\bar{\phi} = \phi$ in (V)).

(XIII) (Least-number operator.) If $\psi(\underline{y}, \mathfrak{a})$ is partial recursive in Θ, so is

$$\phi(\mathfrak{a}) \simeq \mu \underline{y}[\psi(\underline{y}, \mathfrak{a})=0],$$

where $\mu\underline{y}[\psi(\underline{y},\mathfrak{a})=0]$ is the least $\underline{y}$ (if it exists) such that $\psi(0,\mathfrak{a}),\ldots,\psi(\underline{y},\mathfrak{a})$ are all defined and $\psi(\underline{y},\mathfrak{a}) = 0$, and is undefined if such a $\underline{y}$ does not exist.

Proof. Essentially as in IM Example 4 p. 350. Or apply (V) to the example $\phi_1,,,,,\phi_{18}$ in 1.4, thus.

Example 2. To adapt the derivation $\phi_1,\ldots,\phi_{18}$ as given in 1.4 to the present context, first replace its "$\underline{b}$" by "$\mathfrak{a}$" and its "θ" by "ψ". Thereby, $\phi(\mathfrak{a}) \simeq \mu\underline{y}[\psi(\underline{y},\mathfrak{a})=0]$ is partial recursive in ψ. From the (adapted) derivation $\phi_1,\ldots,\phi_{18}$ (of $\phi = \phi_{18}$), it is not hard to see that in the computation of $\phi(\mathfrak{a})$ from ψ for a given assignment to $\mathfrak{a}$ (and a given ψ), E7 will be called upon only to evaluate $\psi(A,\mathfrak{a})$ for the original $\mathfrak{a}$ and a succession of A's which can be evaluated as 0,1,2,... . Therefore we get no fewer values with ψ an assumed function than when, in applying (V), we replace the introduction by S0 of $\phi_2(\underline{y},\mathfrak{a},\underline{d},\underline{e})$ as the assumed function $\psi(\underline{y},\mathfrak{a})$ by the identification of $\phi_2(\underline{y},\mathfrak{a},\underline{d},\underline{e})$ with the last function in a partial recursive derivation from Θ of the same function $\psi(\underline{y},\mathfrak{a})$ expanded by (IV) to $\psi(\underline{y},\mathfrak{a},\underline{d},\underline{e})$. That is, the phenomenon of Example 1 does not arise, so the

---

[6] The presence of S11 in 1.3 made the schema of primitive recursion replaceable by introductions of some particular functions, just as did the μ-schema in μ-recursiveness (cf. IM p. 320, Kleene 1936a).

function given by the new derivation from Θ constructed by (V) is $\mu y[\psi(y,\mathfrak{a})=0]$, not a proper extension of it.

(XIV) (Totally undefined function.) <u>There is a partial recursive function $\phi(\mathfrak{a})$ such that, for each $\mathfrak{a}$, $\phi(\mathfrak{a})$ is undefined</u>.

<u>First proof</u>. As in IM Example 1 p. 350; i.e. $\psi(\eta;\mathfrak{a}) \simeq \eta(\mathfrak{a})$ by S0, $\phi(\mathfrak{a}) \simeq \psi(\phi;\mathfrak{a})$ by S11. <u>Second proof</u>. $\phi(\mathfrak{a}) \simeq \mu y[\psi(y,\mathfrak{a})=0]$ by (XIII), where $\psi(y,\mathfrak{a}) \simeq y'$ by S1.0.

<u>3.2.</u> Our next objective is an enumeration (or parametrization) theorem and its consequences, still taking the ranges of the variables to be types 0, 1, 2, 3. In 1959, while we had to work hard to get a restricted result on substitution of λ-functionals, the enumeration theorem was almost the starting point (S9 p. 13, XII p. 15). It won't be so hard though to get the enumeration theorem here, if we refrain from laboring over the sort of detail with a Gödel numbering which by now has become routine.

We establish a Gödel numbering of the symbols and expressions, as we described them in 2.2, which will apply to any z such that $\mathrm{Ix}(z)$, determining the lists $\theta_1,\ldots,\theta_\ell$ and $\phi_1,\ldots,\phi_p$ (1-fold treatment; cf. 2.5). The variables of type j, assumed to be enumerated as $\beta^j_0, \beta^j_1, \ldots, \beta^j_i, \ldots$, shall have the Gödel numbers $\langle 0,j,i\rangle$ ($j = 0,1,2,3$; $i = 0,1,2,\ldots$). $\theta_1,\ldots,\theta_\ell$ shall have the Gödel numbers $\langle 1,1\rangle,\ldots,\langle 1,\ell\rangle$ where $\ell = \ell h((z)_1)$; $\phi_1,\ldots,\phi_p$ the Gödel numbers $\langle 2,1\rangle,\ldots,\langle 2,p\rangle$ where $p = \ell h(\mathrm{cd}(z))$ (cf. 1.4); and $\lambda, 0, {}', \dot{-}1, \mathrm{cs}$ the Gödel numbers $\langle 3\rangle,\langle 4\rangle,\langle 5\rangle,\langle 6\rangle,\langle 7\rangle$. A composite expression shall have the Gödel number $\langle 8,c_1,\ldots,c_s\rangle$ where $c_1,\ldots,c_s$ are the Gödel numbers of its components. Thus $\phi_i(A_1,\ldots,D_{n_{i3}})$ has $\langle 8,\ulcorner\phi_i\urcorner,\ulcorner A_1\urcorner,\ldots,\ulcorner D_{n_{i3}}\urcorner\rangle$ where "$\ulcorner\ \urcorner$" indicates Gödel numbers, and similarly with $\theta_t(A_1,\ldots,D_{m_{t3}})$; $A(B)$ has $\langle 8,\ulcorner A\urcorner,\ulcorner B\urcorner\rangle$; $\lambda\alpha^{j-1} A$ has $\langle 8,\langle 3\rangle,\ulcorner\alpha^{j-1}\urcorner,\ulcorner A\urcorner\rangle$; A' has $\langle 8,\langle 5\rangle,\ulcorner A\urcorner\rangle$, thinking of the operator symbol $'$ as logically first (as though A' were $'(A)$); etc. If $n_{i0}+n_{i1}+n_{i2}+n_{i3} = 0$,[7] ϕ_i as a 0-expression rather than a function symbol has the Gödel number $\langle 8,\ulcorner\phi_i\urcorner\rangle$; and similarly with θ_t.

[7] This case interests us only marginally. In ORT it was used in the proof of the main theorem ("$\Pi^1_1 \rightarrow$ hyp") of Kleene 1955.

As in 1959 2.1 p. 7, for $\underline{j} > 0$ let $<\alpha^{\underline{j}}_0,\ldots,\alpha^{\underline{j}}_{\underline{n}}> = \lambda\tau^{\underline{j}-1} <\alpha^{\underline{j}}_0(\tau^{\underline{j}-1}),\ldots,\alpha^{\underline{j}}_{\underline{n}}(\tau^{\underline{j}-1})>$ and $(\alpha^{\underline{j}})_{\underline{i}} = \lambda\tau^{\underline{j}-1}(\alpha^{\underline{j}}(\tau^{\underline{j}-1}))_{\underline{i}}$. The scopes of $\lambda\tau^{\underline{j}-1}$ in these are primitive recursive functions of their variables, and, for $0 \leq \underline{i} \leq \underline{n}$, $(<\alpha^{\underline{j}}_0,\ldots,\alpha^{\underline{j}}_{\underline{n}}>)_{\underline{i}} = \alpha^{\underline{j}}_{\underline{i}}$.

(XV) (= (XV.3.3)) For each fixed Θ as described by $\underline{m} = <\underline{m}_1,\ldots,\underline{m}_{\underline{l}}>$ (possibly empty, i.e. with $\underline{m} = 1$), there is a function ϕ^* (of six variables of the types shown below) partial recursive in Θ such that: If $\underline{z}$ is the index of a function ϕ partial recursive in Θ (so $\mathrm{Ix}(\underline{z})$ & $(\underline{z})_1=\underline{m}$), and $\underline{e}$ is the Gödel number of a 0-expression E based (as in 2.2) on $\theta_1,\ldots,\theta_{\underline{l}}$ and the $\phi_1,\ldots,\phi_{\underline{p}}$ determined by $\underline{z}$, then

$$\phi^*(\underline{z},\underline{e},\alpha^0,\alpha^1,\alpha^2,\alpha^3) \simeq E$$

when (for each $\underline{j} = 0,1,2,3$ and each $\underline{i}$) each variable $\beta^{\underline{j}}_{\underline{i}}$ occurring free in E is assigned the value $(\alpha^{\underline{j}})_{\underline{i}}$. Otherwise, $\phi^*(\underline{z},\underline{e},\alpha^0,\alpha^1,\alpha^2,\alpha^3)$ is undefined.

Proof. In the theorem, "Otherwise" means contrary to the specifications from "If" to "then". They include that $\mathrm{Ix}(\underline{z})$ & $(\underline{z})_1=\underline{m}$. That being so, we can tell primitively recursively from $\underline{z},\underline{e}$ whether $\underline{e}$ is the Gödel number of a 0-expression E based on $\theta_1,\ldots,\theta_{\underline{l}},\phi_1,\ldots,\phi_{\underline{p}}$. Thus we get a conjunction of primitive recursive predicates, negating which we get "Otherwise" as a primitive recursive predicate $\underline{O}(\underline{z},\underline{e})$.

We shall assume for the next part of the proof that $\overline{\underline{O}}(\underline{z},\underline{e})$, i.e. not otherwise.

Now consider the cases in the definition of the computation procedure, or the inductive definition of $E \simeq_\Omega \underline{w}$ for a given E and assignment Ω, in 2.4.

E1: E is a type-0 variable $\beta^0_{\underline{i}}$. The case hypothesis is $\underline{e} = <0,0,(\underline{e})_2>$ where $\underline{i} = (\underline{e})_2$, and clearly it will suffice to have

$$\phi^*(\underline{z},\underline{e},\alpha^0,\ldots,\alpha^3) \simeq (\alpha^0)_{(\underline{e})_2} \text{ if } \underline{e}=<0,0,(\underline{e})_2>.$$

E2: E is $\phi_{\underline{i}}(A_1,\ldots,D_{\underline{n}_{\underline{i}3}})$. Given that $\overline{\underline{O}}(\underline{z},\underline{e})$, so $\underline{e}$ is anyway the index of a 0-expression in the language with $\theta_1,\ldots,\theta_{\underline{l}},\phi_1,\ldots,\phi_{\underline{p}}$ based on $\underline{m}_1,\ldots,\underline{m}_{\underline{l}},\underline{z}_1,\ldots,\underline{z}_{\underline{p}}$ ($\underline{z}_{\underline{p}} = \underline{z}$), the remainder of the case hypothesis is simply $(\underline{e})_0=8$ & $(\underline{e})_1=\langle 2,(\underline{e})_{1,1}\rangle$; this says E is of the form $\phi_{\underline{i}}(\ldots)$, and that the arguments of $\phi_{\underline{i}}$ are of the right numbers and types is then necessitated by $\overline{\underline{O}}(\underline{z},\underline{e})$. From $\underline{e}$ with $\underline{z}$ (where $(\underline{e})_{1,1}$ is the $\underline{i}$ of $\phi_{\underline{i}}$; and using 1.4 $(\mathrm{cd}(\underline{z}))_{\underline{i}\dot{-}1}$ is the index $\underline{z}_{\underline{i}}$ of $\phi_{\underline{i}}$; and for $(\underline{z}_{\underline{i}})_0=0$, so S0 applies, $(\mathrm{ps}(\underline{z}))_{\underline{i}\dot{-}1}$ with $(\underline{z}_{\underline{i}})_3 = \underline{t}$ tells whether to use a θ, or to feed back in one of $\phi_2,\ldots,\phi_{\underline{p}}$, and in either case which one), we can primitive recursively (but not altogether simply) find the index $\psi_{E2}(\underline{z},\underline{e})$ of the expression F to which the reduction step by E2 leads. (We operate here without a reserved list $\underline{V}$ of variables.) So we obtain the following specification for ϕ^*:

$$\phi^*(\underline{z},\underline{e},\alpha^0,\ldots,\alpha^3) \simeq \phi^*(\underline{z},\psi_{E2}(\underline{z},\underline{e}),\alpha^0,\ldots,\alpha^3) \text{ if } \overline{O}(\underline{z},\underline{e}) \ \& \ (\underline{e})_0=8 \ \& \ (\underline{e})_1=\langle 2,(\underline{e})_{1,1}\rangle.$$

E4.λ: E is $\{\lambda\beta^{\underline{j}}_{\underline{i}}\ A\}(B)$. Similarly:

$$\phi^*(\underline{z},\underline{e},\alpha^0,\ldots,\alpha^3) \simeq \phi^*(\underline{z},\psi_{E4.\lambda}(\underline{z},\underline{e}),\alpha^0,\ldots,\alpha^3) \text{ if } \overline{O}(\underline{z},\underline{e}) \ \& \ (\underline{e})_0=8 \ \& \ (\underline{e})_1=\langle 3\rangle.$$

E4.1: E is $\beta^1_{\underline{i}}(B)$. Then $\underline{i} = (\underline{e})_{1,2}$ and $\ulcorner B\urcorner = (\underline{e})_2$.

$$\phi^*(\underline{z},\underline{e},\alpha^0,\ldots,\alpha^3) \simeq (\alpha^1(\phi^*(\underline{z},(\underline{e})_2,\alpha^0,\ldots,\alpha^3)))_{(\underline{e})_{1,2}}$$
$$\text{if } \overline{O}(\underline{z},\underline{e}) \ \& \ (\underline{e})_0=8 \ \& \ (\underline{e})_1=\langle 0,1,(\underline{e})_{1,2}\rangle.$$

E4.2: E is $\beta^2_{\underline{i}}(B)$. Say that $\underline{c} = \beta^0_{\underline{k}}$, where $\underline{k} = \psi_{E4.2}(\underline{e})$ with $\psi_{E4.2}$ primitive recursive, is the first number variable not occurring free in B. Then if $\alpha^0 = \langle\alpha^0_0,\alpha^0_1,\ldots,\alpha^0_{\underline{r}}\rangle$ with $\underline{r} \geq \underline{k}$, $[\alpha^0/\underline{p}_{\underline{k}}^{(\alpha^0)_{\underline{k}}}]\cdot\underline{p}_{\underline{k}}^{\underline{s}} = \langle\alpha^0_0,\ldots,\alpha^0_{\underline{k}-1},\underline{s},\alpha^0_{\underline{k}+1},\ldots,\alpha^0_{\underline{r}}\rangle$.

$$\phi^*(\underline{z},\underline{e},\alpha^0,\ldots,\alpha^3) \simeq (\alpha^2(\lambda\underline{s}\ \phi^*(\underline{z},\langle 8,(\underline{e})_2,\langle 0,0,\underline{k}\rangle\rangle,[\alpha^0/\underline{p}_{\underline{k}}^{(\alpha^0)_{\underline{k}}}]\cdot\underline{p}_{\underline{k}}^{\underline{s}},\alpha^1,\alpha^2,\alpha^3)))_{(\underline{e})_{1,2}}$$
$$(\underline{k} = \psi_{E4.2}(\underline{e})) \text{ if } \overline{O}(\underline{z},\underline{e}) \ \& \ (\underline{e})_0=8 \ \& \ (\underline{e})_1=\langle 0,2,(\underline{e})_{1,2}\rangle.$$

E4.3: E is $\beta^3_{\underline{i}}(B)$.

$$\phi^*(\underline{z},\underline{e},\alpha^0,\ldots,\alpha^3) \simeq (\alpha^3(\lambda\sigma\ \phi^*(\underline{z},\langle 8,(\underline{e})_2,\langle 0,1,\underline{k}\rangle\rangle,\alpha^0,\lambda\underline{t}\ [\alpha^1(\underline{t})/\underline{p}_{\underline{k}}^{(\alpha^1(\underline{t}))_{\underline{k}}}]\cdot\underline{p}_{\underline{k}}^{\sigma(\underline{t})},\alpha^2,$$
$$\alpha^3)))_{(\underline{e})_{1,2}} \quad (\underline{k} = \psi_{E4.3}(\underline{e})) \text{ if } \overline{O}(\underline{z},\underline{e}) \ \& \ (\underline{e})_0=8 \ \& \ (\underline{e})_1=\langle 0,3,(\underline{e})_{1,2}\rangle.$$

E5: E is 0. $\phi^*(\underline{z},\underline{e},\alpha^0,\ldots,\alpha^3) \simeq 0$ if $\underline{e} = \langle 4\rangle$.

E6.cs: E is cs(A,B,C). (E6.' and E6.∸1 are similar but simpler.)

$$\phi^*(\underline{z},\underline{e},\alpha^0,\ldots,\alpha^3) \simeq \mathrm{cs}(\phi^*(\underline{z},(\underline{e})_2,\alpha^0,\ldots,\alpha^3),\phi^*(\underline{z},(\underline{e})_3,\alpha^0,\ldots,\alpha^3),$$
$$\phi^*(\underline{z},(\underline{e})_4,\alpha^0,\ldots,\alpha^3)) \text{ if } \overline{O}(\underline{z},\underline{e}) \ \& \ (\underline{e})_0=8 \ \& \ (\underline{e})_1=\langle 7\rangle.$$

E7.$\underline{t}$: E is $\theta_{\underline{t}}(A_1,\ldots,D_{\underline{m}_{\underline{t}3}})$. We make $\underline{l}$ separate cases according as $\underline{t} = 1,\ldots,\underline{l} = \underline{l}h((\underline{z})_1)$. Say e.g. E is $\theta_{\underline{t}}(A,B_1,B_2,C,D)$ ($\underline{m}_{\underline{t}0}=\underline{m}_{\underline{t}2}=\underline{m}_{\underline{t}3}=1$, $\underline{m}_{\underline{t}2}=2$).

$$\begin{aligned}
&\phi^*(\underline{z},\underline{e},\alpha^0,\ldots,\alpha^3) \simeq \\
&\quad\theta_{\underline{t}}(\phi^*(\underline{z},(\underline{e})_2,\alpha^0,\ldots,\alpha^3), \\
&\quad\quad\lambda\underline{s}\ \phi^*(\underline{z},\langle 8,(\underline{e})_3,\langle 0,0,\underline{k}_1\rangle\rangle,[\alpha^0/\underline{p}_{\underline{k}_1}^{(\alpha^0)_{\underline{k}1}}]\cdot\underline{p}_{\underline{k}_1}^{\underline{s}},\alpha^1,\alpha^2,\alpha^3), \\
&\quad\quad\lambda\underline{s}\ \phi^*(\underline{z},\langle 8,(\underline{e})_4,\langle 0,0,\underline{k}_2\rangle\rangle,[\alpha^0/\underline{p}_{\underline{k}_2}^{(\alpha^0)_{\underline{k}2}}]\cdot\underline{p}_{\underline{k}_2}^{\underline{s}},\alpha^1,\alpha^2,\alpha^3), \\
&\quad\quad\lambda\sigma\ \phi^*(\underline{z},\langle 8,(\underline{e})_5,\langle 0,1,\underline{k}_3\rangle\rangle,\alpha^0,\lambda\underline{t}\ [\alpha^1(\underline{t})/\underline{p}_{\underline{k}_3}^{(\alpha^1(\underline{t}))_{\underline{k}3}}]\cdot\underline{p}_{\underline{k}_3}^{\sigma(\underline{t})},\alpha^2,\alpha^3), \\
&\quad\quad\lambda\sigma^2\phi^*(\underline{z},\langle 8,(\underline{e})_6,\langle 0,2,\underline{k}_4\rangle\rangle,\alpha^0,\alpha^1,\lambda\tau\ [\alpha^2(\tau)/\underline{p}_{\underline{k}_4}^{(\alpha^2(\tau))_{\underline{k}4}}]\cdot\underline{p}_{\underline{k}_4}^{\sigma^2(\tau)},\alpha^3))
\end{aligned}$$
$$(\underline{k}_{\underline{i}} = \psi^{\underline{i}}_{E7.\underline{t}}(\underline{e}) \text{ for } \underline{i} = 1,2,3,4) \text{ if } \overline{O}(\underline{z},\underline{e}) \ \& \ (\underline{e})_0=8 \ \& \ (\underline{e})_1=\langle 1,\underline{t}\rangle.$$

Otherwise: Using a totally undefined function $\psi_{\underline{0}}$ (cf. (XIV)):

$$\phi^*(\underline{z},\underline{e},\alpha^0,\ldots,\alpha^3) \simeq \psi_{\underline{0}}(\underline{z},\underline{e},\alpha^0,\ldots,\alpha^3) \text{ if } \underline{0}(\underline{z},\underline{e}).$$

Now all the case specifications, with ϕ^* replaced by η, can be assembled into a definition by cases fitting (XI). For, we can define the χ_0 primitive recursively (in fact independently of $\eta,\Theta,\alpha^0,\ldots,\alpha^3$) so that $\chi_0(\underline{z},\underline{e},\alpha^0,\ldots,\alpha^3) = 0, \ldots, \underline{n}-1$ according as the case hypothesis of E1, ... , E7.*1* or Otherwise applies; and the $\chi_1,\ldots,\chi_{\underline{n}}$ we get by use of (X) for η,Θ as its "Θ". Thus there is $\psi^{\eta}(\underline{z},\underline{e},\alpha^0,\ldots,\alpha^3)$, or more explicitly $\psi^{\eta,\Theta}(\underline{z},\underline{e},\alpha^0,\ldots,\alpha^3)$, representing what is defined by combining the cases (with ϕ^* replaced by η). So using S11, we can get $\phi^*(\underline{z},\underline{e},\alpha^0,\ldots,\alpha^3) \simeq \psi^{\phi^*}(\underline{z},\underline{e},\alpha^0,\ldots,\alpha^3)$ to satisfy all the spececifications.

Now it can be proved under the assumption $\overline{0}(\underline{z},\underline{e})$, by induction over any completed computation tree for E under the assignment Ω extracted from $\alpha^0,\alpha^1,\alpha^2,\alpha^3$ as in the theorem (with given values of Θ), that then $\phi^*(\underline{z},\underline{e},\alpha^0,\ldots,\alpha^3) \simeq \underline{w}$ for $\underline{w}$ the value of E under Ω. Conversely, still assuming $\overline{0}(\underline{z},\underline{e})$, by induction over a completed computation tree for $\phi^*(\underline{z},\underline{e},\alpha^0,\ldots,\alpha^3)$ with result $\underline{w}$ under given values of $\underline{z},\underline{e},\alpha^0,\ldots,\alpha^3$ and Θ, $E \simeq \underline{w}$ under the assignment for E extracted from $\alpha^0,\ldots,\alpha^3$. And of course, $\underline{0}(\underline{z},\underline{e})$ gives that $\phi^*(\underline{z},\underline{e},\alpha^0,\ldots,\alpha^3)$ is undefined.

In these inductions, we will have that the use of S11 to give $\phi_{\underline{p}}(\underline{z},\underline{e},\alpha^0,\ldots,\alpha^3) \simeq \psi^{\phi^*}_{\underline{p}-1}(\underline{z},\underline{e},\alpha^0,\ldots,\alpha^3)$, where $\psi^{\eta}_{\underline{p}-1}(\underline{z},\underline{e},\alpha^0,\ldots,\alpha^3)$ has been defined from η,Θ, for the given Θ and η a variable assumed function, will treat "previous values" of ϕ^* just as η is treated as an assumed function by E7 for the arguments in hand. So, for $\underline{z},\underline{e},\alpha^0,\alpha^1,\alpha^2,\alpha^3$ ranging over our types 0,0,0,1,2,3, ϕ^* defined via S11 is a solution for η of the recursion $\eta(\underline{z},\underline{e},\alpha^0,\ldots,\alpha^3) \simeq \psi^{\eta}_{\underline{p}-1}(\underline{z},\underline{e},\alpha^0,\ldots,\alpha^3)$, just as in the (simpler) case of a primitive recursion (for (XII)). That is, in the computations of $\psi^{\eta}_{\underline{p}-1}(\underline{z},\underline{e},\alpha^0,\ldots,\alpha^3)$ for given type-0,0,0,1,2,3 values of $\underline{z},\underline{e},\alpha^0,\alpha^1,\alpha^2,\alpha^3$, values of η will be called for under E7 only for argument expressions which will all of them be defined in our types (so the phenomenon of Example 1 will not arise). In particular, for values of $\underline{z},\underline{e}$ which, via the evaluations of cs(...) for the application of (XI), give us e.g. Case 4.2 (read with "η" replacing "ϕ^*") with $\lambda\underline{s}\ \eta(\ldots)$ appearing, the part $\alpha^2(\lambda\underline{s}\ \eta(\ldots))$ will eventually be evaluated via a new application of E4.2 with α^2 as the α^2 of E4.2

(as will be seen in Part II). Thereby $\{\lambda \underline{s}\ \eta(\ldots)\}(\underline{s})$ will start subtrees with $\underline{s}$ having the values 0,1,2,..., and after using E4.λ we will come to evaluate by E7 η applied to argument expressions with the new variable $\underline{s}$ in one of them. Similarly with E4.3 and with E7 when the $\theta_{\underline{t}}$ has some arguments of types > 0. All the arguments of η that will arise in the computation will be primitive recursive functions of $\underline{z},\underline{e},\alpha^0,\alpha^1,\alpha^2,\alpha^3$ and new variables like $\underline{s},\sigma,\sigma^2$ introduced into subcomputations for E4.2, E4.3 and E7. ∇

We write (XV) also as (XV.3.3) to express by the first "3" that $\phi_1,\ldots,\phi_{\underline{p}}$ may have variables of each type ≤ 3, and by the second "3" that variables of each type ≤ 3 may occur free in E and may be variables of $\theta_1,\ldots,\theta_{\underline{l}}$. Suppose instead that only variables of types ≤ 1 may occur free in E or be variables of $\theta_1,\ldots,\theta_{\underline{l}}$, but $\phi_1,\ldots,\phi_{\underline{p}}$ may still have variables of types ≤ 3. For example, E might be $\phi_{\underline{p}}(\lambda\beta^2\ \phi_{\underline{q}}(\beta^2,\underline{b},\beta),\underline{b},\beta)$. The following modification (XV.3.1) of (XV) applies. In brief, the use of the schemata with variables of types > 1 does not increase the class of functions of variables of types ≤ 1 partial recursive in functions of variables of types ≤ 1. Similarly, we have (XV.$\underline{j}$.$\underline{k}$) for any $0 \leq \underline{k} \leq \underline{j} \leq 3$. (1959 I pp. 3, 15 doesn't hold in the present theory.)

(XV.3.1) For each fixed Θ (possibly empty) which consists of functions of variables of only types ≤ 1, there is a function ϕ^* (of four variables) partial recursive in Θ in the theory with only types ≤ 1 (thus omitting S4.2, S4.3, S6.2, S6.3, S7.2, S7.3) such that: For $\underline{z},\underline{e}$ as in (XV) from "If" to "then" (in the theory with types 0,1,2,3) but with E containing free only variables of types ≤ 1,

$$\phi^*(\underline{z},\underline{e},\alpha^0,\alpha^1) \simeq E$$

when (for each $\underline{j} = 0,1$ and each $\underline{i}$) each variable $\beta^{\underline{j}}_{\underline{i}}$ occurring free in E is assigned the value $(\alpha^{\underline{j}})_{\underline{i}}$. Otherwise, $\phi^*(\underline{z},\underline{e},\alpha^0,\alpha^1)$ is undefined.

Proof. To adapt the proof of (XV.3.3), we observe that, when the E in its hypothesis contains free only variables of those types,[8] none of the possible computation steps will introduce a higher-type variable. Indeed, E4.2 and E4.3,

[8]If furthermore, the functions Θ have only variables of type 0, then only finitely many branches can issue from any vertex, so any completed computation tree is finite, and our partial recursive functions (functions partial recursive

and E7 with arguments of types > 1, will be inapplicable. So we set out to define a $\phi^*(\underline{z},\underline{e},\alpha^0,\alpha^1)$ instead of $\phi^*(\underline{z},\underline{e},\alpha^0,\ldots,\alpha^3)$. The case hypothesis and the specifications for ϕ^* will then be entirely within the theory for types ≤ 1, in which the constructions used from earlier are all good (as we remarked in 1.1).

(XVI) (Enumeration theorem.) Let $\mathfrak{a}$ be $\underline{n}_0,\underline{n}_1,\underline{n}_2,\underline{n}_3$ variables of types 0,1,2,3; let $\underline{n} = \langle\underline{n}_0,\underline{n}_1,\underline{n}_2,\underline{n}_3\rangle$; and similarly let Θ be characterized by $\underline{m} = \langle\underline{m}_1,\ldots,\underline{m}_{\underline{l}}\rangle$ as in 1.3. There is a function $\Phi_{\underline{m},\underline{n}}(\underline{z},\mathfrak{a})$, also written $\{\underline{z}\}^{\Theta}(\mathfrak{a})$ (for Θ empty, $\{\underline{z}\}(\mathfrak{a})$), partial recursive in Θ such that, if $\phi(\mathfrak{a})$ is partial recursive in Θ with index $\underline{z}$,

$$\{\underline{z}\}^{\Theta}(\mathfrak{a}) \simeq \phi(\mathfrak{a}).$$

Proof. There is no loss of generality in taking $\mathfrak{a}$ to be the first $\underline{n}_0,\underline{n}_1,\underline{n}_2,\underline{n}_3$ variables of the respective lists (preceding (XV)). There is a primitive recursive function $\psi_{\underline{m},\underline{n}}$ such that: if $\mathrm{Ix}(\underline{z})$ & $(\underline{z})_1=\underline{m}$ & $(\underline{z})_2=\underline{n}$, then $\psi_{\underline{m},\underline{n}}(\underline{z})$ is the Gödel number $\underline{e}$ of the 0-expression $\phi_{\underline{p}}(\mathfrak{a})$, where $\phi_1,\ldots,\phi_{\underline{p}}$ is the canonical derivation determined by $\underline{z}$; and $\psi_{\underline{m},\underline{n}}(\underline{z}) \simeq 0$ otherwise. Now put

$$\Phi^{\Theta}_{\underline{m},\underline{n}}(\underline{z},\mathfrak{a}) \simeq \{\underline{z}\}^{\Theta}(\mathfrak{a}) \simeq$$
$$\phi^*(\underline{z},\psi_{\underline{m},\underline{n}}(\underline{z}),\langle\beta^0_0,\ldots,\beta^0_{\underline{n}_0-1}\rangle,\langle\beta^1_0,\ldots,\beta^1_{\underline{n}_1-1}\rangle,\langle\beta^2_0,\ldots,\beta^2_{\underline{n}_2-1}\rangle,\langle\beta^3_0,\ldots,\beta^3_{\underline{n}_3-1}\rangle).$$

(XVII) (S-m-n theorem.) For each $\overline{m} \geq 1$, there is a primitive recursive function $\underline{S}^{\overline{m}}(\underline{z},\underline{y}_1,\ldots,\underline{y}_{\overline{m}})$ such that, if $\phi(\underline{y}_1,\ldots,\underline{y}_{\overline{m}},\mathfrak{b})$ is a function partial recursive in Θ with index $\underline{z}$, then for each fixed values of $\underline{y}_1,\ldots,\underline{y}_{\overline{m}}$, $\underline{S}^{\overline{m}}(\underline{z},\underline{y}_1,\ldots,\underline{y}_{\overline{m}})$ is an index of $\phi(\underline{y}_1,\ldots,\underline{y}_{\overline{m}},\mathfrak{b})$ as a function of $\mathfrak{b}$ partial recursive in Θ. (Cf. 1959 p. 15, IM Theorem XXIII p. 342, Kleene 1938.)[9]

Proof. Since here we chose for conceptual economy to introduce directly only

---

in Θ), whether or not we allow only type 0 and 1 variables in $\phi_1,\ldots,\phi_{\underline{p}}$, are the functions partial recursive (partial recursive in Θ) in the sense of ORT. For, by 3.1 with known results, we get no fewer functions. And the general technique for analyzing finite computations, which was used in Kleene 1936 to prove the normal form theorem with type-0 variables and in IM § 58 was relativized and thereby with uniformity[2] extended to type-1 variables, then applies to our computation trees.

[9] "$\underline{S}$" is for "substitution" of constants $\underline{y}_1,\ldots,\underline{y}_{\overline{m}}$ for the first $\overline{m}$ number variables. In some versions of computability, the function $\underline{S}$ depended on the number $\overline{n}$ of remaining number variables. We are using "$\overline{m}$","$\overline{n}$" in place of the traditional "$\underline{m}$","$\underline{n}$" because the latter have been preempted by 1.3.

the one constant 0, by S2.0, we need first to get the "general" constant function $\kappa_{\underline{y}}(\mathfrak{A}) \simeq \underline{y}$. An index $\iota(\underline{m},\underline{n},\underline{y})$ of it, where $\underline{m},\underline{n}$ describe $\Theta,\mathfrak{A}$, is given by the primitive recursion $\iota(\underline{m},\underline{n},0) = \langle 2,\underline{m},\underline{n}\rangle$, $\iota(\underline{m},\underline{n},\underline{y}') = \langle 4,\underline{m},\underline{n},0,\langle 1,\underline{m},2\underline{n},0\rangle,\iota(\underline{m},\underline{n},\underline{y})\rangle$. Now we use induction on $\overline{\underline{m}}$. Basis. We get $\lambda\mathfrak{b}\,\phi(\underline{y},\mathfrak{b}) = \lambda\mathfrak{b}\,\phi(\kappa_{\underline{y}}(\mathcal{L}),\mathfrak{b})$ by S4.0 with the index $\underline{S}^1(\underline{z},\underline{y}) = \langle 4,(\underline{z})_1,[(\underline{z})_2/2],0,\underline{z},\iota((\underline{z})_1,[(\underline{z})_2/2],\underline{y})\rangle$.

Ind. step. $\underline{S}^{\overline{\underline{m}}+1}(\underline{z},\underline{y}_1,\ldots,\underline{y}_{\overline{\underline{m}}},\underline{y}_{\overline{\underline{m}}+1}) = \underline{S}^1(\underline{S}^{\overline{\underline{m}}}(\underline{z},\underline{y}_1,\ldots,\underline{y}_{\overline{\underline{m}}}),\underline{y}_{\overline{\underline{m}}+1})$.

(XVIII) ((Second) recursion theorem.) Given any function $\psi(\underline{z},\mathfrak{b})$ partial recursive in Θ, an index $\underline{e}$ from Θ of $\lambda\mathfrak{b}\,\psi(\underline{e},\mathfrak{b})$ can be found; thus we can solve for $\underline{z}$ the equation

$$\{\underline{z}\}^{\Theta}(\mathfrak{b}) \simeq \psi(\underline{z},\mathfrak{b}).$$

(Cf. 1959 XIV p. 15, IM pp. 352-353, Kleene 1938.)

Proof, as usual. (Can we afford the space?) Let $\underline{f}$ be an index from Θ of $\lambda\underline{y}\mathfrak{b}\,\psi(\underline{S}^1(\underline{y},\underline{y}),\mathfrak{b})$, and let $\underline{e} = \underline{S}^1(\underline{f},\underline{f})$.

(XIX) (Reflection principle.) There is a function $\phi(\underline{a},\mathcal{L},\mathcal{I})$ partial recursive in Θ such that

$$\phi(\underline{a},\mathcal{L},\mathcal{I}) \simeq \{\underline{a}\}^{\Theta}(\mathcal{L}).$$

(Cf. 1959 S9 p. 13, 3.14 p. 17; of course the indexing is different now.)

Proof. (IV) gives this from (XVI).

REFERENCES

CHURCH, ALONZO

(1936) An unsolvable problem of elementary number theory. Amer. jour. of math., vol. 58, pp. 345-363.

GANDY, R. O.

(1967) Computable functionals of finite type I. Sets, models and recursion theory, Proceedings of The Summer School in Mathematical Logic and Tenth Logic Colloquium, Leicester, Aug.-Sept. 1965, 1967 (North-Holland Pub. Co., Amsterdam), pp. 202-242.

GÖDEL, KURT

(1931) Über formal unentscheidbare Sätze der Principia Mathematica und verwandter Systeme I. Monatsh. für Math. und Physik, vol. 38, pp. 173-198.

KLEENE, STEPHEN COLE

(1936) General recursive functions of natural numbers. Math. Ann., vol. 112, pp. 727-742.

(1936a) A note on recursive functions. Bull. Amer. Math. Soc., vol. 42, pp. 544-546.

(1938) On notation for ordinal numbers. Jour. symbolic logic, vol. 3, pp. 150-155.

(1943) Recursive predicates and quantifiers. Trans. Amer. Math. Soc., vol. 53, pp. 41-73.

(1952) Recursive functions and intuitionistic mathematics. Proceedings of The International Congress of Mathematicians (Cambridge, Mass., Aug.30-Sept. 6, 1950), 1952, vol. 1, pp. 679-685.

(1952a) Introduction to metamathematics. North-Holland Pub. Co. (Amsterdam), P. Noordhoff Ltd. (Groningen), and D. van Nostrand Co. (Princeton, Toronto, New York). Seventh reprint 1974, Wolters-Noordhoff Pub. (Groningen), North-Holland Pub. Co. (Amsterdam, Oxford), American Elsevier Pub. Co. (New York). X + 550 pp. Cited as "IM".

(1955) Hierarchies of number-theoretic predicates. Bull. Amer. Math. Soc., vol. 61, pp. 193-213.

(1959) Recursive functionals and quantifiers of finite types I. Trans. Amer. Math. Soc., vol. 91, pp. 1-52.

(1963) Recursive functionals and quantifiers of finite types II. Trans. Amer. Math. Soc., vol. 108, pp. 106-142.

LEVIN, MICHAEL

(1974) Mathematical logic for computer scientists. Project MAC Technical Report TR-131, M.I.T.

PLATEK, RICHARD A.

(1966) Foundations of recursion theory. Doctoral Dissertation, Stanford University, Jan. 1966, 215 pp., mimeographed.

SCOTT, DANA

(1976) Data types as lattices. SIAM journal on computing, vol. 5, pp. 522-587

TARSKI, ALFRED

(1933) Der Wahrheitsbegriff in den formalisierten Sprachen. Studia philosophica, vol. 1 (1936), pp. 261-405. In Polish, 1933.

TURING, ALAN MATHISON

(1936-7) On computable numbers, with an application to the Entscheidungsproble Proc. London Math. Soc., ser. 2, vol. 42, pp. 230-265.

(1937) Computability and λ-definability. Jour. symbolic logic, vol. 2, pp. 153-163.

(1939) Systems of logic based on ordinals. Proc. London Math. Soc., ser. 2, vol. 45, pp. 161-228.

J.E. Fenstad, R.O. Gandy, G.E. Sacks (Eds.)
GENERALIZED RECURSION THEORY II

LATTICES OF α-RECURSIVELY ENUMERABLE SETS

Manuel Lerman*
Department of Mathematics
University of Connecticut
Storrs, Connecticut 06268

Lattices of α-recursively enumerable sets provide an excellent setting for the study of recursion theory on admissible ordinals. Many theorems of ordinary recursion theory fail in this setting, providing examples which distinguish those properties of ω which are significant for generalized recursion theory from those properties which are particular to ordinary recursion theory. These differences have stimulated the discovery of genuinely new constructions which are not liftings of proofs or constructions from ordinary recursion theory. Yet there is enough similarity with ordinary recursion theory so that the difficult methods of α-recursion theory which were introduced to lift theorems come into play in the study of the lattices. In fact, some of these methods were discovered by considering lattice-theoretic questions.

We will survey lattices of α-r.e. sets, including the case $\alpha = \omega$, in this paper. A few of the easier proofs will be presented in order to provide the reader with the flavor of the methods used in the subject. We will, however, avoid the more difficult proofs. In Section 1, we will present a quick introduction to α-recursion theory and lattices of α-r.e. sets. In Section 2, we will introduce a quotient of the lattice of α-r.e. sets in which many questions of interest are more easily

*The preparation of this paper was partially supported by NSF grant MCS76-07258.

studied. Sections 3 and 4 will deal with structure theory and decidability questions, and Section 5 will deal with algebraic and definability questions.

SECTION 1: PRELIMINARIES

α-recursion theory takes place on fragments of L, Gödel's universe of constructible sets. We define L using the Lévy hierarchy. For sets A and B, we say that B is first-order definable over A if there is a formula R in the language of set theory with one free variable x and parameters (if any) from A such that $B = \{x \in A : R(x)\}$. Let Fodo $(A) = \{B : B$ is first-order definable over $A\}$. We now define L by transfinite induction on the ordinal numbers.

$$L_0 = \emptyset$$
$$L_{\alpha+1} = \text{Fodo}\ (L_\alpha)$$
$$L_\lambda = \bigcup_{\delta<\lambda} L_\delta \ , \quad \lambda \text{ a limit ordinal.}$$

A partial function f on domain L_α is partial α-recursive if its graph is Σ_1-definable over L_α. An α-recursive function is a partial α-recursive function which is total. A subset of L_α is α-r.e. if it is the range of an α-recursive function, and is α-recursive if both it and its complement in L_α are α-r.e. A set is α-finite if it is an element of L_α. As L_α can be put into one-one α-recursive correspondence with α, it suffices to restrict our attention to functions from α^n into α and subsets of α. We will do so, modifying the definition of α-recursive functions and sets to require the properties just on α rather than on all of L_α. α is said to be an admissible ordinal if given any partial α-recursive function f, and any α-finite set $A \subseteq \text{dom}(f)$, then $f(A)$ is α-finite. Those L_α for α an admissible ordinal are the universes with which α-recursion theory concerns itself. For the remainder of the paper, α will always denote an admissible ordinal.

A useful tool for studying lattices of α-r.e. sets is the following enumeration theorem:

THEOREM. There is an α-recursive enumeration $\{W_e : e < \alpha\}$ of all the α-r.e. sets. Furthermore, there is an α-recursive double sequence $\{W_e^\sigma : e < \alpha, \sigma < \alpha\}$ of α-finite sets such that $W_e = \bigcup_\sigma W_e^\sigma$ for all σ, and $W_e^\sigma \subseteq L_\sigma$. (By an α-recursive enumeration above, we mean that there is a partial α-recursive function $f : \alpha^2 \to \alpha$ such that $f(e,x)$ is defined $\leftrightarrow x \in W_e$.)

It is easy to see that the α-r.e. sets are closed under finite unions and intersections, and thus form a lattice, $\mathcal{E}(\alpha)$, known as the lattice of α-r.e. sets.

SECTION 2: GENERALIZED FINITE SETS

The study of $\mathcal{E}(\omega)$ is closely connected with the study of $\mathcal{E}^*(\omega)$, the quotient lattice of $\mathcal{E}(\omega)$ obtained upon factoring by the ideal of finite sets. As "finite" can be generalized in many different ways, we need to find the correct generalization which will give rise to an $\mathcal{E}^*(\alpha)$ which is related to $\mathcal{E}(\alpha)$ in the way that $\mathcal{E}^*(\omega)$ is related to $\mathcal{E}(\omega)$. Let us first list the properties of "finite" which are most important in this relationship.

(1) The finite sets form an ideal of $\mathcal{E}(\omega)$ which

- (a) is definable over $\mathcal{E}(\omega)$,
- (b) contains only ω-recursive sets,
- (c) contains all the finite sets,
- (d) is the largest definable ideal of $\mathcal{E}(\omega)$.

(2) $\mathcal{E}(\omega)$ and $\mathcal{E}^*(\omega)$ are equidecidable; i.e., their elementary theories have the same degree of unsolvability.

We immediately see that "α-finite" is not a good substitute for "finite." For let α^* be the least ordinal such that there is a one-one α-recursive function $f : \alpha \to \alpha^*$. α^* is sometimes less than α (e.g. for $\alpha = \omega_1^{ck}$, the least non-recursive ordinal, $\omega = \alpha^* < \alpha$). If $\alpha^* < \alpha$, then α^* is α-finite but has α-r.e. subsets (e.g. $f[\alpha]$) which are not α-recursive. Hence the α-finite sets do not always form an ideal of $\mathcal{E}(\alpha)$.

The definition which captures more of the above properties than any other is <u>α*-</u>

finite, i.e., α-finite and of ordertype less than α^*. The class of α^*-finite sets does form an ideal of $\mathcal{E}(\alpha)$. This ideal is definable over $\mathcal{E}(\alpha)$ using Lacombe's definition for $\mathcal{E}(\omega)$: A is α^*-finite if and only if every subset of A in $\mathcal{E}(\alpha)$ is α-recursive (the α-recursive sets are those elements of $\mathcal{E}(\alpha)$ which have complements). This ideal contains only α-recursive sets and contains all the finite sets (but not all the α-finite sets when $\alpha^* < \alpha$). It is not always the largest definable ideal of $\mathcal{E}(\alpha)$. That distinction sometimes belongs to the α-bounded sets, i.e., the α-r.e. sets which are subsets of some $\beta < \alpha$. If $\alpha^* = \alpha$, α^*-finite, α-finite, and α-bounded give rise to the same class of α-r.e. sets. If $\alpha^* < \alpha$, we get three distinct classes, but it is not known whether, for all α, the class of α-bounded sets is definable. Lerman [12] shows that the α^*-finite sets and the α-bounded sets are the only ideals of $\mathcal{E}(\alpha)$ which can be definable over $\mathcal{E}(\alpha)$. As property (1b) is more crucial than property (1d), we must lean towards choosing "α^*-finite" as the best generalization of "finite."

Property (2) was established by Lachlan [6], and generalized to all α by Lerman [12], letting $\mathcal{E}^*(\alpha)$ be the quotient of $\mathcal{E}(\alpha)$ obtained upon factoring by the ideal of α^*-finite sets. This is the definition of $\mathcal{E}^*(\alpha)$ which we adopt.

SECTION 3: REDUCTION AND SEPARATION

In this section, we will "lift" a theorem of ordinary recursion theory to α-recursion theory. We choose a theorem involving a simple priority argument, Friedberg's splitting theorem [2] to give the reader a hint of the difficulties involved in lifting theorems. Roughly, the problems encountered are as follows: We must satisfy a sequence of requirements $\{R_e : e < \alpha\}$. For $\alpha = \omega$, it can be shown that after all the requirements $\{R_e : e < \beta < \alpha\}$ are satisfied, only finitely many attempts need to be made to satisfy R_β. An induction then shows that all requirements are satisfied. Once limit ordinals less than α appear, however, we need to show that all $\{R_e : e < \beta < \alpha$, β a limit ordinal$\}$ are satisfied by some stage $\sigma < \alpha$ in order to continue the induction argument. This may be false, however, for a given construction. We circumvent this difficulty by finding a shorter indexing of requirements for which the induction does work. Such indexings were first introduced by Kreisel and Sacks [5] in the case where $\alpha^* = \omega$, and later by Sacks and Simpson [22] for arbitrary α. Other indexings have subsequently been introduced to handle more complicated constructions. The proof below will use α^* to index requirements.

Two disjoint α-r.e. sets A and B are <u>α-recursively inseparable</u> if there is no α-recursive set $R \supseteq A$ such that $R \cap B = \emptyset$. ω-recursively inseparable sets are very useful for the construction of counterexamples in model theory. The existence of α-recursively inseparable sets is a byproduct of the proof of the reduction theorem (Friedberg's splitting theorem) which we prove below. This theorem was first proved by Friedberg [2] for $\alpha = \omega$, and generalized by Machtey [17] for $\alpha^* = \omega$. A proof of the theorem in the generality we give also appears in Lerman [11].

<u>THEOREM 3.1.</u> For any α-r.e. set A which is not α-recursive, there exist disjoint α-r.e. sets A_1 and A_2 with union A such that neither A_1 nor A_2 is α-recursive. Furthermore, we can obtain A_1 and A_2 α-recursively inseparable.

<u>PROOF</u>. Note that if we build A_1 and A_2 α-recursively inseparable, then neither set can be α-recursive.

Let g be a one-one α-recursive function enumerating A. Let $f : \alpha \to \alpha^*$ be a one-one α-recursive function. We will use α^* to assign priorities to requirements. The crucial property of α^* needed is that any α-r.e. subset of $\beta < \alpha^*$ is α-finite. Let $\{R_e : e < \alpha\}$ be a listing of all requirements of the form

$$W_k \cap A_i \neq \emptyset \ , \quad i = 1,2 \ ; \ k < \alpha \ .$$

We say that R_i <u>has higher priority than</u> R_j <u>at stage</u> σ if $i,j < \sigma$ and $f(i) < f(j)$. We say that $R_j \equiv W_k \cap A_i \neq \emptyset$ <u>requires attention at stage</u> σ if $j = f(x)$ for some $x < \sigma$, $W_k \cap A_i = \emptyset$, and $g(\sigma) \in W_k^\sigma$.

<u>The construction; stage σ</u>: Let $R_j \equiv W_k \cap A_i \neq \emptyset$ be the requirement of highest

priority which requires attention at stage σ . If such a j exists, place $g(\sigma) \in A_i$ and say that R_j <u>receives attention at stage</u> σ . Otherwise, place $g(\sigma) \in A_1$.

This completes the construction. Clearly A_1 and A_2 are disjoint α-r.e. sets with union A .

<u>LEMMA 3.2</u>. For all $i < \alpha^*$, there is a $\sigma < \alpha$ such that for all $\tau \geq \sigma$ and $j < \alpha$, if $f(j) < i$ then R_j does not require attention at stage τ .

<u>PROOF</u>. Let $F = \{k < i : (\exists j)(f(j) = k$ and $(\exists \sigma)(R_k$ receives attention at stage $\sigma))\}$. F is an α-r.e. subset of $i < \alpha^*$, hence is α-finite. For $k \in F$, let $h(k)$ be the least stage σ such that $R_{f^{-1}(k)}$ receives attention at stage σ . h is partial α-recursive, so $\sup(\{h(k) : k \in F\}) = \sigma < \alpha$ by the admissibility of α . σ is the desired stage since a requirement never requires attention after it has received attention. ⊠

<u>LEMMA 3.3</u>. There is no α-recursive set R such that $A_1 \subseteq R$ and $A_2 \cap R = \emptyset$.

<u>PROOF</u>. Assume that such an R exists. We obtain a contradiction by showing that A must be α-recursive. Let $R = W_k$, and let $R_j \equiv A_2 \cap W_k \neq \emptyset$. By Lemma 3.2, there is a stage $\sigma > j$ such that no R_n for which $f(n) < f(j)$ receives attention at any stage $\tau \geq \sigma$. If $\tau \geq \sigma$ and R_j requires attention at stage τ , then R_j would receive attention at stage τ , and so $A_2 \cap W_k \neq \emptyset$, an impossibility. Hence R_j does not require attention at any stage $\tau \geq \sigma$. We now see that A_1 is α-recursive. For $x \in A_1 \leftrightarrow x$ is placed in A_1 before stage σ or $x \in R = W_k$ and if τ is the least ordinal such that $x \in W_k^\tau$, then x is placed in A_1 before stage τ . A similar argument shows that A_2 is α-recursive. Hence $A = A_1 \cup A_2$ is α-recursive. ⊠

The theorem now follows easily from Lemma 3.3.

<u>SECTION 4: SIMPLE SETS</u>

Let $\langle L, \vee, \wedge, 0, 1\rangle$ be a lattice with least element 0 and greatest element 1.

We say that $a \in L$ is <u>L-simple</u> if for all $b \in L$, $b \wedge a = 0$ implies $b = 0$. If L is a quotient lattice of $\mathcal{E}(\alpha)$ and $A \in \mathcal{E}(\alpha)$ we say that A <u>is simple for L</u> if the congruence class of A corresponding to L is L-simple. If $L = \mathcal{E}^*(\alpha)$, we call sets which are simple for $\mathcal{E}^*(\alpha)$ <u>simple α-r.e. sets</u>.

Simple ω-r.e. sets were defined by Post [19], who also constructed such sets which are not ω-recursive. Post's definition is normally generalized to: A is simple if $\alpha - A$ is not α-finite and for any α-r.e. set $B \subseteq \alpha - A$, B is α-finite. It is straightforward to see that our class of simple α-r.e. sets are just those given by the generalization of Post's definition together with the sets with α*-finite complements. This equivalence follows from the fact that if A is α-finite but not α*-finite, then A has an α-r.e. subset which is not α-recursive.

Post used a construction by diagonalization to construct his simple ω-r.e. sets. The requirements for that construction are those of Theorem 3.1 together with the requirements that for all n, the complement of the simple set has more than n elements. He handled the latter requirements by guaranteeing that for all n, $S \cap \{x : x < 2n\}$ has cardinality at most n, where S is the simple set constructed. This part of the construction fails for arbitrary α, but a priority argument along the lines of the proof of Theorem 3.1 can be used to generalize this theorem. One merely adds requirements to keep exactly α^* many elements out of S. A proof can be found in Simpson [25].

Various types of simple sets play an important role in decision procedures for fragments of the elementary theory of $\mathcal{E}^*(\alpha)$. In order to discuss these simple sets, we need the following definitions.

A function $f : \beta \to \gamma$ is a <u>cofinality function</u> if the image of f is cofinal with γ. A partial function $f : \beta \to \gamma$ is a <u>projection</u> if the image of f is $\{\delta : \delta < \gamma\}$. The projection f is <u>tame</u> if f is total on β and for all $\delta < \beta$, $\{f(x) : x < \delta\}$ is α-finite.

We say that $\lim_\sigma f(\sigma,x) = y$ if $f(\tau,x) = y$ for all sufficiently large $\tau < \alpha$. $f : \beta \to \gamma$ is an S_n function if there is an α-recursive function f' of n variables such that for all $x < \beta$, $f(x) = \lim_{\sigma_1} \ldots \lim_{\sigma_{n-1}} f'(\sigma_1,\ldots,\sigma_{n-1},x)$.

If Γ is a level of the arithmetical or S_n hierarchy, we call β the Γ-projectum of α if β is the least ordinal such that there is a projection $f : \beta \to \alpha$ of class Γ. The tame Δ_2-projectum of α is the least $\beta \leq \alpha$ such that there is a tame Δ_2-projection of $f : \beta \to \alpha$. The Γ-cofinality of α is the least $\beta \leq \alpha$ such that there is a cofinality function $f : \beta \to \alpha$ of class Γ.

An α-r.e. set M is said to be maximal if $\alpha - M$ is not α^*-finite and for all α-r.e. sets W, either $W \cap (\alpha - M)$ or $(\alpha - W) \cap (\alpha - M)$ is α^*-finite. Thus maximal sets are exactly those sets whose congruence classes in $\mathcal{E}^*(\alpha)$ are maximal elements. Friedberg [2] first constructed maximal ω-r.e. sets. Sacks [21] showed that there are no maximal $\aleph_1^L$-r.e. sets, where $\aleph_1^L$ is the smallest uncountable cardinal of L. We present Simpson's modification of Sacks' proof as an example of a theorem and proof in α-recursion theory which has no counterpart in ω-recursion theory.

THEOREM 4.1. There are no maximal $\aleph_1^L$-r.e. sets.

PROOF. We note that Gödel [3] shows that all subsets of ω which lie in L are elements of $L_{\aleph_1^L}$. From Gödel's proof, there is an $\aleph_1^L$-recursive function $f : \omega \times \aleph_1^L \to \{0,1\}$ such that for all $\beta \neq \gamma$, $\lambda x f(x,\beta) \neq \lambda x f(x,\gamma)$; i.e., f gives a one-one listing of $\aleph_1^L$-many subsets of ω.

Assume that M is a maximal $\aleph_1^L$-r.e. set. We will show that $\aleph_1^L$ has cofinality ω in L, and so obtain a contradiction. For each $x < \omega$, let $g(x)$ be the least $\sigma < \aleph_1^L$ such that either

$$(\tau > \sigma)(\tau \notin M \to f(x,\tau) = 0)$$

or

$$(\tau > \sigma)(\tau \notin M \to f(x,\tau) = 1) .$$

Such a σ must exist since M is maximal. The theorem will follow once we show that g is a cofinality function. If g were not a cofinality function, choose $\sigma > g(x)$ for all $x < \omega$. As $\aleph_1^L - M$ must be $\aleph_1^L$-unbounded, there are $\tau > \rho > \sigma$ with $\tau, \rho \in \aleph_1^L - M$. But then $\lambda x f(x,\tau) = \lambda x f(x,\rho)$, an impossibility. ⊠

The admissible ordinals for which maximal α-r.e. sets exist have been classified by Lerman [13]. Partial results were previously obtained by Kreisel and Sacks [5], Sacks [21] and Lerman and Simpson [16].

<u>THEOREM 4.2</u>. Maximal α-r.e. sets exist if and only if the S_3-projectum of α is ω .

Maximal α-r.e. sets with α-bounded complements have been studied by Kreisel and Sacks [5], Owings [18] and Leggett [8]. Leggett has classified the admissible ordinals for which such sets exist.

<u>THEOREM 4.3</u>. Maximal α-r.e. sets with α-bounded complements exist if and only if $\alpha^* = \omega$.

An α-r.e. set H is said to be <u>hyperhypersimple (hhs)</u> if $\alpha - H$ is not α^*-finite and the lattice of supersets of H in $\mathcal{E}(\alpha)$ forms a boolean algebra. Post [19] originally defined hhs ω-r.e. sets differently, and Lachlan [6] discovered the above definition and showed it equivalent to Post's definition for $\alpha = \omega$. Maximal sets are hhs, so hhs ω-r.e. sets exist. For $\alpha = \omega$, Lachlan [6] has classified the boolean algebras which can occur as lattices of supersets of hhs sets. From Theorem 4.2, hhs α-r.e. sets exist if the S_3-projectum of α is ω . For such α , Cooper (unpublished) has constructed hhs sets with no maximal supersets. Further results obtained by Chong and Lerman [1] are now summarized.

<u>THEOREM 4.4</u>. Hyperhypersimple α-r.e. sets exist if either the S_3-projectum of α is ω , or if the S_2-cofinality of α is less than the tame S_2-projectum of α . In the latter case, the hhs sets are exactly those sets whose complements have order-type $< \alpha^*$ with a final segment of order-type less than the tame S_2-

projectum of α . Hyperhypersimple α-r.e. sets fail to exist if the S_3-cofinality of α is greater than ω .

An α-r.e. set R is said to be *r-maximal* if $\alpha - H$ is not α^*-finite and for every α-recursive set W either $W \cap (\alpha - R)$ or $(\alpha - W) \cap (\alpha - R)$ is α^*-finite. It is easy to see that any maximal set is r-maximal. Robinson [20] and Lachlan [6] have studied r-maximal ω-r.e. sets. Lerman and Simpson [16] use the construction of Theorem 4.1 to rule out the existence of r-maximal sets for certain α . The known results are summarized in the following theorem.

THEOREM 4.5. r-maximal α-r.e. sets exist if the S_3-projectum of α is ω . They fail to exist if α^* is not a limit of α-cardinals and the Σ_3-cofinality of α is α^* .

If $A \subseteq B$, we call A a *major subset* of B if $B - A$ is not α^*-finite and for every α-r.e. set W , if $\alpha - (W \cup B)$ is α^*-finite then $\alpha - (W \cup A)$ is α^*-finite. Lachlan [6] has shown that every ω-r.e. set B which is not α-recursive has a major subset, and that if B is r-maximal and $A \subseteq B$ then A is r-maximal if and only if $B - A$ is finite or A is a major subset of B . In this way, he obtains r-maximal ω-r.e. sets which are not maximal. The following theorem of Leggett and Shore [9] summarizes the known results and subsumes earlier results of Lerman [14].

THEOREM 4.6. If the Σ_2-projectum of α equals the Σ_2-cofinality of α , then every α-r.e. set which is not α-recursive has a major subset.

A natural question to ask at this point is whether, for all α , there is a formula of the language of lattice theory satisfied by some, but not all, simple α-r.e. sets. An affirmative answer is given by

THEOREM 4.7. For all α , there is a formula of the language of lattice theory with one free variable satisfied by some, but not all, simple α-r.e. sets.

PROOF. If the S_2-cofinality of α is less than the tame S_2-projectum of α , then Chong and Lerman [1] show that the formula "S is hyperhypersimple" differentiates between simple sets. Otherwise, Leggett and Shore [9] show that the formula "S is a major subset of some α-r.e. set" differentiates between simple sets. (This latter formula was previously used by Lerman [14] to obtain the result in the special case when α is a regular cardinal of L .) ⊠

An important open question of recursion theory is to determine the degree of unsolvability of the elementary theory of $\mathcal{E}(\alpha)$. Lachlan [7] showed that $\mathcal{E}(\omega)$ and $\mathcal{E}^*(\omega)$ are equidecidable, a result generalized by Lerman [12] to all α . Let $\mathcal{L}$ be the language of the pure predicate calculus with equality, binary relation symbols to be interpreted as union and intersection, a unary function symbol to be interpreted as complementation, and a unary relation symbol to be interpreted as distinguishing the α-r.e. sets. The language $\mathcal{L}$ applied to the boolean algebra generated by the α-r.e. sets is equivalent to the usual language of lattice theory applied to the r.e. sets, and is useful for studying decidability questions.

THEOREM 4.8. The $\forall\exists$ theory of $\mathcal{E}^*(\alpha)$ in the language $\mathcal{L}$ is decidable if either the S_2-cofinality of α and the tame S_2-projectum of α are ω and $\alpha^* = \alpha$, or if the S_3-cofinality of α and the Σ_3-projectum of α are both α and there is a greatest α-cardinal.

The decidability for $\alpha = \omega$ was obtained by Lachlan [7]. The remaining cases were obtained by Lerman [11] and include the case where α is a regular cardinal of L .

The types of simple sets considered in this section play an important role in the decision procedures. A complete classification of those α for which such sets exist would be valuable for extending Theorem 4.8 to all α . Progress has been made recently by Lerman and Soare towards obtaining a decision procedure for the $\forall\exists$ theory of $\mathcal{E}^*(\omega)$ in the language $\mathcal{L}'$ obtained from $\mathcal{L}$ by adjoining a unary relation symbol to be interpreted as distinguishing the maximal sets.

SECTION 5: DEFINABILITY AND AUTOMORPHISMS

The first question which we consider in this section is the characterization of all definable ideals, filters, and congruence relations of $\mathcal{E}(\alpha)$. One obstacle towards obtaining such a characterization is the determination of whether "α-bounded" or equivalently "α-finite" is definable. If $\alpha^* = \alpha$ then "α-finite", "α-bounded" and "α^*-finite" are all equivalent over $\mathcal{E}(\alpha)$, and so all are definable. "α-finite" is known to be definable in other cases, e.g., for $\omega = \alpha^* < \alpha$, Owings [18] shows that A is α-finite if and only if $(\exists M_1)(\exists M_2)(M_2 \subseteq M_1 \subseteq A$ & A is α-recursive & $M_1 \cup (\alpha - A)$ is maximal & M_2 is maximal in M_1 (i.e., $M_1 - M_2$ cannot be split into two pieces, each non-α^*-finite, by any α-r.e. set) & $(B)(M_2 \neq B \cap M_1))$. A summary of the cases where α-finite is known to be definable can be found in Lerman [10].

The definable ideals of $\mathcal{E}(\alpha)$ are characterized in Lerman [12]. The ideal of α^*-finite sets is one such ideal. There is at most one additional definable ideal, the ideal of α-bounded sets, but this ideal is definable only when "α-bounded" is definable.

Several definable filters are known to exist, but some become trivial (equal to the filter of sets with α^*-finite complements) for various choices of α . The filter of simple sets is always definable, and the filter of sets simple for $\mathcal{E}_b(\alpha)$ (the quotient of $\mathcal{E}(\alpha)$ obtained upon factoring by the ideal of bounded sets) is definable exactly when "α-bounded" is definable. One of these filters will always be the largest definable filter (Lerman [10]). Other definable filters which are sometimes non-trivial are the filter of hhs sets and sets with α^*-finite complements, the filter of sets with no maximal supersets, the filter of sets with no r-maximal supersets, the filter of sets with no hhs supersets, and the filter of sets with no r-maximal or hhs supersets. All these filters are non-trivial and different over $\mathcal{E}(\omega)$. For example, to obtain a set with no r-maximal or hhs superset, we start with a maximal set M and let A be a major subset of M . Let f be a one-one ω-recursive function enumerating M , and let

$B = f^{-1}(A)$. B is the desired set. Since A is r-maximal, it follows that B has no hhs superset, else by Lachlan [6], there would be a recursive set R such that $A \cup R = M$ contradicting the r-maximality of A . Owings' splitting theorem [18] implies that B has no r-maximal superset. It is unknown whether there are infinitely many filters which are definable over $\mathcal{E}(\alpha)$.

Various definable congruence relations not corresponding to filters or ideals have been identified. One such is: A is simple with B if $(A - B) \cup (B - A)$ contains no α-r.e. set which is not α*-finite. This will be the largest definable congruence relation of $\mathcal{E}(\alpha)$ unless "α-bounded" is definable. In the latter case, the largest definable congruence relation is given by: A is simple with B for $\mathcal{E}_b(\alpha)$ if $(A - B) \cup (B - A)$ contains no α-r.e. set which is not α-bounded. Another definable congruence relation which is sometimes non-trivial is: A is major with B if for all α-r.e. sets W , $\alpha - (W \cup A)$ is α*-finite if and only if $\alpha - (W \cup B)$ is α*-finite. It is unknown whether there are infinitely many definable congruence relations in $\mathcal{E}(\alpha)$. One way to try to construct infinitely many might be to iterate mixtures of the above congruence relations and filters to successive quotients of $\mathcal{E}(\alpha)$ and to show that the procedure does not terminate. A detailed discussion of the situation can be found in Lerman [10].

The other topic which we consider in this section deals with automorphisms of $\mathcal{E}(\alpha)$ and $\mathcal{E}^*(\alpha)$. Little is known for $\alpha \neq \omega$. A detailed summary for $\alpha = \omega$ can be found in Soare [26]. It is easy to see that every automorphism of $\mathcal{E}(\alpha)$ is determined by a permutation of α , but that not every permutation of α gives rise to an automorphism of $\mathcal{E}(\alpha)$. Furthermore, every automorphism of $\mathcal{E}(\alpha)$ determines an automorphism of $\mathcal{E}^*(\alpha)$. Using maximal sets, Kent [4] showed that there are $2^{\aleph_0}$ automorphisms of $\mathcal{E}(\omega)$. All these automorphisms give rise to the same automorphism of $\mathcal{E}^*(\omega)$. Lachlan used a different method to show that there are $2^{\aleph_0}$ automorphisms of $\mathcal{E}^*(\omega)$. The number of automorphisms of $\mathcal{E}(\alpha)$ and $\mathcal{E}^*(\alpha)$ for arbitrary α remains to be determined. Soare [26] shows that every automorphism of $\mathcal{E}^*(\omega)$ comes from some automorphism of $\mathcal{E}(\omega)$, so the determination of the automorphisms of $\mathcal{E}(\omega)$ and of $\mathcal{E}^*(\omega)$ are closely related. The existence of such a relationship for arbitrary α has yet to be determined.

A subset $\mathcal{S}$ of $\mathcal{E}(\alpha)$ is an <u>α-orbit</u> if for any $A \in \mathcal{S}$, $\mathcal{S} = \{B \in \mathcal{E}(\alpha)$: there is an automorphism of $\mathcal{E}(\alpha)$ carrying A to B$\}$. It is easy to see that for all α {B : B is α-recursive and neither B nor $\alpha - B$ is α-finite} is an α-orbit as is {B : B is α-finite and has α-cardinality κ}. Soare [26] has shown that {B : B is maximal} and {B : B is the intersection of exactly n distinct maximal sets} are ω-orbits for each n. These, essentially, are the only ω-orbits which have been classified. Maximal sets do not form an α-orbit for certain α as was demonstrated by Leggett [8] by producing two maximal sets whose complements have different order-types which do not allow an automorphism. It would be interesting to determine whether the α-orbit of a maximal set M is determined just by properties of the order-type and the boundedness of the complement of M. It seemed natural following Soare's classification of the ω-orbit of a maximal set to conjecture that the class of hhs sets with no maximal supersets also forms an ω-orbit. Unfortunately, this is not so, as was recently shown by Lerman, Shore, and Soare [15].

Another interesting class of problems deals with bases for automorphisms of $\mathcal{E}(\alpha)$ and extendability of automorphisms from sublattices of $\mathcal{E}(\alpha)$ to all of $\mathcal{E}(\alpha)$. This question has been studied by Shore and Soare, and is summarized in Shore [23]. Such problems led Shore to the discovery of a new definable class in $\mathcal{E}(\alpha)$, the nowhere simple sets [24].

References

[1] C.T. Chong and M. Lerman: Hyperhypersimple α-r.e. sets, Ann. of Math. Logic 9 (1976) 1-48.

[2] R.M. Friedberg: Three theorems on recursive enumeration, J. Symbolic Logic 23 (1958) 309-316.

[3] K. Gödel: Consistency proof for the general continuum hypothesis, Proc. Nat.

Acad. Sci. U.S.A. 25 (1935) 220-224.

[4] C.F. Kent: Constructive analogues of the group of permutations of the natural numbers, Trans. Amer. Math. Soc. 104 (1962) 347-362.

[5] G. Kreisel and G.E. Sacks: Metarecursive sets, J. Symbolic Logic 31 (1966) 1-21.

[6] A.H. Lachlan: On the lattice of recursively enumerable sets, Trans. Amer. Math. Soc. 130 (1968) 1-37.

[7] ____________: The elementary theory of recursively enumerable sets, Duke Math. J. 35 (1968) 123-146.

[8] A. Leggett: Maximal α-r.e. sets and their complements, Ann. of Math. Logic 6 (1974) 293-357.

[9] A. Leggett and R.A. Shore: Types of simple α-recursively enumerable sets, J. Symbolic Logic 41 (1976) 681-694.

[10] M. Lerman: Congruence relations, filters, ideals and definability in lattices of α-recursively enumerable sets, J. Symbolic Logic 41 (1976) 405-418.

[11] ____________: On elementary theories of some lattices of α-recursively enumerable sets, to appear.

[12] ____________: Ideals of generalized finite sets in lattices of α-recursively enumerable sets, to appear.

[13] ____________: Maximal α-r.e. sets, Trans. Amer. Math. Soc. 188 (1974) 341-386.

[14] ____________: Types of simple α-recursively enumerable sets, J. Symbolic Logic 41 (1976) 419-426.

[15] ____________, R.A. Shore and R.I. Soare: R-maximal major subsets, in preparation.

[16] ____________ and S.G. Simpson: Maximal sets in α-recursion theory, Israel J. Math. 14 (1973) 236-247.

[17] M. Machtey: Admissible ordinals and lattices of α-r.e. sets, Ann. of Math. Logic 2 (1971) 379-417.

[18] J.C. Owings: Recursion, metarecursion, and inclusion, J. Symbolic Logic 32 (1967) 173-179.

[19] E.L. Post: Recursively enumerable sets of positive integers and their decision problems, Bull. Amer. Math. Soc. 50 (1944) 284-316.

[20] R.W. Robinson: Two theorems on hyperhypersimple sets, Trans. Amer. Math. Soc. 128 (1967) 531-538.

[21] G.E. Sacks: Post's problem, admissible ordinals, and regularity, Trans. Amer. Math. Soc. 124 (1966) 1-23.

[22] G.E. Sacks and S.G. Simpson: The α-finite injury method, Ann. of Math. Logic 4 (1972) 343-367.

[23] R.A. Shore: Determining automorphisms of the recursively enumerable sets, to appear.

[24] ____________: Nowhere simple sets and the lattice of recursively enumerable sets, to appear.

[25] S.G. Simpson: Recursion theory over admissible structures, Ω-series, Springer-Verlag, Heidelberg, in preparation.

[26] R.I. Soare: Automorphisms of the lattice of recursively enumerable sets Part I: Maximal sets, Ann. of Math. 100 (1974) 80-120.

J.E. Fenstad, R.O. Gandy, G.E. Sacks (Eds.)
GENERALIZED RECURSION THEORY II

HIGH α-RECURSIVELY ENUMERABLE DEGREES

Wolfgang Maass

Mathematisches Institut der Universität München

A degree $\underset{\sim}{a}$ is said to be high if $\underset{\sim}{a}' = 0''$ where $\underset{\sim}{a}'$ is the jump of $\underset{\sim}{a}$ and 0 is the degree of the empty set. Thus $0'$ is a high degree but in ordinary recursion theory (ORT) there exist as well high recursively enumerable (r.e.) degrees below $0'$ according to a theorem of Sacks [12]. The proof of this result is a very nice application of the infinite injury priority method.

It follows from the theorem of Sacks that the notion high is not trivial. Further results show that the notions high and low ($\underset{\sim}{a}$ is low if $\underset{\sim}{a}' = 0'$) are in fact important for the study of the fine structure of the r.e. degrees in ORT. The intuitive meaning is that $\underset{\sim}{a}$ is high if $\underset{\sim}{a}$ is near to $0'$ and $\underset{\sim}{a}$ is low if $\underset{\sim}{a}$ is near to 0 in the upper semilattice of the r.e. degrees. Therefore these notions are useful for the study of non-uniformity effects in this structure where one looks for theorems which hold in some regions of this semilattice but not everywhere (see e.g. Lachlan [4]).

In addition high degrees are interesting for technical reasons. Some results have been proved for high degrees and it is not yet known whether they are true for all r.e. degrees (see e.g. Cooper [1]).

Finally high degrees are a link between the structure of r.e. degrees and the structure of r.e. sets according to a theorem of Martin (see [15]): A degree contains a maximal r.e. set if and only if it is a high r.e. degree.

In α-recursion theory for admissible ordinals α the deeper properties of r.e. degrees and r.e. sets are explored in a general setting and one tries to find out which assumptions are really needed in order to do certain constructions. We refer the reader to the survey papers by Lerman and Shore in this volume for more information.

It turned out that in fact several priority arguments can be transferred to α-recursion theory (see e.g. Sacks-Simpson [14], Shore [16], Shore [18]). Other results of ORT have been proved for many admissible α but it is still open whether they hold for all admissible α(e.g. the existence of minimal pairs of α-r.e. degrees [6] ,[21] and the existence of minimal α-degrees [17],[7]). Lerman [5] closed the gap between provable existence and provable non-existence in the case of maximal α-r.e. sets.

For some time one thought that the existence of high α-r.e. degrees below $0'$ was as well completely settled by Shore [20], but an error was found in the proof of Theorem 2.3. in [20]*. The problem was then open again except for Σ_2-admissible α where the existence proof from ORT works and for α such that $0'$ is the only non-hyperregular α-r.e. degree where every α-r.e. degree below $0'$ is low according to [20] (these are the types (1) and (4) in our characterization in §3).

We close the gap in this paper by proving that high α-r.e. degrees below $0'$ exist if and only if $\sigma 2cf\alpha \geq \sigma 2p\alpha$. This result was not expected and is different from the result in [20]. We think that the new result is a lucky circumstance for α-recursion theory since it was thought in [20] that the situation is somewhat trivial (every non-hyperregular α-r.e. degree is high). Now it turns out that inadmissibility (in form of non-hyperregularity) influences the behaviour of the jump of an α-r.e. degree but is

*I would like to thank R.A. Shore for informing me about this.

not so strong that it overruns everything (this will become even clearer in our forthcoming paper [11]).

The plan of this paper is as follows:

§0. contains some basic definitions and facts.

In §1. we construct high α-r.e. degrees below $0'$ for the case $\alpha > \sigma 2\mathrm{cf}\,\alpha \geq \sigma 2\mathrm{p}\,\alpha$. We give some motivation for the construction so that this chapter should be readable for anyone who has seen before an infinite injury priority argument in ORT (e.g.[23]). The construction reflects several typical features of α-recursion theory and uses strategies which would not work in ORT.

In §2. we prove that there exist no high α-r.e. degrees below $0'$ in the case $\sigma 2\mathrm{cf}\,\alpha < \sigma 2\mathrm{p}\,\alpha$ by using some basic properties of strongly inadmissible structures. Along the way some first results are proved about a distinguished degree between $0'$ and $0''$ for which we write $0^{3/2}$.

A summary is given in §3. . Four types of admissible ordinals have to be distinguished as far as the behaviour of the jump of r.e. degrees is concerned.

§0. Preliminaries

Lowcase greek letters are always ordinals, β and λ are always limit ordinals and α is always admissible in this paper. We consider only structures $\mathfrak{B} = \langle L_\beta, B \rangle$ where $B \subseteq L_\beta$ and B is <u>regular</u> over L_β, i.e. $\forall \gamma < \beta (L_\gamma \cap B \in L_\beta)$. We say that a set $D \subseteq L_\beta$ is $\Sigma_n \mathfrak{B}$ if D is definable by some Σ_n formula (which may contain elements of L_β as parameters) over the structure $\mathfrak{B}$.

For $\lambda \leq \beta$ one writes $\sigma \mathrm{ncf}^{\mathfrak{B}} \lambda$ for the least $\delta \leq \lambda$ such that some $\Sigma_n \mathfrak{B}$ function maps δ cofinally into λ and one writes $\sigma \mathrm{np}^{\mathfrak{B}} \beta$ for the least $\delta \leq \beta$ such that some $\Sigma_n \mathfrak{B}$ function p projects β into δ (i.e. p maps β 1-1 into δ). We write <u>$\sigma \mathrm{ncf}\, \alpha$</u> instead of $\sigma \mathrm{ncf}^{L_\alpha} \alpha$ and <u>$\sigma \mathrm{np}\, \alpha$</u> instead of $\sigma \mathrm{np}^{L_\alpha} \alpha$.

A set $D \subseteq L_\beta$ is called <u>$\mathfrak{B}$-r.e.</u> (<u>$\mathfrak{B}$-recursive</u>) if D is $\Sigma_1 \mathfrak{B}$ ($\Delta_1 \mathfrak{B}$). We say that a set D is <u>tame-$\Sigma_n \mathfrak{B}$</u> if the set of "positive neighborhoods" $\{K \in L_\beta \mid K \subseteq D\}$ is $\Sigma_n \mathfrak{B}$. A set $K \subseteq L_\beta$ is called <u>β-finite</u> if $K \in L_\beta$.

An ordinal $\delta < \beta$ is called a <u>(regular) β-cardinal</u> if $L_\beta \models [\delta$ is a (regular) cardinal$]$.

We fix for the following universal $\Sigma_n \mathfrak{B}$ sets $U_n^{\mathfrak{B}}$ (i.e. for every set $D \subseteq L_\beta$: D is $\Sigma_n \mathfrak{B}$ if and only if $D = \{x \mid \langle e,x \rangle \in U_n^{\mathfrak{B}}\}$ for some $e \in \beta$) which are given by some $\Sigma_n \mathfrak{B}$ definition. In the special case $n = 1$ we write $W_e^{\mathfrak{B}}$ for $\{x \mid \langle e,x \rangle \in U_1^{\mathfrak{B}}\}$.

For sets $A, D \subseteq L_\beta$ one says that <u>A is $\mathfrak{B}$-reducible to D</u> (written $A \leq_{\mathfrak{B}} D$) if there is some index $e \in \beta$ such that for all $K \in L_\beta$

$$K \subseteq A \leftrightarrow \exists H_1 H_2 \in L_\beta (\langle K,0,H_1,H_2 \rangle \in W_e^{\mathfrak{B}} \wedge H_1 \subseteq D \wedge H_2 \subseteq L_\beta - D)$$

and

$$K \subseteq L_\beta - A \leftrightarrow \exists H_1 H_2 \in L_\beta (\langle K,1,H_1,H_2 \rangle \in W_e^{\mathfrak{B}} \wedge H_1 \subseteq D \wedge H_2 \subseteq L_\beta - D).$$

The index e can be communicated by writing $A \leq^{e}_{\mathcal{L}} D$.

One further defines that A is weakly $\mathcal{L}$-reducible to D in the same way but with the sets $K \in L_\beta$ restricted to singletons $\{x\}$ (written $A \leq_{w\mathcal{L}} D$).

An equivalence relation $A =_{\mathcal{L}} D$ is defined by $A \leq_{\mathcal{L}} D \wedge D \leq_{\mathcal{L}} A$ and the equivalence classes are called $\mathcal{L}$-degrees . One says that a degree $\underset{\sim}{a}$ has certain properties if there exist a set $A \in \underset{\sim}{a}$ which has all these properties.

We study in this paper the α-jump operator (see Shore [20] for a discussion of the definition) :

$A' := \{\langle e,x\rangle \mid \exists\ H_1\ H_2 \in L_\alpha\ (\langle x,H_1,H_2\rangle \in W_e \wedge H_1 \subseteq A \wedge H_2 \subseteq L_\alpha - A)$

is the jump of a set $A \subseteq L_\alpha$ in α-recursion theory (we always write W_e instead of $W_e^{L_\alpha}$). Since we have $A \leq_\alpha D \rightarrow A' \leq_\alpha D'$ this definition gives rise to the definition of the α-jump operator $\underset{\sim}{a} \mapsto \underset{\sim}{a}'$ for α-degrees $\underset{\sim}{a}$.

We write 0 for the α-degree of the empty set and 0'' instead of (0')' . Observe that $U_1^{L_\alpha} \in 0'$ and (using the admissibility) $U_2^{L_\alpha} \in 0''$. Furthermore we have for regular sets A that $U_1^{\langle L_\alpha, A\rangle} =_\alpha A'$.

One says that an α-r.e. set A is complete if $A \in 0'$; otherwise A is called incomplete.

We often use without further mentioning the regular set theorem of Sacks which says that every α-r.e. degree contains a regular α-r.e. set (see [13], [22],[8] for proofs).

For a set $A \subseteq L_\alpha$ one writes rcf A for the least $\delta \leq \alpha$ such that a cofinal function $f : \delta \rightarrow \alpha$ exists which is weakly α-reducible to A . The set A is called hyperregular if rcf A = α , otherwise A is called non-hyperregular.

Observe that we have for regular A $\operatorname{rcf} A = \sigma 1\mathrm{cf}^{\langle L_\alpha, A\rangle}\alpha$, in particular A is non-hyperregular iff $\langle L_\alpha, A\rangle$ is inadmissible. Hyperregularity is -contrary to regularity- a property of degrees rather than of single representativs : if $\underset{\sim}{a}$ is an α-degree then $\operatorname{rcf} A$ is the same for every $A \in \underset{\sim}{a}$.

Simpson proved in his thesis [22] that for any $\gamma \in \alpha$ we have that $\gamma = \operatorname{rcf} A$ for some α-r.e. A iff γ is a regular α-cardinal and $\sigma 2\mathrm{cf}^{L_\alpha}\gamma = \sigma 2\mathrm{cf}\,\alpha$. The following Lemma combines in b) Simpson's result with Theorem 2.1. of Shore [19] . The proofs of a) and c) are straightforward (consider a Σ_2 projection from κ into $\sigma 2\mathrm{p}\alpha$ for c)).

<u>Lemma 1 :</u>

a) $0'$ is a non-hyperregular α-degree iff $\sigma 2\mathrm{cf}\,\alpha < \alpha$.

b) There exists an incomplete non-hyperregular α-r.e. degree iff either $\sigma 2\mathrm{p}\alpha \leq \sigma 2\mathrm{cf}\,\alpha < \alpha$ or $\sigma 2\mathrm{cf}\,\alpha < \sigma 2\mathrm{p}\alpha < \alpha$ and there is a regular α-cardinal $\kappa \geq \sigma 2\mathrm{p}\alpha$ such that $\sigma 2\mathrm{cf}^{L_\alpha}\kappa = \sigma 2\mathrm{cf}\,\alpha$.

c) We have $\sigma 2\mathrm{cf}^{L_\alpha}\kappa = \sigma 2\mathrm{cf}\,\alpha$ for every regular α-cardinal κ such that $\sigma 2\mathrm{p}\alpha < \kappa$ and $\sigma 2\mathrm{cf}\,\alpha < \kappa$.

Finally define for any structure $\mathcal{L} = \langle L_\beta, B\rangle$

$\rho_{n,\beta}^{\mathcal{L}} := \mu\delta \leq \beta($ a $\Sigma_n\,\mathcal{L}$ set $M \subseteq \delta$ exists such that $M \notin L_\beta)$

(we write $\rho_{n,\beta}$ instead of $\rho_{n,\beta}^{L_\beta}$).

According to Jensen's Uniformization Theorem [2] we have $\rho_{n,\beta} = \sigma n\mathrm{p}\beta$ for every $n > 0$ and every limit ordinal β.

We will often use without further mentioning the equalities $\rho_{n,\alpha} = \sigma n\mathrm{p}\alpha$ for $n = 1,2$ which are easier to show because Σ_2 - uniformization is trivial for admissible α .

We refer the reader to Devlin [2] for all details concerning constructibility.

§1. Construction of high α-r.e. degrees

At first we sketch the construction of incomplete high r.e. sets in ORT. The original proof is due to Sacks [12]. Additional ideas of Lachlan and Soare are used in the very perspicuous version of the construction as it is presented in Soare[23] (we refer the reader to this paper for more motivation and details concerning the proof in ORT).

In order to bring the requirement $A' \in 0''$ in the reach of a recursive construction we associate with a fixed Σ_2 set $S \in 0''$ a r.e. set B_S which is defined by

$\langle e,y\rangle \in B_S \leftrightarrow \forall y' \leq y \exists z \neg\phi(e,y',z)$

where $\exists y \forall z \phi(\cdot,y,z)$ is a fixed Σ_2 definition of S over L_ω. Then we have for every $e \in \omega$ $1\text{-}S(e) = \lim_{y\to\omega} B_S(\langle e,y\rangle)$ and it is enough to insure that for all $e \in \omega$ $\lim_{y\to\omega} A(\langle e,y\rangle) \simeq \lim_{y\to\omega} B(\langle e,y\rangle)$ in order to get $A' \in 0''$. So for every e we set up a positive requirement $P_e : \lim_{y\to\omega} A(\langle e,y\rangle) = \lim_{y\to\omega} B(\langle e,y\rangle)$. P_e is a requirement which is hard to satisfy if $e \notin S$ since in this case we have to put all but finitely many elements of $\{\langle e,y\rangle \mid y \in \omega\}$ into A.

A conflict arises because we have to satisfy as well for all $e \in \omega$ $N_e : \neg C = \Phi_e(A)$ where $C \in 0'$ is a fixed r.e. set. The requirements N_e are satisfied by preserving a disagreement between $C_t(x)$ and $\Phi_{e,t}(A_t,x)$ for a suitable argument x and by forcing the appearance of such a disagreement (respectively of an argument x such that $\Phi_e(A,x)\uparrow$) on the way of preserving as well agreements between $C_t(x)$ and $\Phi_{e,t}(A_t,x)$ for all x out of an initial segment γ of ω which is chosen as long as possible ($\gamma \leq \omega$).(We have written $\Phi_e(A,\cdot)$ for the function which is partially recursive in A with index e.)

This strategy of preservation in order to get $\neg C \leq^e A$ is on first sight contrary to intuition. But if we preserve -as soon as it appears during the construction- agreement between $C_t(x)$ and $\phi_{e,t}(A_t,x)$ for every $x \in \gamma$ we actually try to make $C \upharpoonright \gamma$ recursive. Since $C = C \upharpoonright \omega$ is not recursive we must then have $\gamma < \omega$ and therefore $\neg \phi_e(A,\gamma) \simeq C(\gamma)$.

We write I_e for the injury set of N_e which is the set of all elements x that are put into A -as demanded by some positive requirement $P_{e'}$ with $e' < e$ - although they destroy a computation in A which should be preserved in order to satisfy N_e . Then we can be a little more exact in our description of the preservation strategy and say that although we try to make $C \upharpoonright \gamma$ recursive we only succeed in making $C \upharpoonright \gamma$ recursive in I_e (γ as before). But we can still get the wanted conclusion $\gamma < \omega$ even if I_e is infinite since we need only $C \not\leq I_e$ for the argument above. Since I_e is recursive in $\{\langle e',y\rangle \in A \mid e' < e \wedge y \in \omega\}$ we can prove $C \not\leq I_e$ for every $e \in \omega$ during the inductive argument where one shows that for every $e \in \omega$ $\neg C \leq^e A$ and $\lim_{y \to \omega} A(\langle e,y\rangle) = \lim_{y \to \omega} B(\langle e,y\rangle)$ (we use here that $C \not\leq \{\langle e',y\rangle \in B \mid e' < e \wedge y \in \omega\}$ for every e).

Observe that in writing $C(\gamma)$ etc. we have followed the usual convention to identify a set with its characteristic function.

The construction from ORT works as well for Σ_2 admissible α (Shore [20]). But there are several reasons why this construction does not work for the other admissible α . We discuss five of these problems in the following and we simultaneously try to motivate the new features of the subsequent construction for the case $\alpha > \sigma 2cf\alpha \geq \sigma 2p\alpha$.

<u>1)</u> Assume that we succeed in constructing the set A in such

a way that $\forall e \in \alpha (A^{(e)} =^* B_S{}^{(e)})$ with S and B_S as before (define for any set M : $M^{(e)} := M \cap (\{e\} \times L_\alpha)$; $M_1 =^* M_2$ means that $M_1 - M_2 \in L_\alpha$ and $M_2 - M_1 \in L_\alpha$).

This doesn't imply in general that $S \leq_\alpha A'$ if $\sigma 2\mathrm{cf}\,\alpha < \alpha$. We have of course for every $e \in \alpha$ that $e \in S \leftrightarrow$
$\exists y_e \forall y \geq y_e (\neg \langle e,y\rangle \in A) \leftrightarrow \exists y_e (\neg \langle p,e,y_e\rangle \in A')$
for some fixed parameter p. But if we want to reduce in the same way questions "$K \subseteq S$" to questions about A' we need the existence of a bound for the set $\{y_e \mid e \in K\}$ of witnesses. Since $S \in 0''$ can't be tame-Σ_2 L_α if α is not Σ_2 admissible (see §2.) we can hardly expect that this bound exists for all α-finite K such that $K \subseteq S$.

We overcome this difficulty by using in a positive way that α is not Σ_2 admissible. For these α there exist non-hyperregular α-r.e. sets and in the case $\alpha > \sigma 2\mathrm{cf}\,\alpha \geq \sigma 2\mathrm{p}\,\alpha$ there exist even incomplete non-hyperregular α-r.e. sets according to Shore [19]. But for non-hyperregular A we can avoid the search for witnesses y_e : Take a cofinal function $f : \mathrm{rcf}\,A \to \alpha$ which is weakly α-recursive in A. Then we have

$e \in S \leftrightarrow \forall x \in \mathrm{rcf}\,A\, \exists y\, z(y = f(x) \wedge z \geq y \wedge \neg\langle e,z\rangle \in A) \leftrightarrow$
$\{p\} \times \mathrm{rcf}\,A \times \{e\} \subseteq A'$

for some fixed parameter p which implies that for every α-finite K we have

$K \subseteq S \leftrightarrow \{p\} \times \mathrm{rcf}\,A \times K \subseteq A'$.

<u>Convention:</u> We say "α-recursive in" and "weakly α-recursive in" for "$\leq_\alpha$" respectively "$\leq_{w\alpha}$" as usual. But there is a problem with this interpretation, see[9].

<u>2)</u> For the considered α where $\alpha > \sigma 2\mathrm{cf}\,\alpha \geq \sigma 2\mathrm{p}\,\alpha$ it can happen that $0''$ does not contain a regular Σ_2 L_α set. According to [11] this occurs if and only if $\sigma 3\mathrm{cf}\,\alpha < \sigma 3\mathrm{p}\,\alpha$. We will con-

struct in [11] an α where $\sigma 3cf\alpha < \sigma 3p\alpha \leq \sigma 2p\alpha < \sigma 2cf\alpha < \alpha$. This example is the most difficult one with respect to our construction of incomplete high α-r.e. degrees since $0''$ does not contain a regular Σ_2 set and we have $\sigma 2p\alpha < t\sigma 2p\alpha$ (see [6] for the definition of the tame Σ_2 projectum $t\sigma 2p\alpha$).

3) In consequence of the preceding the plan for our construction is as follows : We take a fixed incomplete non-hyperregular α-r.e. set D and make sure that $A^{(0)} =^* D$ in order to make A non-hyperregular. Further for $e > 0$ we want to have that $A^{(e)} =^* B_S{}^{(e)}$. As before we set up for every $e \in \alpha$ a positive requirement P_e which tries to satisfy this condition concerning $A^{(e)}$.

It is crucial for the infinite injury argument that the set of those elements which should be put into A in order to satisfy all requirements out of an initial segment of the priority list is not too complicated. According to point 2) this forces us to make our priority list no longer than $\sigma 2p\alpha$ because only for α-finite sets K of α-cardinality less than $\sigma 2p\alpha$ it is guaranteed that $B_S \cap K \times L_\alpha$ is α-recursive. It is not easy to work with such a short priority list in an infinite injury construction since the α-recursive approximation to this list is very weak if $\sigma 2p\alpha < \sigma 2cf\alpha$. We introduce a clause b) in the construction which makes it possible to control in many situations those unwanted injuries which are merely due to bad guessing of priorities.

4) We want to prove by induction on the priority $p(e)$ that for every e we have $A^{(e)} =^* B^{(e)}$. There is a problem in the case that $p(e)$ is a limit ordinal since the induction hypothesis doesn't imply then that $\bigcup\{A^{(i)} \mid p(i) < p(e)\} =^* \bigcup\{B^{(i)} \mid p(i) < p(e)\}$ and we can't control the degree of the injury set

I_e . We use the fact that this situation is only possible if $\sigma 2cf\alpha > \omega$ since $\sigma 2cf\alpha \geq \sigma 2p\alpha$. $\sigma 2cf\alpha > \omega$ implies that there are enough fixpoint stages in the construction so that it is in fact not necessary to determine the degree of the injury set I_e .

5) There is a problem with the preservation strategy of Sacks in the case that there are non-hyperregular injury sets I_e (which will occur in our construction since $A^{(0)}$ is non-hyperregular). If we want to preserve then agreements $C(x) = \phi_e(A,x)$ for $x < \gamma$ these computations may altogether use an unbounded part of A even if $\gamma < \alpha$. Since this would endanger the positive requirements of lower priority we have to be much more careful with preservations. For this sake we introduce "e-fixpoints" in the case $\sigma 2cf\alpha > \omega$. In the case $\sigma 2cf\alpha = \omega$ we divide α into rcf D many blocks as in Shore [18] (doing the same thing in the case $\sigma 2cf\alpha \geq \sigma 2p\alpha > \omega$ would be troublesome because of limit points in the priority list).

Theorem 1 : Assume that $\alpha > \sigma 2cf\alpha \geq \sigma 2p\alpha$. If C and D are α-r.e. sets such that $C \not\leq_\alpha D$ and D is non-hyperregular then there exists an α-r.e. set A such that $D \leq_\alpha A$, $C \not\leq_\alpha A$ and $A' =_\alpha 0''$.

The rest of this chapter is devoted to the proof of this theorem. After some preparations we will describe the construction of the set A for the case $\sigma 2cf\alpha > \omega$. We will show in the Lemmata 3,4,5 that this set A has the properties we want. The construction for the case $\sigma 2cf\alpha = \omega$ is rather close to the construction in ORT and will be discussed briefly afterwards.

We fix for the following regular α-r.e. sets $C, D \subseteq \alpha$ such that $C \not\leq_{\alpha} D$ and D is non-hyperregular. $(C_{\sigma})_{\sigma<\alpha}$ and $(D_{\sigma})_{\sigma<\alpha}$ are in the following fixed α-recursive enumerations of these sets.

Take a Σ_2 L_α set $S \subseteq \alpha$ such that $S \in 0''$ and fix a Δ_0 formula Ψ such that $\beta \in S \leftrightarrow L_\alpha \models \exists \gamma \forall x \Psi(\beta,\gamma,x)$. Define the α-r.e. set $B \subseteq \alpha \times \alpha$ as follows :

$$\langle\beta,\gamma\rangle \in B :\leftrightarrow ((\beta = 0 \wedge \gamma \in D) \vee (\beta > 0 \wedge L_\alpha \models \forall \gamma' \leq \gamma \exists x \neg\Psi(\beta,\gamma',x))) .$$

Then we have for $\beta > 0$:

$\beta \in S \rightarrow \{\gamma | \langle\beta,\gamma\rangle \in B\} = \mu\delta(L_\alpha \models \forall x \Psi(\beta,\delta,x)) < \alpha$ and

$\neg\beta \in S \rightarrow \{\gamma | \langle\beta,\gamma\rangle \in B\} = \alpha$.

Fix an α-recursive enumeration $(B_\sigma)_{\sigma<\alpha}$ of B for the following.

For any set M and any $x \in L_\alpha$ we will use in the following $M^{(x)}$ as an abbreviation for $M \cap (\{x\} \times L_\alpha)$.

A_σ will be the set of elements which have been put into A before stage σ .

Lemma 2 : Assume that K is an α-finite set of α-cardinality less than $\sigma 2\mathrm{cf}\,\alpha$. Further assume that W is an α-r.e. set such that $W^{(x)}$ is regular for every $x \in K$. Then $\bigcup \{W^{(x)} | x \in K\}$ is regular.

Proof : Fix an enumeration $(W_\sigma)_{\sigma<\alpha}$ of W . For given $\beta < \alpha$ define a Σ_2 function $f : K \rightarrow \alpha$ by $f(x) := \mu\sigma(W_\sigma \cap W^{(x)} \cap L_\beta = W^{(x)} \cap L_\beta)$. There exists a bound σ_0 for $\mathrm{Rg}\, f$ and we have $\bigcup \{W^{(x)} | x \in K\} \cap L_\beta \subseteq W_{\sigma_0}$.

In the following we will write $x \in W_{e,\sigma}$ for $L_\sigma \models \phi(\langle e,x\rangle)$ where ϕ is a fixed Σ_1 L_α definition of $U_1^{L_\alpha}$.

For the considered α-r.e. sets A and C and their enumerations $(A_\tau)_{\tau<\alpha}$ and $(C_\tau)_{\tau<\alpha}$ we will say that <u>at stage σ there exists a computation of "$C \leq^e A$" for "$K \subseteq L_\alpha - C$" with negative neighborhood H</u> if
$\exists H' \in L_\sigma(\langle K,1,H',H\rangle \in W_{e,\sigma} \wedge H' \subseteq A_\sigma \wedge H \subseteq L_\sigma - A_\sigma)$.

<u>Case i) : $\alpha > \sigma 2\mathrm{cf}\alpha \geq \sigma 2\mathrm{p}\alpha$ and $\sigma 2\mathrm{cf}\alpha > \omega$.</u>

The next definition is the fixpoint device which was mentioned in point 5) of the motivation.

λ is an <u>e-fixpoint</u> at stage $\beta \geq \lambda$:$\leftrightarrow$
for every $\tau < \lambda$ there is a τ' such that $\tau \leq \tau' < \lambda$ and there is a stage $\sigma < \lambda$ such that at stage σ there exists a computation of "$C \leq^e A$" for "$\tau' - C_\lambda \subseteq L_\alpha - C$" with negative neighborhood H and we have $H \subseteq L_\alpha - A_\beta$.

We say that this e-fixpoint λ is <u>inactive</u> at stage β if $C_\lambda \cap \lambda \neq C_\beta \cap \lambda$.

The <u>"restraint function" $r : \alpha \times \alpha \to \alpha$</u> will play a similar role as in Soare [23] and is defined by cases :
<u>Case 1):</u> There exists a stage $\sigma \leq \beta$ such that some $\lambda < \sigma$ is an inactive e-fixpoint at all stages in $[\sigma,\beta] := \{\tau \mid \sigma \leq \tau \leq \beta\}$. Take the least such σ . Define $r(e,\beta)$ to be the least $\lambda < \sigma$ which is an inactive e-fixpoint at all stages in $[\sigma,\beta]$.
<u>Case 2):</u> Define $r(e,\beta)$ to be the union of all e-fixpoints λ at β otherwise.

We fix an 1-1 Σ_2 L_α function g which maps $\sigma 2\mathrm{p}\alpha$ partially onto α . $g^{-1}(e)$ will be the priority of the requirements P_e, N_e for $e \in \alpha$. Using the assumption $\sigma 2\mathrm{p}\alpha \leq \sigma 2\mathrm{cf}\alpha$ it is easy to see that $g \upharpoonright (\gamma \cap \mathrm{dom}\ g)$ is α-finite and $B^{(<\gamma)} \leq_\alpha D$ for every $\gamma < \sigma 2\mathrm{p}\alpha$ where $B^{(<\gamma)} := \bigcup \{ B^{(e)} \mid g^{-1}(e) < \gamma \}$.

We further need an α-recursive approximation function $g^{\cdot}(\cdot)$ (of two arguments) with α-recursive domain which has the property that for all $\gamma < \sigma 2p\alpha$ there exists an ordinal $\tau_\gamma < \alpha$ such that for all $x \in \gamma \cap \mathrm{dom}\, g$ and all $\tau \geq \tau_\gamma$ we have $g^\tau(x) \simeq g(x)$.
In addition we want to have that

(1) $g(x)\downarrow \leftrightarrow \exists \sigma_0 \forall \sigma \geq \sigma_0 (g^\sigma(x)\downarrow)$ and

(2) $\forall$ limits $\lambda < \alpha (g^\lambda(x)\downarrow \leftrightarrow \exists \sigma_0 < \lambda \forall \sigma (\sigma_0 \leq \sigma < \lambda \rightarrow g^\sigma(x)\downarrow))$

and that $g^\sigma(\cdot)$ is 1-1 for every $\sigma < \alpha$.
Because of the distinguished role of the requirement P_0 we further need that $g(0) \simeq 0$ and $g^\sigma(0) \simeq 0$ for all $\sigma < \alpha$.

The definition of an approximation function $g^{\cdot}(\cdot)$ with these properties is routine.

Observe that in general we can't get the following property which one would really like to have :

$\forall \gamma < \sigma 2p\alpha \exists \sigma_0 \forall z \leq \gamma \forall \sigma \geq \sigma_0 (g^\sigma(z) \simeq g(z))$

(see the points 2) and 3)).

Construction :

At stage σ we consider every $\langle\beta,\gamma\rangle \in B_{\sigma+1}$ such that $g^\sigma(z) \simeq \beta$ for some $z < \sigma 2p\alpha$.

If $\langle\beta,\gamma\rangle$ is not already an element of A_σ we put $\langle\beta,\gamma\rangle$ into A at stage σ if

a) $\langle\beta,\gamma\rangle \geq r(g^\sigma(z'),\sigma)$ for all $z' \in (z+1) \cap \mathrm{dom}\, g^\sigma$ and

b) $\langle\beta,\gamma\rangle \geq \tau$ for all $\tau < \sigma$ such that not $(z+1) \cap \mathrm{dom}\, g^\sigma \subseteq (z+1) \cap \mathrm{dom}\, g^\tau$.

End of construction.

For a negative requirement N_e there are in general unboundedly many stages σ where some positive requirement $P_{e'}$ of truly lower priority (i.e. $g^{-1}(e) \leq g^{-1}(e')$) thinks that it may injure N_e because of the weak approximation property of $g^{\cdot}(\cdot)$. The following Lemma shows that in some special situations these unwanted injuries will not occur because of clause b) in the construction.

In the following we always write τ_γ for the least τ such that $\forall \sigma \geq \tau \; \forall x \in \gamma \cap \text{dom } g \; (g^\sigma(x) \simeq g(x))$.

Lemma 3 : Assume that $\gamma < \sigma 2p\alpha$, $\tau_\gamma \leq \sigma \leq \tau$, $\gamma \cap \text{dom } g^\sigma = \gamma \cap \text{dom } g$ and $z \in \gamma \cap \text{dom } g$. If at stage τ an element $\langle \beta, \delta \rangle$ is put into A such that $\langle \beta, \delta \rangle < \sigma$ and $\langle \beta, \delta \rangle < r(g^\tau(z), \tau)$ then there exists a $z' < z$ such that $z' \in \gamma \cap \text{dom } g$ and $g(z') \simeq \beta$.

Proof : According to the construction there exists a z' such that $g^\tau(z') \simeq \beta$. Since clause b) does not restrain $\langle \beta, \delta \rangle$ at stage τ we have $(z'+1) \cap \text{dom } g^\tau \subseteq (z'+1) \cap \text{dom } g^\sigma$. We further have $z' < z$ because $\langle \beta, \delta \rangle < r(g^\tau(z), \tau)$ and $\langle \beta, \delta \rangle$ is not restrained by clause a) at stage τ. Since $\tau \geq \sigma \geq \tau_\gamma$ it follows that $(z'+1) \cap \text{dom } g^\tau = (z'+1) \cap \text{dom } g$ and $g^\tau \restriction ((z'+1) \cap \text{dom } g) = g \restriction ((z'+1) \cap \text{dom } g)$. In particular we have that $z' \in \text{dom } g$ and $g(z') \simeq g^\tau(z') \simeq \beta$.

The following Lemma will solve the problem which was described in point 3) of the motivation : In the case where the priority $g^{-1}(e)$ of some negative requirement N_e is a limit ordinal we have problems to control $\bigcup \{ A^{(i)} \mid g^{-1}(i) < g^{-1}(e) \}$ and the injury set I_e. Lemma 4 gives a sufficient condition for a stage σ that some computation which exists at stage σ will not be destroyed later. It is important that this condition can be expressed by using just $\bigcup\{B^{(i)} \mid g^{-1}(i) < g^{-1}(e)\}$, not $\bigcup\{A^{(i)} \mid g^{-1}(i) < g^{-1}(e)\}$. A fixpoint argument in Lemma 5 will show that this condition will be met by an unbounded set of stages.

The properties of $\sigma 2p\alpha$ and the assumption $\sigma 2p\alpha \leq \sigma 2cf\alpha$ imply that for every $\gamma < \sigma 2p\alpha$ there exists a stage $\tau \geq \tau_\gamma$ such that for every $z \in \gamma \cap \text{dom } g$

$\exists \delta \exists \lambda$(λ is an inactive g(z)-fixpoint at all stages $\sigma \geq \delta$) $\rightarrow$

$\exists \lambda$(λ is an inactive g(z)-fixpoint at all stages $\sigma \geq \tau$) .

In the following we write τ_γ' for the least such $\tau \geq \tau_\gamma$.

We further define $B^{(<\gamma)} := \bigcup\{B^{(e)} \mid g^{-1}(e) < \gamma\}$ and $B_\sigma^{(<\gamma)} := B^{(<\gamma)} \cap B_\sigma$.

<u>Lemma 4 :</u> Assume that $\sigma \geq \tau_\gamma'$ is a stage such that $\gamma \cap \text{dom } g^\sigma = \gamma \cap \text{dom } g$, $B_\sigma^{(<\gamma)} \cap \sigma = B^{(<\gamma)} \cap \sigma$ and no element $x < \sup\{r(g(z),\sigma) \mid z \in \gamma \cap \text{dom } g\}$ is put into A at stage σ . Then we have for every $z \in \gamma \cap \text{dom } g$ and for every stage $\tau \geq \sigma$: $r(g(z),\tau) \geq r(g(z),\sigma)$ and no element $x < \sigma$ with $x < r(g(z),\tau)$ is put into A at stage τ .

<u>Proof :</u> Induction on $z \in \gamma \cap \text{dom } g$.

Assume for a contradiction that some $x < \sigma$ with $x < r(g(z), \sigma_0)$ is put into A at stage σ_0 .

By Lemma 3 there is some $z' < z$ such that $z' \in \text{dom } g$ and $g(z') \simeq \beta$ where $x = \langle\beta, \delta\rangle$ for some δ . Therefore we have $x \in B^{(<\gamma)} \cap \sigma = B_\sigma^{(<\gamma)} \cap \sigma$. We consider two cases :

<u>a)</u> x was not put into A at stage σ since $x < r(g(z''),\sigma)$ where $z'' \in (z'+1) \cap \text{dom } g$. By our induction hypothesis we have $r(g(z''),\sigma) \leq r(g(z''),\sigma_0) = r(g^{\sigma_0}(z''), \sigma_0)$ and x will not be put into A at stage σ_0 either.

<u>b)</u> x was not put into A at stage σ since there exists $\sigma' < \sigma$ such that $\sigma' > x$ and not $(z'+1) \cap \text{dom } g^\sigma \subseteq (z'+1) \cap \text{dom } g^{\sigma'}$.

Since $(z'+1) \cap \operatorname{dom} g^{\sigma} \subseteq (z'+1) \cap \operatorname{dom} g^{\sigma_0}$ x is not put into A at stage σ_0 because of clause b) in the construction.

It remains to prove that $r(g(z),\tau) \geq r(g(z),\sigma)$ for all $\tau \geq \sigma$. Assume that there is a minimal stage $\sigma_0 > \sigma$ such that $r(g(z),\sigma_0) < r(g(z),\sigma)$. By the preceding no element $y < r(g(z),\sigma)$ will be put into A at some stage τ where $\sigma \leq \tau \leq \sigma_0$. $r(g(z),\sigma_0) < r(g(z),\sigma)$ is therefore only possible if $r(g(z),\sigma_0)$ is defined according to case 1) of the definition of r whereas $r(g(z),\sigma)$ is defined according to case 2). Since $r(g(z),\sigma_0) < \sigma$ no element $y < r(g(z),\sigma_0)$ will be put into A at any stage $\tau \geq \sigma_0$: Otherwise assume that σ_1 is the minimal such τ. Since $r(g(z),\sigma_0)$ is defined according to case 1) we have $r(g^{\sigma_1}(z),\sigma_1)$ $= r(g(z),\sigma_0) > y$ and y can't be put into A at stage σ_1 as it was shown in the first part of this proof. Thus we have proved that some $\lambda < \sigma_0$ is an inactive g(z)-fixpoint at all stages in $[\sigma_0,\alpha)$ whereas there is no inactive g(z)-fixpoint at stage σ. Since we have $\tau_\gamma' \leq \sigma$ this gives a contradiction to the definition of τ_γ' and we have proved that $r(g(z),\tau) \geq r(g(z),\sigma)$ for all $\tau \geq \sigma$.

<u>Remark:</u> If σ satisfies the assumptions of Lemma 4 then no element $x < \sup\{r(g(z),\sigma) \mid z \in \gamma \cap \operatorname{dom} g\}$ is put into A at any stage $\tau \geq \sigma$. Therefore these stages σ play a role in this proof which is similar to the role of "true stages" (see Soare[23]) in the proof in ORT.

<u>Lemma 5 :</u> For every $e \in \alpha$ we have

a) $\neg C \leq_\alpha^e A$ and

b) $A^{(e)} =^* B^{(e)}$.

<u>Proof :</u> For convenience we prove a) and b) simultaneously by

induction on $g^{-1}(e)$. Assume for the following that $g^{-1}(e) = z$ and that a) and b) are true for all e' such that $g^{-1}(e') < z$. Observe that this assumption does in general not imply that $\bigcup\{A^{(e')} \mid g^{-1}(e') < z\} =^* \bigcup\{B^{(e')} \mid g^{-1}(e') < z\}$ if we have $\sigma 3\mathrm{cf}\,\alpha \leq z$, which is of course possible (see point 3) of the motivation). But we get the information that $\bigcup\{A^{(e')} \mid g^{-1}(e') < z\}$ is regular : Since every $B^{(e')}$ is regular we get from $A^{(e')} =^* B^{(e')}$ that every $A^{(e')}$ is regular as well. Then Lemma 2 implies that $\bigcup\{A^{(e')} \mid g^{-1}(e') < z\}$ is regular. This is the only fact which we use from our induction hypothesis so that in the case $\sigma 1\mathrm{p}\,\alpha = \alpha$ we don't need an induction at all (this is rather surprising if compared with the situation in ORT , see[23]).

For $\gamma := z+1$ we write M for the set of those stages σ where the assumptions of Lemma 4 are satisfied. We want to prove that M is unbounded in α by using the regularity of $A^{(<z)} := \bigcup\{A^{(e')} \mid g^{-1}(e') < z\}$.

For $\lambda_n < \alpha$ define $\lambda_{n+1} := \mu\tau > \lambda_n\ (\forall y \in (((z+1) \cap \mathrm{dom}\, g^{\lambda_n}) - \mathrm{dom}\, g)\ \exists \tau' \leq \tau\ (g^{\tau'}(y)\uparrow)\ \wedge\ B_\tau^{(<\gamma)} \cap \lambda_n = B^{(<\gamma)} \cap \lambda_n\ \wedge\ A_\tau^{(<z)} \cap \lambda_n = A^{(<z)} \cap \lambda_n)$.

By using property (1) of $g^{\cdot}(\cdot)$ and the fact that $B^{(<\gamma)}$ and $A^{(<z)}$ are regular α-r.e. sets it is easy to see that $\lambda_{n+1} < \alpha$ exists. For every given $\lambda_0 < \alpha$ with $\lambda_0 > \tau_\gamma'$ define then $\lambda := \sup\{\lambda_n \mid n \in \omega\}$. We have $\lambda < \alpha$ since the function $n \mapsto \lambda_n$ is $\Sigma_2\ L_\alpha$. It follows from property (2) of the appoximating function and Lemma 3 that $\lambda \in M$.

We write IN for the α-finite set of all $z' \leq z$ such that some $\lambda < \alpha$ is an inactive $g(z')$-fixpoint in $[\tau,\alpha)$ for some $\tau < \alpha$ $([\tau,\alpha) := \{\tau' \mid \tau \leq \tau' < \alpha\})$. Then we have that $r(g(z'),\cdot)$ is constant in $[\sigma_0,\alpha)$ for every $z' \in$ IN according to Lemma 4 , where σ_0 is the least element of M. Therefore it is enough to show $\sup\{r(g(z'),\sigma) \mid \sigma \in M \wedge z' \in ((z+1) \cap \text{dom } g - \text{IN})\} < \alpha$ in order to prove that $\sup\{r(g(z'),\sigma) \mid \sigma \in M \wedge z' \in (z+1) \cap \text{dom } g\} < \alpha$.

Thus assume for a contradiction that

$\forall \tau < \alpha \exists \sigma \in M \exists z' \in ((z+1) \cap \text{dom } g - \text{IN})(\tau < r(g(z'),\sigma))$.

This implies that for every $K \in L_\alpha$

$$K \subseteq L_\alpha - C \leftrightarrow \exists \sigma \in M \exists z' \in ((z+1) \cap \text{dom } g - \text{IN}) (\sup K < r(g(z'),\sigma) \wedge K \subseteq L_\alpha - C_\sigma).$$

The part "$\rightarrow$" of this equivalence is obvious from our assumptions. For a proof of "$\leftarrow$" we assume that σ, z', K do satisfy the right side. By Lemma 4 we have that $r(g(z'),\sigma)$ is defined according to case 2) of the definition of r since $\neg z' \in \text{IN}$. Therefore there is at σ some $g(z')$-fixpoint $\lambda \leq r(g(z'),\sigma)$ such that $\sup K < \lambda$ and λ is not an inactive $g(z')$-fixpoint at σ which means that $C_\sigma \cap \lambda = C_\lambda \cap \lambda$. By Lemma 4 there is no stage $\tau \geq \sigma$ such that an element $y < \lambda$ is put into A at stage τ. Therefore there is no $y < \lambda$ and $\tau \geq \sigma$ such that $y \in C_{\tau+1} - C_\tau$ since otherwise some $\lambda' \leq \lambda$ would be an inactive $g(z')$-fixpoint in $[\tau,\alpha)$, contradicting $\neg z' \in \text{IN}$. Thus we have proved that $C_\sigma \cap \lambda = C \cap \lambda$ which shows that $K \subseteq L_\alpha - C$.

The equivalence which was just proved implies that $C \leq_\alpha B^{(<\gamma)}$ since "$\sigma \in M$" can be expressed α-recursively in $B^{(<\gamma)}$. But this is absurd because we have $B^{(<\gamma)} \leq_\alpha D$. Thus we have proved that $S := \sup\{r(g(z'),\sigma) \mid \sigma \in M \wedge z' \in (z+1) \cap \text{dom } g\} < \alpha$.

In order to prove a) assume for a contradiction that $C \leq_\alpha^e A$. For $\lambda'_n < \alpha$ we define $\lambda'_{n+1} < \alpha$ by

$\lambda'_{n+1} := \mu\tau > \lambda'_n$((the same as in the definition of λ_{n+1}) $\wedge$

$C_\tau \cap \lambda'_n = C \cap \lambda'_n \wedge$ (at stage τ there exists a computation of

"$C \leq^e_\alpha A$" for "$\lambda'_n - C \subseteq L_\alpha - C$" with negative neighborhood H such that $H \subseteq L_\alpha - A$)) .

We have again that $\lambda' := \sup\{\lambda'_n \mid n \in \omega\} < \alpha$ for every given $\lambda'_0 < \alpha$ since the function $n \mapsto \lambda'_n$ is Σ_2 definable and if we start with some $\lambda'_0 > \tau_\gamma'$ it is obvious that $\lambda' \in M$ and λ' is a g(z)-fixpoint at λ' . Now it can't be the case that $r(g(z),\sigma)$ is defined for some $\sigma \in M$ according to case 1) of the definition of r because this contradicts $C \leq^e_\alpha A$ (use Lemma 4) . This implies that $r(g(z), \lambda') = \lambda'$ for all these stages $\lambda' \in M$ which is absurd since we have proved just before that $S < \alpha$.

For the proof of b) we choose $\sigma_1 \in M$ such that $\sigma_1 > S$. It follows from Lemma 4 that $A^{(e)} \cap \sigma_1 = A^{(e)}_{\sigma_1+1} \cap \sigma_1$. Further we have $A^{(e)} - \sigma_1 = B^{(e)} - \sigma_1$ by the definition of S which together shows that $A^{(e)} =^* B^{(e)}$.

The proof of Theorem 1 is now very easy. We get $\neg C \leq_\alpha A$ and $D = B^{(0)} =_\alpha A^{(0)} \leq_\alpha A$ from Lemma 5 . In order to show $S \leq_\alpha A'$ we fix a cofinal function $f : \mathrm{rcf}\, A \to \alpha$ which is weakly α-recursive in A (A is non-hyperregular since D is non-hyperregular and $D \leq_\alpha A$). Lemma 5 b) implies that for every $\beta \in \alpha$

$\beta \in S \leftrightarrow \forall x < \mathrm{rcf}\, A\ \exists \delta > f(x)\ (\neg \langle \beta, \delta \rangle \in A)$

(we may assume without loss of generality that $0 \in S$) .

There is a parameter $p \in \alpha$ such that for all β and x

$x < \mathrm{rcf}\, A \wedge \exists \delta > f(x)\ (\neg \langle \beta, \delta \rangle \in A) \leftrightarrow \langle p, x, \beta \rangle \in A'$. Then we have $K \subseteq S \leftrightarrow \{p\} \times \mathrm{rcf}\, A \times K \subseteq A'$.

Concerning the computation for "$K \subseteq L_\alpha - S$" we observe that "$K \subseteq L_\alpha - S$" can be written as a Π_2 formula. Since we have $U_2^{L_\alpha} \leq_{w\alpha} A'$ (see the first part of the proof of Theorem 2 b) in §2.) this Π_2 fact can be expressed α-recursively in A' .

Case ii) : $\alpha > \sigma 2\mathrm{cf}\alpha = \sigma 2\mathrm{p}\alpha = \omega$.

The proof of Theorem 1 is simpler in this case since the problem at limits of the priority list doesn't occur (see point 4) of the motivation). The construction is closer to the one in ORT [23] but we have to be aware of the other points in the motivation and the fact that we can't use the regularity of A as it is done in ORT ("true stages").

According to point 5) of the motivation we fix a strictly increasing cofinal function $f : \mathrm{rcf}\ D \to \alpha$ which is weakly α-recursive in $B^{(0)}$ with an index e . We define then

$f^{\sigma}(x)\downarrow \;:\leftrightarrow\; \exists \tau \leq \sigma\ \exists y\ \exists H\ (\langle x,y,H\rangle \in W_{e,\tau} \wedge H \subseteq \{0\} \times L_{\alpha} - A_{\sigma}^{(0)})$.

If $f^{\sigma}(x)\downarrow$ we go to the least such $\tau \leq \sigma$ and choose $\langle x,\hat{y},\hat{H}\rangle$ minimal (with respect to a fixed canonical Δ_1 L_{α} well ordering $\prec_{\alpha}$ of L_{α}) such that $\langle x,\hat{y},\hat{H}\rangle \in W_{e,\tau} \wedge \hat{H} \subseteq L_{\alpha} - A_{\sigma}^{(0)}$. We then say that $f^{\sigma}(x) \simeq \hat{y}$ and $\hat{H}$ is the negative neighborhood of this computation.

Further we fix a Σ_2 L_{α} function g such that dom g $= \omega$ and g maps ω 1-1 onto α . We have in this case a very nice approximation $g^{\cdot}(\cdot)$ to g where dom $g^{\cdot}(\cdot) = \alpha \times \omega$ and $\forall n < \omega\ \exists \sigma\ \forall m \leq n\ \forall \tau \geq \sigma\ (g^{\tau}(m) \simeq g(m))$. We require further that $g^{\sigma}(\cdot)$ is 1-1 for every σ and that $g(0) \simeq g^{\sigma}(0) \simeq 0$ for all σ .

Analogously as in Soare [23] we define functions l and r relative to fixed enumerations $(C_{\sigma})_{\sigma<\alpha}$ and $(A_{\sigma})_{\sigma<\alpha}$.

For $e,\sigma \in \alpha$ choose $l(e,\sigma) \leq \mathrm{rcf}\ D$ maximal such that for all $x < l(e,\sigma)$ the following holds :
There is a stage $\tau \leq \sigma$ such that $f^{\tau}(x)\downarrow$ and the negative neighborhood $\hat{H}$ of this computation satisfies $\hat{H} \subseteq L_{\alpha} - A_{\sigma}$ and at stage τ there exists a computation of "$C \leq_{\alpha}^{e} A$" for

"$K_{x,\tau} := f^{\tau}(x) - C_{\tau} \subseteq L_{\alpha} - C$" with negative neighborhood H and H satisfies $H \subseteq L_{\alpha} - A_{\sigma}$ (we then write $\tilde{\tau}$ for the minimal such $\tau \leq \sigma$ and $\tilde{H}$ for the minimal such H).

For $u(e,x,\sigma) := \mu\gamma(\hat{H} \subseteq \gamma \wedge \tilde{H} \subseteq \gamma)$ we then demand in the case that σ is a successor stage that no $y < u(e,x,\sigma)$ was put into A at stage $\sigma-1$.

Finally we demand (for any σ) that $C_{\tilde{\tau}} \cap f^{\sigma}(x) = C_{\sigma} \cap f^{\sigma}(x)$.

If we then have for this $l(e,\sigma)$ that $l(e,\sigma) < \text{rcf } D$ and for $x = l(e,\sigma)$ all the conditions in the definition are satisfied except the last one (i.e. $C_{\tilde{\tau}} \cap f^{\sigma}(x) = C_{\sigma} \cap f^{\sigma}(x)$) we say that e is inactive at stage σ and define

$r(e,\sigma) := \sup\{u(e,x,\sigma) \mid x \leq l(e,\sigma)\}$.

Otherwise we define

$r(e,\sigma) := \sup\{u(e,x,\sigma) \mid x < l(e,\sigma)\}$.

It is convenient to choose the universal enumeration $(W_e)_{e<\alpha}$ in such a way that $W_0 = \phi$ so that we have $l(0,\sigma) = r(0,\sigma) = 0$ for all σ .

Construction :

At stage σ we consider every $\langle\beta,\gamma\rangle \in B_{\sigma+1}$ such that $\beta \in \text{Rg } g^{\sigma}(\cdot)$. If $\langle\beta,\gamma\rangle$ is not already an element of A_{σ} we put $\langle\beta,\gamma\rangle$ into A at stage σ if

$\langle\beta,\gamma\rangle \geq r(g^{\sigma}(m),\sigma)$ for all $m \leq n$ where $g^{\sigma}(n) \simeq \beta$.

End of construction.

The claims of Theorem 1 follow as in case i) from the following Lemma .

Lemma 6 : For every $n \in \omega$ we have

a) $A^{(g(n))} =^* B^{(g(n))}$ and

b) $\neg C \leq_\alpha^{g(n)} A$.

Proof: Induction on n . a) and b) are trivial for $n = 0$ since $W_0 = \phi$ and for all σ $g^\sigma(0) \simeq 0$. Assume for the following that $n > 0$.

a) We get from the properties of B and the induction hypothesis that $A^{(<n)} := \bigcup \{A^{(e)} \mid g(e) < n\}$ is regular. Choose σ_0 such that $\forall \sigma \geq \sigma_0 \ \forall m \leq n \ (g^\sigma(m) \simeq g(m))$ and define

$T_n := \{\sigma > \sigma_0 \mid \sigma$ is a successor stage and an element y is put into $A^{(<n)}$ at stage $\sigma-1$ such that $y \cap A^{(<n)} = y \cap A_\sigma^{(<n)}\}$.

Define $I := \{m \leq n \mid \exists \sigma \in T_n$ ($g(m)$ is inactive at σ)$\}$.

Then there is a stage $\sigma_1 \geq \sigma_0$ such that $\forall \sigma \geq \sigma_1 \ \forall m \in I \ (r(g(m),\sigma) = r(g(m),\sigma_1))$.

Take further any $m \in (n+1)-I$ and assume that $\sup\{r(g(m),\sigma) \mid \sigma \in T_n\} = \alpha$. Then we have $\sup\{l(g(m),\sigma) \mid \sigma \in T_n\} = \mathrm{rcf}\ D$ (by the definition of rcf D) which implies the contradiction $C \leq_\alpha A^{(<n)} \leq_\alpha D$. Thus we have shown that $\sup\{r(g(m),\sigma) \mid m \leq n \wedge \sigma \in T_n\} < \alpha$ which is used for the proof of $A^{(g(n))} =^* B^{(g(n))}$ as usual.

b) $C \leq_\alpha^{g(n)} A$ implies that $\sup\{l(g(n),\sigma) \mid \sigma \in T_n\} = \mathrm{rcf}\ D$ which is absurd according to the preceding.

The proof of Theorem 1 is now finished. We have proved Theorem 1 in order to get the following corollary :

Corollary : Assume that $\sigma 2\mathrm{cf}\,\alpha \geq \sigma 2\mathrm{p}\,\alpha$. Then there exist incomplete high α-r.e. degrees.

Proof of the corollary: The case $\alpha = \sigma 2\mathrm{cf}\,\alpha$ is proved in Shore [20]. For the other admissible α there exist incomplete non-hyperregular α-r.e. sets D if $\sigma 2\mathrm{cf}\,\alpha \geq \sigma 2\mathrm{p}\,\alpha$ according to Shore [19] (see also [11] for another proof of this fact). Apply Theorem 1 to this set D and an α-r.e. set $C \in 0'$.

§2. The degree $0^{3/2}$

For those α where incomplete non-hyperregular α-r.e. degrees exist there exists a distinguished α-degree between $0'$ and $0''$ for which we write $0^{3/2}$. We will show in the following and in [11] that there is a close connection between $0^{3/2}$ and the jump of non-hyperregular α-r.e. degrees.

Lemma 7: Assume α is such that incomplete non-hyperregular α-r.e. degrees exist. Then there is an α-degree $0^{3/2}$ such that

a) $0' <_\alpha 0^{3/2} <_\alpha 0''$

b) $0^{3/2}$ is the greatest $\Delta_2 L_\alpha$ degree (i.e. $0^{3/2}$ contains a $\Delta_2 L_\alpha$ set and $D \leq_\alpha 0^{3/2}$ for every $\Delta_2 L_\alpha$ set D)

c) $0^{3/2}$ is the greatest tame-$\Sigma_2 L_\alpha$ degree (i.e. $0^{3/2}$ contains a set S such that $\{K \in L_\alpha \mid K \subseteq S\}$ is $\Sigma_2 L_\alpha$ and we have $D \leq_\alpha 0^{3/2}$ for every set D with this property)

d) $U_2^{L_\alpha} \leq_{w\alpha} \underset{\sim}{a} \leftrightarrow 0^{3/2} \leq_\alpha \underset{\sim}{a}$ for the set $U_2^{L_\alpha} \in 0''$ and any $\underset{\sim}{a}$.

Remark: If α is Σ_2 admissible then $0'$ is the greatest $\Delta_2 L_\alpha$ degree and $0''$ is the greatest tame Σ_2 L degree. Thus for the α of the Lemma they meet together in the middle, one coming from below, the other coming from above.

<u>Proof:</u> $\mathcal{L} := \langle L_\alpha, C\rangle$ with $C \in 0'$ regular and α-r.e. is inadmissible. A set $S \subseteq L_\alpha$ is $\Delta_2 L_\alpha$ (tame-$\Sigma_2\, L_\alpha$) if and only if S is $\Delta_1 \mathcal{L}$ (tame-$\Sigma_1 \mathcal{L}$). Friedman [3] observed that for inadmissible β a greatest $\Delta_1\, L_\beta$ β-degree exists which lies strictly between 0 and $0'$ and which is an upper bound for the tame-$\Sigma_1\, L_\beta$ degrees. This result can't be generalized to all inadmissible structures $\langle L_\beta, D\rangle$ even if D is regular over L_β: The structure $\mathcal{L} = \langle L_{\aleph_\omega^L}, C\rangle$ with $C \in 0'$ $\aleph_\omega^L$-r.e. and regular is inadmissible (we have $\omega = \sigma 1\mathrm{cf}^{\mathcal{L}} \aleph_\omega^L < \sigma 1\mathrm{p}^{\mathcal{L}} \aleph_\omega^L = \aleph_\omega^L$) but $0'$ is the greatest $\Delta_1 \mathcal{L}$ degree. But Friedman's argument works as well for those inadmissible structures $\mathcal{L} = \langle L_\beta, B\rangle$ where $\sigma 1\mathrm{p}^{\mathcal{L}} \beta < \beta$. According to Lemma 1 we have $\sigma 2\mathrm{p}\,\alpha < \alpha$ for those α where incomplete non-hyperregular α-r.e. degrees exist. Since we have $\sigma 1\mathrm{p}^{\mathcal{L}} \alpha = \sigma 2\mathrm{p}\,\alpha$ for the considered structure $\mathcal{L} = \langle L_\alpha, C\rangle$ there is no problem with the additional assumption in this case.

Take a $\Delta_1 \mathcal{L}$ set $M \subseteq \alpha$ out of the greatest $\Delta_1 \mathcal{L}$ $\mathcal{L}$-degree $\underset{\sim}{r}$ and define $0^{3/2}$ to be the α-degree of the $\Delta_2\, L_\alpha$ set $C \vee M := \{2x \mid x \in C\} \cup \{2x+1 \mid x \in M\}$. Then we have for every set $S \subseteq L_\alpha$ that S is (weakly) $\mathcal{L}$-recursive in M if and only if S is (weakly) α-recursive in $C \vee M$. Therefore we can prove a) and b) for the so defined α-degree $0^{3/2}$ by using the corresponding properties of the $\mathcal{L}$-degree $\underset{\sim}{r}$.

In order to prove c) it remains to show that $\underset{\sim}{r}$ contains a tame-$\Sigma_1 \mathcal{L}$ set. In the case $\sigma 2\mathrm{cf}\,\alpha \geq \sigma 2\mathrm{p}\,\alpha$ this follows from Theorem 1 in [9]. If we have $\sigma 2\mathrm{cf}\,\alpha < \sigma 2\mathrm{p}\,\alpha$ then $\mathcal{L}$ is strongly inadmissible and tame-$\Sigma_1 \mathcal{L}$ sets which are not of degree 0 may or may not exist for these $\mathcal{L}$, depending on the fine structure of $\mathcal{L}$ as it is shown in §2 of [9]. However in our situation where incomplete non-hyperregular α-r.e. degrees exist we have an α-cardinal $\varkappa \geq \sigma 2\mathrm{p}\,\alpha$ such that $\sigma 2\mathrm{cf}^{L_\alpha} \varkappa = \sigma 2\mathrm{cf}\,\alpha$

according to Lemma 1 . Therefore we can apply the construction of Lemma 5 in [9] and get a tame-$\Sigma_1\mathcal{L}$ set of degree $\underset{\sim}{r}$.

Property d) follows from Theorem 2 in [9] .

Remark : The greatest $\Delta_2\ L_\alpha$ and the greatest tame-$\Sigma_2\ L_\alpha$ degree can be determined for the other admissible α as well. The results might be useful for the study of $\Sigma_2\ L_\alpha$ degrees.

For α with $\sigma 2cf\alpha < \sigma 2p\alpha = \alpha$ we have that the greatest $\Delta_2\ L_\alpha$ degree is equal to $0''$ and the greatest tame-$\Sigma_2\ L_\alpha$ degree is equal to $0'$ (thus these two degrees have switched their places compared with Σ_2 admissible α).

For the other α with the property that $0'$ is the only non-hyperregular α-r.e. degree we have that $\sigma 2cf\alpha < \sigma 2p\alpha < \alpha$ and in this case there is a greatest Δ_2 L degree strictly between $0'$ and $0''$ whereas the greatest tame-Σ_2 degree is either equal to the greatest Δ_2 degree (if $\sigma 2cf^{L_\alpha}(\sigma 2p\alpha) = \sigma 2cf\alpha$) or is equal to $0'$ (otherwise) as one can see by using Lemma 1 and arguments of §2 in [9] .

For all α which are not Σ_2 admissible we have that the greatest $\Delta_2\ L_\alpha$ degree $\underset{\sim}{r}$ has the property that $U_2^{L_\alpha} \leq_{w\alpha} \underset{\approx}{a} \leftrightarrow r \leq_\alpha \underset{\approx}{a}$ for every α-degree $\underset{\approx}{a}$.

The following rather technical Lemma will be the heart of the proof of Theorem 2 . It generalizes an observation of Shore (Lemma 3.3 in [18]) which also has important applications in β-recursion theory (see Lemma 3 , §2 in [9]).

Lemma 8 : Consider a structure $\mathcal{L} = \langle L_\beta, B \rangle$ and a limit ordinal $\lambda \leq \beta$ such that $\sigma 1 cf^{\mathcal{L}} \beta < \rho^{\mathcal{L}}_{1,\beta}$ and $\sigma 1 cf^{\mathcal{L}} \lambda < \rho^{\mathcal{L}}_{1,\beta}$ (see §0. for definitions).

If $D \subseteq L_\lambda$ is regular over L_λ and $\{K \in L_\lambda \mid K \subseteq D\}$ is $\Sigma_1 \mathcal{L}$ then $\{K \in L_\lambda \mid K \subseteq L_\lambda - D\}$ is $\Sigma_1 \mathcal{L}$ as well.

Proof: The same trick as in Shore [18] is used. Fix a $\Sigma_1 \mathcal{L}$ definition Ψ of the set $\{K \in L_\lambda \mid K \subseteq D\}$, a cofinal $\Sigma_1 \mathcal{L}$ function $p : \sigma 1 cf^{\mathcal{L}} \lambda \to \lambda$ and a cofinal $\Sigma_1 \mathcal{L}$ function $q : \sigma 1 cf^{\mathcal{L}} \beta \to \beta$. Define a $\Pi_1 \mathcal{L}$ set $M \subseteq \sigma 1 cf^{\mathcal{L}} \lambda \times \sigma 1 cf^{\mathcal{L}} \beta$ by

$$\langle \gamma, \delta \rangle \in M \;:\leftrightarrow\; \forall x \in L_{p(\gamma)} \,(x \in D \to \langle L_{q(\delta)}, L_{q(\delta)} \cap B \rangle \models [\exists K\,(x \in K \wedge \Psi(K))]).$$

Then we have in fact $M \in L_\beta$ and thus get a $\Sigma_1 \mathcal{L}$ definition of $\{K \in L_\lambda \mid K \subseteq L_\lambda - D\}$ as follows :

$$K \in L_\lambda \wedge K \subseteq L_\lambda - D \leftrightarrow \exists \gamma\, \delta (\langle \gamma, \delta \rangle \in M \wedge K \subseteq L_{p(\gamma)} \wedge K \in L_{q(\delta)} \wedge \langle L_{q(\delta)}, L_{q(\delta)} \cap B \rangle \models [\forall x \in K \neg \exists K' (x \in K' \wedge \Psi(K'))]).$$

Theorem 2 : Assume that $\sigma 2 cf \alpha < \sigma 2 p \alpha$ and $\underset{\sim}{a}$ is an incomplete α-r.e. degree. Then we have

a) $\underset{\sim}{a}' = 0'$ if $\underset{\sim}{a}$ is hyperregular (Shore) and
b) $\underset{\sim}{a}' = 0^{3/2}$ if $\underset{\sim}{a}$ is non-hyperregular .

Proof : a) is contained in Shore [20] . It follows immediately from Lemma 8 : Choose $\mathcal{L} := \langle L_\alpha, C \rangle$ with $C \in 0'$ α-r.e. and regular, $\lambda := \alpha$, $D \in \underset{\sim}{a}'$ regular and $\Sigma_1 \langle L_\alpha, A \rangle$ where $A \in \underset{\sim}{a}$ is α-r.e. and regular .

b) Assume that $A \in \underset{\sim}{a}$ is α-r.e., incomplete, regular and non-hyperregular. Then we have $\alpha > \sigma 1 cf^{\langle L_\alpha, A \rangle} \alpha \geq \sigma 1 p^{\langle L_\alpha, A \rangle} \alpha$ according to Shore [18] (this fact follows immediately from Lemma 8). For $\pi := \sigma 1 cf^{\langle L_\alpha, A \rangle} \alpha$ we can find a $\Sigma_1 \langle L_\alpha, A \rangle$ function g

which maps α 1-1 onto τ . Take any set S which is defined by a Σ_2 formula $\exists y \forall z \phi(x,y,z)$ over L_α . Then we have
$x \notin S \leftrightarrow \forall y \exists z \neg\phi(x,y,z) \leftrightarrow \forall y \in \tau \exists \tilde{y} \exists z\ (g(\tilde{y}) = y \wedge$
$\neg\phi(x,\tilde{y},z)\ ) \leftrightarrow \{e\} \times \tau \times \{x\} \subseteq A'$
for some fixed index e . This implies $S \leq_{w\alpha} A'$ (it is this fact which is actually proved in Theorem 2.3. in [20]). We get then $0^{3/2} \leq_\alpha A'$ from Lemma 7 d).

In order to get $0^{3/2} =_\alpha A'$ we show that A' is $\Delta_2\ L_\alpha$ (this implies $A' \leq_\alpha 0^{3/2}$ by Lemma 7 b) . Since A' is obviously $\Sigma_2\ L_\alpha$ it is enough to show that A' is $\Pi_2\ L_\alpha$. We do this by showing that $\tilde{A} := f[A']$ is $\Pi_2\ L_\alpha$ where $f : \alpha \to \sigma 1 p^{\langle L_\alpha, A\rangle}\alpha$ is a 1-1 $\Sigma_1 \langle L_\alpha, A\rangle$ map. We apply Lemma 8 to the structure
$\mathcal{L} := \langle L_\alpha, C\rangle$ with $C \in 0'$ α-r.e. and regular,
$\lambda := \sigma 1 p^{\langle L_\alpha, A\rangle}\alpha$ and $D := \tilde{A}$. The assumptions of the Lemma are all satisfied in this situation :
We have $\sigma 1 cf^{\mathcal{L}}\alpha = \sigma 2 cf\alpha < \sigma 2 p\alpha = \rho^{\mathcal{L}}_{1,\alpha}$, $\sigma 1 cf^{\mathcal{L}}\lambda \leq \sigma 2 cf\alpha < \rho^{\mathcal{L}}_{1,\alpha}$
(take a cofinal $\Sigma_2\ L_\alpha$ function $q : \sigma 2 cf\alpha \to \alpha$; $f \circ q$ is then cofinal in λ because according to Shore [18] we have
$\rho^{\langle L_\alpha, A\rangle}_{1,\alpha} = \sigma 1 p^{\langle L_\alpha, A\rangle}\alpha \leq \sigma 1 cf^{\langle L_\alpha, A\rangle}\alpha$, therefore $f^{-1}[Rg\ f \cap \gamma]$ is bounded for every $\gamma < \lambda$),
$\tilde{A}$ is regular over L_λ (because $\tilde{A}$ is $\Sigma_1 \langle L_\alpha, A\rangle$) and
$\{K \in L_\lambda \mid K \subseteq \tilde{A}\}$ is $\Sigma_1 \langle L_\alpha, A\rangle$ (since $\lambda \leq \sigma 1 cf^{\langle L_\alpha, A\rangle}\alpha$).

Therefore $\{K \in L_\lambda \mid K \subseteq L_\lambda - \tilde{A}\}$ is $\Sigma_1 \mathcal{L}$ according to Lemma 8 which implies that $\tilde{A}$ is $\Pi_2\ L_\alpha$.

---

§3. Summary

Two factors determine the results about the jump of α-r.e. degrees : the relative size of $\sigma 2\mathrm{cf}\,\alpha$ and $\sigma 2\mathrm{p}\,\alpha$ and the existence of incomplete non-hyperregular α-r.e. degrees.

Therefore we distinguish four different types of admissible ordinals α :

(1) $\sigma 2\mathrm{cf}\,\alpha \geq \sigma 2\mathrm{p}\,\alpha$ and there exist no incomplete non-hyperregular α-r.e. degrees
(these are exactly those α which are Σ_2 admissible)

(2) $\sigma 2\mathrm{cf}\,\alpha \geq \sigma 2\mathrm{p}\,\alpha$ and there exist incomplete non-hyperregular α-r.e. degrees
(these are exactly those α which satisfy $\alpha > \sigma 2\mathrm{cf}\,\alpha \geq \sigma 2\mathrm{p}\,\alpha$)

(3) $\sigma 2\mathrm{cf}\,\alpha < \sigma 2\mathrm{p}\,\alpha$ and there exist incomplete non-hyperregular α-r.e. degrees

(4) $\sigma 2\mathrm{cf}\,\alpha < \sigma 2\mathrm{p}\,\alpha$ and there exist no incomplete non-hyperregular α-r.e. degrees .

For the types (2) and (3) there exists the distinguished degree $0^{3/2}$ between $0'$ and $0''$ with the properties that have been described in Lemma 7 .

For α of type (4) we have $\underset{\sim}{a}' = 0'$ for every incomplete α-r.e. degree $\underset{\sim}{a}$ (Shore [20]).

For α of type (3) we have for incomplete α-r.e. degrees $\underset{\sim}{a}$ that $\underset{\sim}{a}' = 0'$ if $\underset{\sim}{a}$ is hyperregular respectively $\underset{\sim}{a}' = 0^{3/2}$ if $\underset{\sim}{a}$ is non-hyperregular according to Theorem 2 .

For α of type (1) and (2) there exist incomplete α-r.e. degrees $\underset{\sim}{a}$ such that $\underset{\sim}{a}' = 0''$ according to §1. (see Shore [20] for type (1)).

In particular we have thus shown the following :

Corollary: Assume that α is admissible. Then there exist high incomplete α-r.e. degrees if and only if $\sigma 2cf\alpha \geq \sigma 2p\alpha$.

We will continue the study of type (1) and (2) in [11]. It turns out that (2) is the most interesting type as far as results about the jump of α-r.e. degrees are concerned.

REFERENCES :

[1] S.B. Cooper, Minimal pairs and high recursively enumerable degrees, J.Symb.Logic 39 (1974), 655-660

[2] K.J. Devlin, Aspects of constructibility, Springer Lecture Note 354 (1973)

[3] S.D. Friedman, β-Recursion Theory, to appear

[4] A.H. Lachlan, A recursively enumerable degree which will not split over all lesser ones, Ann.Math.Logic 9 (1975), 307-365

[5] M. Lerman, Maximal α-r.e. sets, Trans.Am.Math.Soc. 188 (1974), 341-386

[6] M.Lerman and G.E. Sacks, Some minimal pairs of α-recursively enumerable degrees, Ann.Math.Logic 4 (1972), 415-422

[7] W. Maass, On minimal pairs and minimal degrees in higher recursion theory, Arch.math.Logik 18 (1977), 169-186

[8] W. Maass, The uniform regular set theorem in α-recursion theory, to appear in J.Symb.Logic

[9] W. Maass, Inadmissibility, tame RE sets and the admissible collapse, to appear in Ann.Math.Logic

[10] W. Maass, Fine structure theory of the constructible universe in α- and β-recursion theory, to appear in the

Proceedings of "Definability in Set Theory" (Oberwolfach 1977), Springer Lecture Note

[11] W. Maass, On α- and β-recursively enumerable degrees, in preparation

[12] G.E. Sacks, Recursive enumerability and the jump operator, Trans.Am.Math.Soc. 108 (1963), 223-239

[13] G.E. Sacks, Post's problem, admissible ordinals and regularity, Trans.Am.Math.Soc. (1966), 1-23

[14] G.E. Sacks and S.G. Simpson, The α-finite injury method, Ann. Math.Logic 4 (1972), 323-367

[15] J.R. Shoenfield, Degrees of Unsolvability, North Holland/ American Elsevier, Amsterdam/New York (1971)

[16] R.A. Shore, Splitting an α-recursively enumerable set, Trans.Am.Math.Soc. 204 (1975), 65-77

[17] R.A. Shore,Minimal α-degrees,Ann.Math.Logic 4(1972),393-414

[18] R.A. Shore, The recursively enumerable α-degrees are dense, Ann.Math.Logic 9 (1976), 123-155

[19] R.A. Shore, The irregular and non-hyperregular α-r.e. degrees, Israel J.Math. 22, No.1,(1975), 28-41

[20] R.A. Shore, On the jump of an α-recursively enumerable set, Trans. Am.Math.Soc. 217 (1976), 351-363

[21] R.A. Shore, Some more minimal pairs of α-recursively enumerable degrees, to appear

[22] S.G. Simpson, Admissible ordinals and recursion theory, Ph.d.dissertation, M.I.T. (1971)

[23] R.I. Soare, The infinite injury priority method, J.Symb. Logic 41 (1976), 513-529

J.E. Fenstad, R.O. Gandy, G.E. Sacks (Eds.)
GENERALIZED RECURSION THEORY II

Extendability of ZF Models in the von Neuman Hierarchy to Models of KM theory of classes

by

W.Marek (Warszawa) and A.M.Nyberg (Bø i Telemark)

0. Introduction.

The problem of extendability of models of ZF set theory to models of KM theory of classes is as follows:

If $\langle M,E\rangle$ is a model of ZF, when can we find a family $\mathfrak{F} \subseteq \mathfrak{P}(M)$ such that $\langle M \cup \mathfrak{F}, M, E \cup \in \restriction (M \times \mathfrak{F})\rangle$ is a model of KM (Kelley-Morse theory of classes)? The problem is particulary interesting when M is a transitive set and E is the membership relation, $\in$, on this set.

As pointed out in Marek-Mostowski [4] in fact at least two notions of extendability are involved. Namely, apart from the above mentioned notion there is a more restrictive notion, for transitive $\langle M,\in\rangle$ only, namely, one says that $\langle M,\in\rangle$ is β-extendable if and only if there is an $\mathfrak{F}$ such that $\langle M \cup \mathfrak{F}, M, \in\rangle$ is a β-model of KM, i.e. one for which the notion of wellordering, W.O. $(\cdot)$, is absolute.

In Marek-Mostowski [4] it is proved that the notions of extendability and β-extendability do not coincide on the class of denumerable transitive models of set theory:

0.1. Theorem. If α_0 is the least ordinal ρ such that L_ρ is extendable and α_1 is the least ordinal ν such that L_ν is β-extendable, then

(1) $\alpha_0 < \alpha_1$

(2) α_0 is denumerable in $\langle L_{\alpha_1},\in\rangle$

(3) $\langle L_{\alpha_1},\in\rangle \models$ "L_{α_0} is extendable"

(4) Both α_0 and α_1 are denumerable.

On the other hand it is easy to prove:

0.2. Theorem. If M is a transitive model of ZF set theory such that $\mathrm{cf}(On \cap M) > \omega$ then

$\langle M,\in\rangle$ is extendable implies that $\langle M,\in\rangle$ is β-extendable, moreover, for any family $\mathcal{F}$ such that $\langle M\cap\mathcal{F},M,\in\rangle\models KM$ this structure is a β-model (in fact the notion of wellordering is elementary over M).

Thus extendability and β-extendability coincide on a class of transitive models with cofinality of height at least ω_1. This leaves open the case of cofinality character ω.

In this paper we are going to treat that case and we show that the notions of extendability and β-extendability do not coincide for the class of uncountable models of ZF with cofinality of height equal to ω. To be more specific we are going to prove:

<u>0.3. Theorem.</u> If α_0 is the least ρ such that V_ρ is extendable and α_1 the least ρ such that V_ρ is β-extendable, then

(1) $\alpha_0 < \alpha_1$

(2) $\langle V_{\alpha_1},\in\rangle\models$ "$cf\alpha_0 = \omega$"

(3) $\langle V_{\alpha_1},\in\rangle\models$ "$\langle V_{\alpha_0},\in\rangle$ is extendable"

(4) Both $cf(\alpha_0)$ and $cf(\alpha_1)$ are ω.

(Note that (2) and (3) are obvious since we treat V_α's.)

This settles the problem. In order to get this solution we tried to mimick the idea behind the proof of theorem 0.1. (This is in fact Barwise's idea from Barwise [1].) However, it turned out that the completeness result which is necessary could not be found in either Nyberg [8] or Green [3]. Fortunately (see paragraph 2), the second author of this paper had handy a completeness theorem for a version of M-logic (a generalization of ω-logic) for apropriate M's of cofinality ω. This turned out to be exactly the kind of result we needed to carry through the argument.

In order to prove our theorem we show how to pass from the structures for theory of classes to the structures of set theory.

Let ZFC^- be the ZF set theory without the powerset axiom, in which we add the following scheme of choice:

$$(x)_a(Ey)\Phi(x,y) \to (Ef)(Func(f)\wedge Domf = a\wedge(x)_a\Phi(x,fx))$$

Let Inacc(x) be the following formula:

$$Trans(x)\wedge\omega\in x\wedge(z)_x(\mathcal{P}(z)\in x)\wedge(f)(Func(f)\wedge f\subseteq x\to (a)_x(f*(a)\in x))$$

The following two results characterize the notions of extendability and β-extendability:

0.4. Theorem. Let M be a transitive model of ZFC. Then M is extendable if and only if there exists a (possibly non-standard) model $\langle N,E \rangle$ of ZFC^- such that:

(1) $M \in Sp(N)$ (thus Sp(N) is an end-extension of M)

(2) $\langle N,E \rangle \models$ "Inacc(M)"

In the case of β-extendability the characterization is as follows:

0.5. Theorem. Let M be a transitive model of ZFC. Then M is β-extendable if and only if there exists a transitive model N of ZFC^- such that

(1) $M \in N$

(2) $\langle N, \in \rangle \models$ "Inacc (M)"

(There is a characterization of extendability for arbitrary models of ZFC (not only standard) in Marek-Srebrny [5].)

There is an alternative characterization of the notion of β-extendability in terms of Ramified Analysis over M (for the definition see Marek-Mostowski [4] or Moschovakis [7]). Namely let $R.A.^M$ be the family of all ramified analytical subsets of M.

0.6. Theorem. Let M be a transitive model of ZFC. Then M is β-extendable if and only if $\langle R.A.^M \cup M, M, \in \rangle$ satisfies the second order replacement axiom i.e. if and only if

$$(X)_{R.A.^M} \ (\mathrm{Func}(X) \Rightarrow (x)_M \ (X * x \in M))$$

The ramified Analysis over a transitive model of set theory is the least β-extension satisfying the comprehension scheme (though not necessarily replacement) with "nice" reflection properties. This implies that we have the following:

0.7. Theorem. If M is a transitive model of ZFC then the cofinality character of the closure ordinal of Ramified Analysis is ω.

I. Set Theory.

The following lemma will be useful in the sequel.

1.1. Lemma. If V_α is β-extendable then there is a $\xi < \alpha$ such that

(1) $cf(\xi) = \omega$

(2) $\langle V_\alpha^+, \in, V_\alpha\rangle \equiv \langle V_\xi^+, \in, V_\xi\rangle$

(3) $\langle V_\xi, \in\rangle \prec \langle V_\alpha, \in\rangle$

Proof: By theorem 0.5 there is a transitive model N of ZFC^- such that $V_\alpha \in N$ and $\langle N, \in\rangle \models$"Inacc$(V_\alpha)$". Thus $V_\alpha^+ \in N$ (since N is recursively inaccessible). We reason now in N:

We are going to construct (within N) sequences $\{S_n\}_{n\in\omega}$, $\{\eta_n\}_{n\in\omega}$, $\{S_n'\}_{n\in\omega}$ such that:

i) $\langle S_n, \in\rangle \prec \langle V_\alpha^+, \in\rangle$

ii) $S_n' = S_n \cap V_\alpha$

iii) $S_n' \subseteq V_{\eta_n}$

iv) $V_{\eta_n} \subseteq S_{n+1}$

v) $\{S_n'\}_{n\in\omega} \in V_\alpha$

Indeed, pick $S_0 \in N$ denumerable such that $\langle S_0, \in\rangle \prec \langle V_\alpha^+, \in\rangle$, set $S_0' = S_0 \cap V_\alpha$ and η_0 = the least ρ such that $S_0' \subseteq V_\rho$.

Given S_n, S_n', η_n, pick $S_{n+1} \supset V_{\eta_n}$ such that $\langle S_{n+1}, \in\rangle \prec \langle V_\alpha^+, \in\rangle$, S_{n+1} of least possible rank and power, set $S_{n+1} = S_{n+1} \cap V_\alpha$ and η_{n+1} = the least ρ such that $S_{n+1}' \subseteq V_\rho$. Since V_α is inaccessible in N we have that $\eta_{n+1} < \alpha$.

The use of the principle of dependent choices can be eliminated and we get a sequence $\{S_n'\}_{n\in\omega} \in N$. By the inaccessibility of V_α in N, $\{S_n'\}_{n\in\omega} \in V_\alpha$ and thus $\bigcup S_n' \in V_\alpha$. Since $\bigcup_{n\in\omega} V_{\eta_n} = \bigcup_{n\in\omega} S_n'$, the ordinal $\xi = \bigcup_{n\in\omega} \eta_n$ has the properties:

(a) $\xi < \alpha$

(b) $\langle V_\xi, \in\rangle \prec \langle V_\alpha, \in\rangle$.

(a) is obvious and (b) follows from the fact that for all n $\langle S_n', \in\rangle \prec \langle V_\alpha, \in\rangle$.

Set $S = \bigcup_{n\in\omega} S_n$. Clearly $\langle S,\in\rangle \prec \langle V_\alpha^+,\in\rangle$
Since $S_n' = S_n \cap V_\alpha$ we have that $S \cap V_\alpha = \bigcup_{n\in\omega} S_n' = V_\xi$.
The set S is not transitive but $S \cap V_\alpha$ is.

The contraction function π for S is constant on $S \cap V_\alpha = V_\xi$ and so $\pi(V_\alpha) = V_\xi$. Thus $\langle \pi * S,\in, V_\xi\rangle \equiv \langle V_\alpha^+,\in, V_\alpha\rangle$ and so $\pi(S) = V_\xi^+$. By our construction $\xi = \bigcup_n \eta_n$ and $\{\eta_n\}_{n\in\omega}$ is increasing. Thus $cf(\xi) = \omega$ which completes the proof of the lemma. ⊠

We recall the following well known fact about models of ZF in the von Neuman hierarcy.

<u>1.2. Theorem.</u> If $\langle V_\alpha,\in\rangle \models$ ZF and $cf(\alpha) > \omega$, then there is $\gamma<\alpha$ such that $\langle V\gamma,\in\rangle \prec \langle V_\alpha,\in\rangle$

Proof: Using any fixed enumeration $\{\phi_n\}_{n\in\omega}$ of formulae of the language of set theory we find - by reflection principle - an increasing sequence $\{\xi_n\}_{n\in\omega}$ such that $\langle V_{\xi_m},\in\rangle$ reflects the first m formulae of $\{\phi_n\}_{n\in\omega}$. But then, setting $\gamma = \bigcup_{n\in\omega}\xi_n$, we have $\langle V\gamma,\in\rangle \prec \langle V_\alpha,\in\rangle$. Since $cf(\gamma) = \omega$ and $cf(\alpha)>\omega$ we are done. ⊠

Corollary: The least α such that $\langle V_\alpha,\in\rangle \models$ ZF has cofinality character ω.

We generalize the above theorem and corollary as follows:

<u>1.3. Theorem.</u> If $cf(\alpha)>\omega$ and V_α is β-extendable then there is $\gamma<\alpha$ such that

i) $\langle V\gamma,\in\rangle \prec \langle V_\alpha,\in\rangle$

ii) V_γ is β-extendable

iii) $cf(\gamma) = \omega$

Proof: Since V_α is β-extendable, the structure $\langle R.A.^{V_\alpha}, V_\alpha,\in\rangle$ is the least β-extension of V_α. Now, pick $\{\nu_n\}_{n\in\omega}$ such that $\langle R.A.^{V_\alpha}_{\nu_n}, V_\alpha,\in\rangle \prec_n' \langle R.A.^{V_\alpha}, V_\alpha,\in\rangle$. Using coding within $R.A.^{V_\alpha}$ and a cross-breading of reasoning from the above lemma and theorem, we construct sequences $\{S_n\}_{n\in\omega}$, $\{S_n'\}_{n\in\omega}$, $\{\eta_n\}_{n\in\omega}$ such that:

i) $\langle S_n, V_\alpha,\in\rangle \prec \langle R.A.^{V_\alpha}_{\nu_n}, V_\alpha,\in\rangle$

ii) $S_n' = S_n \cap V_\alpha$

iii) η_n = the least ξ such that $S_n' \subseteq V_\xi$

iv) $V_{\eta n} \subseteq S_{n+1}$

Clearly $< \bigcup_{n\in\omega} S_n, V_\alpha, \in > \prec < R.A.^{V_\alpha}, V_\alpha, \in >$, $\bigcup_{n\in\omega} V_{\eta_n} = \bigcup_{n\in\omega} S'_n$.
Putting $\gamma = \bigcup_{n\in\omega} \eta_n$ we have, using the fact that for every n $<S'_n, \in> \prec <V_\alpha, \in>$, that $<V_\gamma, \in> \prec <V_\alpha, \in>$.
Set $S = \bigcup_{n\in\omega} S_n$. Then $<S, V_\alpha, \in> \prec <R.A.^{V_\alpha}, V_\alpha, \in>$ and $S \cap V_\alpha = V_\gamma$.
If we look closely at the contraction function on S, we see that: on $V_\gamma = S \cap V_\alpha$ it is the identity, whereas on elements of $S - V_\alpha$ it is just the restriction of those sets to V_γ !! Thus the structure we obtain must be a β-model, since if $<\pi * S, V_\gamma, \in> \models WO[X]$ then there must be a X_1 such that $X = X_1 \upharpoonright V_\gamma$ and $<S, V_\alpha, \in> \models WO[X_1]$.
But then $<R.A.^{V_\alpha}, V_\alpha, \in> \models WO[X_1]$ and so X_1 is really a wellordering. So $X = X_1 \upharpoonright V_\gamma$ is also a wellordering. This completes the proof of the theorem. ⊠

1.4. Corollary. The least α such that V_α is β-extendable has cofinality character ω.

Proof: Immediate from the theorem.

II. Logic and Definability.

The aim of this section is to provide enough of infinitary logic to be able to mimick the argument of Marek-Mostowski [4] even in this uncountable setting. Since we are interested in a particular application we will try to restraint our selves from the temptation to give the most general results possible. The more general setting including details and proofs not provided here, will be found in Nyberg [9].

The main result of this section reads as follows:

2.1. Theorem. (Consistency lemma) There is a first order formula $con_{\overline{M}}(T)$ in the language $L = \{\varepsilon, \overline{M}, T\}$ (ε binary, $\overline{M}$ and T unary predicate symbols) such that whenever $M = V_\alpha$, where α is a limit ordinal of cofinality ω, and whenever T is a first order definable set of sentences over the structure $\mathfrak{M} = < M, \in >$. Then

$< \mathbb{HYP}_{\mathfrak{M}}, T > \models con_{\overline{M}}(T)$ if and only if T has an $\mathfrak{M}$-model.

We can if needed allow parameters from M to occur in the formulas in T as well as in the first order definition of (the codes of) the sentences in T. The terminology used follows essentially that of Barwise [2]. Note that in this case $\mathbb{HYP}_{\mathfrak{M}}$ is really the structure $\langle V_\alpha^+, V_\alpha, \in\rangle$. In fact we make the convention that any $\mathfrak{M}$-model $\langle A,M,E\rangle$ is such that $\langle A,E\rangle$ is an end-extension of $\langle M,\in\rangle$.

It is worth pointing out that the crucial point of this theorem is that the formula $\mathrm{con}_{\overline{M}}(T)$ is independent of the particular structure $\mathfrak{M}$. It is to obtain this we have to use $\mathfrak{M}$-logic and the completeness theorem for $\mathfrak{M}$-logic, obtained in Nyberg [9]. In Nyberg [8] a kind of completeness theorem for the infinitary language $\mathcal{L}_{\mathbb{HYP}_{\mathfrak{M}}}$ was obtained. To be precise it was shown that $\mathbb{HYP}_{\mathfrak{M}}$ is "uniformly Σ_1 complete." This actually yields a first order formula say con(T), such that on $\mathbb{HYP}_{\mathfrak{M}}$, this formula is true for any Σ_1 theory $T \subseteq \mathcal{L}_{\mathbb{HYP}_{\mathfrak{M}}}$ if and only if T has a model. It turns out, however, that this formula might depend on the particular structure $\mathfrak{M}$ in question, so there is no direct way of applying that result in this context. Using the results of Green [3] it is likely that one could obtain a model independent formula con(T) with the desired properties. The problem is that in order to obtain the completeness theorem there, a ω-cofinal sequence in α would have to be a member of $\mathbb{HYP}_{\mathfrak{M}}$. This will certainly not be the case in our application. (It won't even appear in the first Σ_∞^0-admissible set over $\mathfrak{M}$.) We hope that the reader now will understand why we are introducing $\mathfrak{M}$-logic into this picture.

Since the paper of Nyberg [9] is not yet readily available to everyone we will give an outline of the results to be used here. We will not involve ourselves in the most general setting so let us just stick to the case where $\mathfrak{M} = \langle M, \in\rangle$ and $M=V_\alpha$, for some limit ordinal $\alpha>\omega$ of cofinality ω.

By $\mathfrak{M}$-logic in this setting we follow Barwise [2] except for the end-extension condition, mentioned earlier in this chapter, and some new axioms in our definition of the syntactical $\mathfrak{M}$-consequense relation $\vdash_M$.

That is, $\vdash_M$ is defined by means of the M-rule modus ponens, and generalization as in Barwise [2]. We take all the axioms there as axioms for $\mathfrak{M}$-logic but we add a few more due to $\mathfrak{M}$ being uncountable.

The first new axiom has nothing to do with uncountability, however, it just takes care of the end-extension condition:

i) $\forall x \forall y(\overline{M}(y) \wedge x \; \varepsilon y \rightarrow \overline{M}(x))$

The next two schemes, although they look like reflection principles, actually work as some kind of distributive laws in the proofs of the completeness theorem for $\mathfrak{M}$-logic:

For each $m \in M$ and each formula $\phi \in L_{\omega\omega}$ the following are axioms for $\mathfrak{M}$-logic.

ii) $(Ez)(z \in \overline{\text{power}(m)} \wedge (y)(y \; \varepsilon z \leftrightarrow y \; \varepsilon \; \overline{m} \wedge \phi))$

iii) $(Ez)\Big[(\overline{M}(z) \wedge \text{"Func}(z)\text{"} \wedge \text{"}z: \overline{\omega} \rightarrow \overline{\text{power}(m)}\text{"} \wedge \text{"}\overline{m} = \bigcup\{z(i) \mid i \; \varepsilon \; \overline{\omega}\}\text{"}$
$\wedge (k)_{\overline{\omega}} \; ((x)_{z(k)} (Ey)(\overline{M}(y) \wedge \phi(x,y)) \rightarrow (Ev)(\overline{M}(v)$
$\wedge (x)_{z(k)} (Ey)_v \phi(x,y)))\Big]$

It is of course assumed that z does not occur free in ϕ.

The language L is here the language of $\mathfrak{M}$ together with a unary predicate symbol $\overline{M}$ and constant symbols $\overline{m}$ for each $m \in M$. The abbreviations like "Func(z)" should be taken as their usual definition in terms of ε.

In Nyberg [9] there is a proof of the completeness theorem for $\mathfrak{M}$-logic:

<u>2.2. Completeness theorem for $\mathfrak{M}$-logic.</u> T is consistent in $\mathfrak{M}$-logic (i.e. for no sentence ϕ, $T \vdash_M \phi \wedge \neg\phi$) if and only if T has an $\mathfrak{M}$-model.

Accepting this result we can now turn to the proof of the main result of this paragraph:

<u>Proof of theorem 2.1.</u> The consequence relation of $\mathfrak{M}$-logic is inductively defined. In fact with a coding of formulas of $L_{\omega\omega}$ as members of M there is a first order formula $\Psi(x,S,T)$, with S and T occuring positively, such that on $\mathfrak{M}$, $x \in I_\Psi(T)$ if and only if x is a code of some formula χ such that $T \vdash_M \chi$.

There is then a Σ_1 formula $\phi(x,T)$ such that whenever T is Σ_1 definable on $\mathbb{HYP}_{\mathfrak{M}}$ we have that $\langle \mathbb{HYP}_{\mathfrak{M}}, T \rangle \models \phi(x,T)$ if and only if $x \in I_\Psi(T)$. At this point it is important to notice that we can make the choice of ϕ dependent of just Ψ and not the particular structure $\mathfrak{M}$. For more details the reader can consult Barwise [2] or Nyberg [8].

To conclude the proof, let $\mathrm{con}_{\overline{M}}(T)$ be the formula $\neg\phi(\ulcorner\chi\wedge\neg\chi\urcorner, T)$ where $\ulcorner\chi\wedge\neg\chi\urcorner(\in V_\omega)$ is a code and χ is some sentence in $L_{\omega\omega}$. If T is a first order definable set of sentences over $\mathfrak{M}$ we then have:

$$\mathbb{HYP}_{\mathfrak{M}} \models \mathrm{con}_{\overline{M}}(T)$$

By definition of $\mathrm{con}_{\overline{M}}$. $\Updownarrow$

$$\mathbb{HYP}_{\mathfrak{M}} \models \neg\phi(\ulcorner\chi\wedge\neg\chi\urcorner, T)$$

Since T will be Σ_1 (in fact Δ_0) over $\mathbb{HYP}_{\mathfrak{M}}$. $\Updownarrow$

$$\ulcorner\chi\wedge\neg\chi\urcorner \notin I_\Psi(T)$$

By definition of Ψ $\Updownarrow$

$$T \not\models_{\mathfrak{m}} \chi\wedge\neg\chi$$

By the completeness theorem 2.2.

T has an $\mathfrak{M}$-model.

But this is exactly the claim of theorem 2.1. ⊠

III. The Result.

We will now prove the theorem which is our purpose for this paper, namely:

3.1. Theorem. Let α_0 be the least ρ such that $\langle V_\rho, \in\rangle \models ZFC$ and V_ρ is extendable and let α_1 be the least ρ such that $\langle V_\rho, \in\rangle \models ZFC$ and V_ρ is β-extendable, then:

(a) $\alpha_0 < \alpha_1$

(b) $cf(\alpha_0) = cf(\alpha_1) = \omega$

(c) V_{α_0} is extendable within V_{α_1}.

Proof: It is enough to show (a). (b) follows from the proof of (a) and considerations of I. (c) is obvious by (a) since we are dealing with the von Neuman hierarchy and any extension of V_{α_0} belongs to V_{α_0+2} and α_1 is limit. In order to prove (a) we proceed as follows:

By the Corollary 1.4 α_1 has cofinality character ω. Take as $\mathfrak{M}$ (in paragraph II) the structure $\langle V_{\alpha_1}, \in\rangle$. Consider $\mathfrak{M}$-logic and the structure $\mathbb{HYP}_{\mathfrak{M}}$ which - by our remarks - we can identify with $\langle V^+_{\alpha_1}, V_{\alpha_1}, \in\rangle$. In the language L of this structure consider the following theory T: ZFC^- + "M is inaccessible." This theory has an $\mathfrak{M}$-model (by theorems 0.3 or 0.4). Thus by the Consistency theorem 2.1 $\langle V_\alpha^+, V_\alpha, \in\rangle \models con_{\overline{M}}(T)$. The formula $con_{\overline{M}}(T)$ is a first order formula (it is actually Π_1). (We tacitly assume that T is replaced by its definition in $con_{\overline{M}}(T)$). Now, using the β-extendability of M, we apply lemma 1.1 getting the structure $\langle V_\xi^+, V_\xi, \in\rangle$ such that $\langle V_\xi, \in\rangle \prec \langle V_{\alpha_1}, \in\rangle$, $cf(\xi) = \omega$ and $\langle V_\xi^+, V_\xi, \in\rangle \equiv \langle V_\alpha^+, V_\alpha, \in\rangle$. Thus in particular $\langle V_\xi^+, V_\xi, \in\rangle \models con_{\overline{M}}(T)$. Again, using the consistency theorem this time taking $\langle V_\xi, \in\rangle$ as $\mathfrak{M}$, we have a $\mathfrak{M}$-model of ZFC^- + "M is inaccessible." This model will be an end-extension of $\langle V_\xi \cup \{V_\xi\}, \in\rangle$ thus of $\langle V_\xi^+, \in\rangle$. Hence V_ξ is extendable and so $\alpha_0 < \alpha_1$. ⊠

One can ask if V_ξ as obtained in the previous proof, actually is V_{ξ_0}. The answere is that it is definitly not V_{ξ_0}. Indeed, take $T = Th(V_{\alpha_1})$. $T \in V_{\omega+1}$ being a set of natural numbers and so we have:

$\langle V_{\alpha_1}, \in\rangle \models$ "There is a model N of T of the form $\langle V_\nu \in\rangle$ such that N is extendable."

The same must hold in $\langle V_\xi, \in\rangle$ and so, using the absoluteness of the V-function with respect to V_ξ we find that there must be a $\nu < \xi$ such that V_ν is extendable and $\langle V_\nu, \in\rangle \equiv \langle V_{\alpha_1}, \in\rangle$. (We could even find an extension of V_ν which is elementary equivalent to a β-extension of V_{α_1}.)

Now $\langle V_{\alpha_1}, \in\rangle \models$ "There exists extendable models of the form V_ξ"

The same must then hold in $\langle V_\nu, \in\rangle$. Thus $\alpha_0 < \nu < \xi < \alpha_1$.

Another corollary to the proof can be obtained by slightly modifying lemma 1.1. We can add the clause that V_ξ contains a given prescribed element $\xi < \alpha_1$. Thus we get the following:

3.2. Theorem. Let α_1 be as in theorem 3.1. There exists a cofinal elementary tower of extendable, non β-extendable models of the form V_ξ.

Also since the length of the tower of elementary subsystems (of form V_ξ) of V_{α_1} is α_1 and the tower is continuous, we have elements of the tower such that the cofinality character of them must be $\geq \omega_1$. Thus there are ξ's less than α_1 such that $\langle V_\xi, \in\rangle \prec \langle V_{\alpha_1}, \in\rangle$ and V_ξ is not extendable. (Recall theorem 0.2.)

There are versions of the results of part I for models of ZF which are extendable not only to models of KM but also to models of KM_n (higher order class theories). Analogous results about the difference between extendability and β-extendability of V_α's to models of such theorems can be obtained.

This paper was initiated in Warsaw due to the Oslo-Warsaw University Exchange program and completed at this Conference.

We express our thanks to J.E.Fenstad for making this possible.

Institute of Mathematics
University
00-901 WARSZAWA, Poland

and

Telemark Regional College
3800 Bø i Telemark
NORWAY

References:

[1] Barwise, J. Infinitary Methods in the Model Theory of Set Theory; in: Logic Colloquium 69, pp 53-66, Amsterdam, North-Holland 1972.

[2] Barwise, J. Admissible Sets and Structures; Berlin, Heidelberg, New York, Springer, 1975.

[3] Green, J. Σ_1-compactness for next admissible sets; JSL, vol 39, 1974, pp 105-116.

[4] Marek, W., Mostovski, A. On extendability of models of ZF set theory to the models of Kelley-Morse theory of classes; in: Logic Conference Kiel 1974, pp 460-542. Springer Lecture Notes 499, Berlin, Heidelberg New-York, Springer 1975.

[5] Marek, W., Srebrny, M. Urelements and extendability; in: Set Theory and Hierarchy Theory, A Memorial Tribute to Andrzej Mostovski, pp 203-219. Springer Lecture Notes 537, Berlin Heidelberg New York, Springer 1976.

[6] Marek, W., Zbierski, P. On higher order set theories; Bull. Acad. Pol. Sci., Ser. Sci. Math. Astron. Phys. XXI (1973) pp 97-101.

[7] Moschovakis, Y.N. Elementary Induction on Abstract Structures; Amsterdam, North-Holland 1974.

[8] Nyberg, A.M. Uniform inductive definability and infinitary languages. JSL, vol 41, 1976 pp 109-120.

[9] Nyberg, A.M. Applications of model theory to recursion theory on structures of strong cofinality ω, Preprint series, Institute of Mathematics, University of Oslo, no.17. 1974 (To appear in revised version.)

J.E. Fenstad, R.O. Gandy, G.E. Sacks (Eds.)
GENERALIZED RECURSION THEORY II

ON THE ROLE OF THE SUCCESSOR FUNCTION IN RECURSION THEORY

by

Johan Moldestad
University of Oslo

The successor function $s(x) = x+1$, where x is a variable ranging through the natural numbers, is a basic function in ordinary recursion theory. All the recursive functions can be generated from s and some other simple functions (projections and constant functions) by a few schemes. It is also chosen as a basic function in several settings of abstract recursion theory. In [3] Moschovakis defines the functions on an abstract domain by schemes, and s is needed already to formulate the scheme of primitive recursion. For the same reason it is needed in the axiomatic approach in [4]. [1] is also an axiomatic approach to recursion theory, and there Fenstad has proved some simple results from the axioms alone without using s (in chapter 1, section 2), but s is brought in in the proofs of more advanced results, for instance the simple representation theorem 1.6.3: Let Θ be a precomputation theory. Then there exists a Θ-computable function f such that Θ is equivalent to PR[f]. Here PR[f] is a computation theory which is inductively defined by an operator Γ. The n-th level of Γ, denoted by Γ^n, is defined from Γ^{n-1}. The following result is an important part of the proof: The function p defined by $p(n,\langle a,\sigma\rangle) \simeq z$ iff $(a,\sigma,z) \in \Gamma^n$ is Θ-computable. This is proved by an application of the recursion theorem, since $\lambda u.\ p(n,u)$ can be defined from $\lambda u.\ p(n-1,u)$ in the same way as Γ^n is defined from Γ^{n-1}. So it is needed to pass from n to $n-1$. This shift operator can easily be constructed from s

(1.4.2 in [1]).

What happens if we leave out s? Below we give an example of a "recursion theory" where s is not recursive (quotation marks because one can argue whether such a class of functions should be called a recursion theory). It has other properties than recursion theories with s. One of the main results in Platek's thesis [5] (First Recursion Theorem 5.3.2) fails in the example. In Platek's proof s is assumed to be recursive. In [2] I gave a proof of Platek's theorem Omitting this assumption (theorem 29). There is a mistake in that proof, and by the example below the "theorem" is false.

Let $HC(\sigma)$ denote the hereditarily consistent objects of type σ over the natural numbers N. Let $HC = \cup\{HC(\sigma): \sigma$ is a type symbol$\}$ $HC^{\ell} = \cup\{HC(\sigma): \sigma$ is a type symbol of level $\leq \ell\}$. See [5] or §12 in [2] for notations and definitions. $HC(0) = N$. $HC(1) =$ the set of partial functions $f:N \to N$. If $B \subseteq HC$ let $R_{\omega}(B)$ be the least set $X \subseteq HC$ such that $B \subseteq X$, X contains the fixed point operators $FP^{(\sigma\to\sigma)\to\sigma}$ for all types $\sigma \neq 0$, some combinatorial functions, and DC (definition by cases), and X is closed under composition, i.e. if $f,g \in X$ and the types of f,g are $\sigma \to \tau$, σ respectively, then $f(g)$ (which is of type τ) is in X. f is recursive from B if $f \in R_{\omega}(B)$. We let $R_{\ell}(B)$ be defined as $R_{\omega}(B)$ with the exception that $FP^{(\sigma\to\sigma)\to\sigma}$ is assumed to be in X only when the level of σ is at most ℓ. Let $R_{\omega}(B)^{\ell} = R_{\omega}(B) \cap HC^{\ell}$. The function DC mentioned above is of type $0 \to 0 \to 1 \to 1 \to 0 \to 0$ and it is defined by:

$$DC(x,y,f,g,v) = \begin{cases} f(v) & \text{if } x = y \\ g(v) & \text{if } x \neq y . \end{cases}$$

Platek's First Recursion Theorem (5.3.2 in [5]):

Suppose $B \subseteq HC^{\ell+2}$, and B contains the number 0 and the successor function s. Then $R_\omega(B)^{\ell+3} = R_{\ell+1}(B)^{\ell+3}$.

(In [5] B is supposed to contain pairing and depairing functions for $HC(0)$. When $HC(0) = N$ there are pairing and depairing functions which are recursive in $0,s$, hence this assumption can be dropped.)

Let $\ell = 0$. By Platek's theorem: If $0,s \in B$, $B \subseteq HC^2$, then $R_\omega(B)^3 = R_1(B)^3$. In the example below $B \subseteq HC^2$, $s \notin B$, and $R_2(B)^2 \neq R_1(B)^2$. So the conclusion in Platek's theorem fails for that particular B.

The example:

Let f_i, g_i $(i \in N)$, S,T (of types $1,1$, $1 \to 1$, $1 \to (1 \to 1)$ respectively) be defined by:

$$f_i(x) = \begin{cases} 0 & \text{if } x = 0,2,4,\ldots,2i \\ x & \text{otherwise} \end{cases}$$

$$g_i(x) = \begin{cases} 0 & \text{if } x = 1,3,5,\ldots,2i+1 \\ x & \text{otherwise} \end{cases}$$

$$S(f) = \begin{cases} f_i & \text{if } f = f_{i+1} \\ \uparrow & \text{otherwise} \end{cases}$$

$$T(f,g) = \begin{cases} g_0 & \text{if } f = f_0 \\ g_{i+1} & \text{if } f = f_{i+1} \text{ and } g = g_i \\ \uparrow & \text{otherwise} \end{cases}$$

where $\uparrow$ denotes the partial function of type 1 which is totally undefined. Let

$$B = \{0,1,2,\ldots,f_0,f_1,f_2,\ldots,g_0,g_1,g_2,\ldots,S,T\}.$$

Then $B \subseteq HC^2$.

Theorem: $R_1(B)^2 \neq R_2(B)^2$.

Proof: Let $U \in HC(1\to 1)$ be defined by:

$$U(f) = \begin{cases} g_i & \text{if } f = f_i \\ \uparrow & \text{otherwise} \end{cases}$$

Then $U = FP^{(\sigma\to\sigma)\to\sigma}(\lambda U^\sigma f^1 . T(f,U(Sf)))$, where $\sigma = 1 \to 1$. The level of σ is 2, hence $U \in R_2(B)^2$. To prove that $U \notin R_1(B)^2$ it suffices to prove that if $H \in R_1(B) \cap HC(1\to 1)$ then there is an $H' \in R_0(B) \cap HC(1\to 1)$ such that $Hf_i = H'f_i$ for $i = 0,1,2,\ldots$. Then the set $\{Hf_i : i \in N\}$ $(= \{H'f_i : i \in N\})$ will contain at most finitely many g_i's because $H' \in R_0(B)$. Hence $U \neq H$.

Proposition: If $H \in R_1(B) \cap HC(1\to 1)$, then there is an $H' \in R_0(B) \cap HC(1\to 1)$ such that $Hf_i = H'f_i$ for all $i \in N$.

Before going into the details of a complete proof we present a characteristic example of how to get H', given H. Let

$HZ = \lambda x.\ FP[\lambda p\ uv.\ t_1](4,x)$, where

$t_1(u,v) = DC(u,v,\lambda w.\ 0,\lambda w.\ t_2,0)$,

$t_2(u,v) = DC(f_1(u),f_3(u),\lambda w'.\ p(f_2(v),f_4(v)),\lambda w'.\ t_3,0)$,

$t_3(v) \quad = DC(p(q(1,1,0,v),0),0,\ \lambda w''.\ 0,\ \lambda w''.\ 1,0)$,

$q = FP[\lambda q\ abcd.\ DC(a,0,Z,\lambda w'''.\ t_4,d)]$,

$t_4(a,b,c,d) = DC(p(1,8),0,\lambda r.\ S(\lambda s.\ q(b,c,a,s))(d),\lambda r.\ 0,0)$

$x,u,v,w,w',w'',w''',a,b,c,d,r$ are variables of type 0 , Z,p,q are variables of type 1 , $0\to 1$, $0\to(0\to(0\to 1))$ respectively.

Let $Z = f_Q$ for some number Q. Let $p = FP[\lambda p\ uv.\ t_1]$. To see what $p(u,v)$ and $q(a,b,c,d)$ are, regard the following diagrams p(u,v):

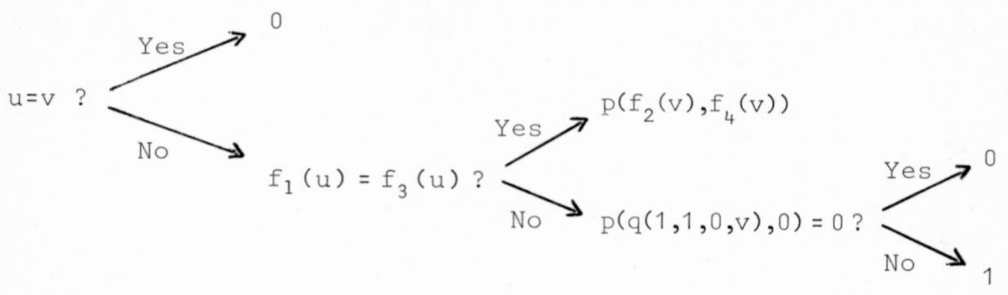

q(a,b,c,d):

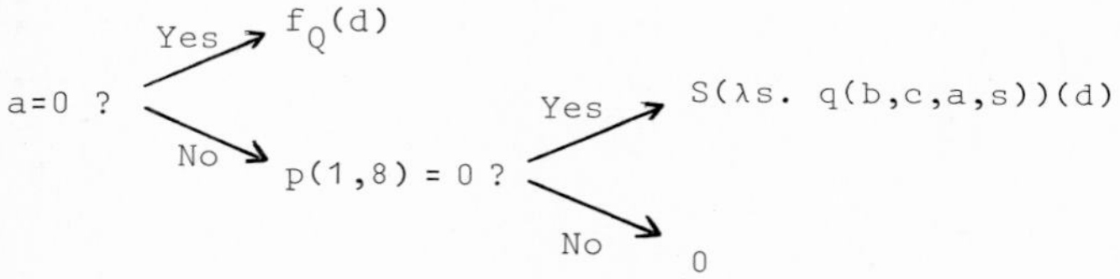

The values of p(u,v) and q(a,b,c,d) depend on the answers to questions, and can be found to the right of the arrows when no more questions are asked. p is inductively defined as follows: p^0 is totally undefined. $p^{\alpha+1} = \lambda uv.\ t_1$ (with p^α substituted for p in t_1). For each p^α q is inductively defined as follows: $q^{\alpha,0}$ is totally undefined, $q^{\alpha,\beta+1} = \lambda abcd.\ DC(a,0,f_Q,\lambda w'''.\ t_4,d)$ (with p^α, $q^{\alpha,\beta}$ substituted for p,q respectively in t_4). Let $q^\alpha = q^{\alpha,\infty}$.

By the diagram:

$$
\begin{aligned}
p^{\alpha+1}(u,v) &= 0 \quad \text{if } u = v \\
&= p^\alpha(f_2(v),f_4(v)) \quad \text{if } u \neq v \ \& \ f_1(u) = f_3(u) \\
&= 0 \quad \text{if } u \neq v,\ f_1(u) \neq f_3(u),\ p^\alpha(q^\alpha(1,1,0,v),0) = 0 \\
&= 1 \quad \text{if } u \neq v,\ f_1(u) \neq f_3(u),\ p^\alpha(q^\alpha(1,1,0,v),0) \neq 0
\end{aligned}
$$

$$q^{\alpha,\beta+1}(a,b,c,d) = f_Q(d) \quad \text{if} \quad a = 0$$
$$= S(\lambda s.\ q^{\alpha,\beta}(b,c,a,s))(d) \quad \text{if} \quad a \neq 0,\ p^{\alpha}(1,8) = 0$$
$$= 0 \quad \text{if} \quad a \neq 0,\ p^{\alpha}(1,8) \neq 0$$

By some calculations one can prove:

$q^{0,1}(a,b,c,d) = f_Q(d)$ if $a = 0$, $q^0 = q^{0,1}$. $p^1(u,v) = 0$ if $u = v$, $q^1 = q^0$ because $p^1(1,8)$ is undefined.
$p^2(u,v) = 0$ if $u \neq v$, $f_1(u) = f_3(u)$, $f_2(v) = f_4(v)$. $q^2 = q^1$ because $p^2(1,8)$ is undefined.
$p^3(u,v) = 0$ if $u \neq v$, $f_1(u) = f_3(u)$, $f_2(v) \neq f_4(v)$, $f_2(v) = f_3(v)$. $p^3(1,8) = 0$. If $Q = 0$ then $q^3 = q^2$. If $Q = 1$ then $q^{3,1} = q^2$, $q^{3,2}(a,b,c,d) = f_0(u)$ if $a \neq 0$, $b = 0$. $q^3 = q^{3,2}$. If $Q > 1$ then $q^{3,2}(a,b,c,d) = f_{Q-1}(d)$ if $a \neq 0$, $b = 0$. $q^{3,3}(a,b,c,d) = f_{Q-2}(d)$ if $a \neq 0$, $b \neq 0$, $c = 0$. $q^3 = q^{3,3}$. If $Q = 0$ or 1 then $p^4 = p^3$. If $Q = 2$ or 3 then $p^4(u,v) = 0$ if $u \neq v$, $f_1(u) \neq f_3(u)$, $f_1(v) = f_3(v)$. If $Q = 4$ then $p^4(u,v) = 0$ if $u \neq v$, $f_1(u) \neq f_3(u)$, $f_2(v) = f_3(v)$. If $Q > 4$ then $p^4(u,v) = 0$ if $u \neq v$, $f_1(u) \neq f_3(u)$. $q^4 = q^3$ for any Q. $p^5(u,v) = 0$ for all u,v if $q > 4$. $q^5 = q^4$ for any Q. $p^{\infty} = p^5$ when $Q > 4$, $= p^4$ when $2 \leq Q \leq 4$, $= p^3$ when $Q < 2$.

Define H' as follows: Substitute $FP[\lambda q\ abcd.\ DC(a,0,Z,\lambda w'''.\ t_4,d)]$ by the term q_3, where $q_0 = \lambda abcd.\ \uparrow$ ($\uparrow$ is the symbol for undefined), $q_{n+1} = \lambda abcd.\ DC(a,0,Z,\lambda w'''.\ t_4,d)$ with q_n substituted for q in t_4. Three iterations are sufficient because $q^{\alpha} = q^{\alpha,3}$ for any α. Then replace $FP[\lambda puv.\ t_1]$ by p_5, where the term p_5 is defined similarly.

Each p^{α}, $q^{\alpha,\beta}$ is defined by cases involving $f_1, f_2, f_3, f_4, f_Q, f_{Q-1}, f_{Q-2}$ and the numbers $0,1$. Only finitely many cases can be defined by these functions, as compositions of them can be reduced, i.e., $f_m(f_n(x)) = f_k(x)$, where $k = \max(m,n)$. This is why the inductions stop after finitely many iterations, and the number of iterations is independent of Q and x. If a skift operator for numbers for instance $s(n) = n+1$, things would have been different, as $s(s(n)) \neq s(n)$. In that case there might be no uniform upper bound to the number of iterations.

Proof of the proposition: Suppose $H \in R_1(B) \cap HC(1\to1)$. H can be expressed as a composition of combinators, DC, fixed point operators $FP^{(\sigma\to\sigma)\to\sigma}$ where the level of σ is 1, and objects from B. We replace the combinators by the corresponding λ-terms, i.e. $I^{\sigma\to\sigma}$ by $\lambda u^{\sigma}.u$, $K^{\sigma\to\tau\to\sigma}$ by $\lambda u^{\sigma} v^{\tau}.\ u$, and S by $\lambda uvw.\ u(w)(v(w))$ (the superscripts denote the types). Then we perform the following reductions in the λ-term for H thus obtained: Replace a term $(\lambda u^{\sigma}.\ s)t^{\sigma}$, where s and t are λ-terms, by the λ-term s' obtained from s by substituting t in all free occurrences of u in s. Only finitely many reductions can be done because the terms are typed. Let E be the expression thus obtained. So $H = \lambda Z\,x.\ E$.

Let $\vec{c},\vec{f},\vec{g}$ be the objects from B (except S and T) which occur in E.

Suppose t is a subterm of E of type 0, with free variables among $x,\vec{u}$. Then t is one of the following expressions:

I	u	$u \in \vec{u}$
II	x	
III	n	$n \in \vec{c}$
IV	$f_j(s)$	$f_j \in \vec{f}$
V	$g_j(s)$	$g_j \in \vec{g}$
VI	$Z(s)$	
VII	$S(\lambda v.\ s_1)(s_2)$	

VIII $T(\lambda v.s_1, \lambda v.s_2)(s_3)$

IX $DC(s_1, s_2, \lambda v.s_3, \lambda v.s_4, s_5)$

X $FP(\lambda p\vec{v}.s)(s_1 \ldots s_k)$ (the level of p is 1)

XI $p(s_1 \ldots s_k)$

where $s, s_1, \ldots, s_k$ are subterms of t of type 0 .

We will associate a tree Y to the term E. To each node in Y we assign a term t as follows:
Assign E to the top node. Suppose t is assigned to a node. If t is of type I, II or III there are no nodes below. If t is of type IV, V or VI there is one node just below. Assign the term s to this node. If t is of type VII there are two nodes just below, one node for s_1 and one for s_2. If t is of type VIII there are three nodes just below. Assign s_1, s_2, s_3 respectively to these nodes. If t is of type IX there are five nodes just below. Assign s_1, s_2, s_3, s_4 s_5 respectively to these nodes. If t is of type X there are $k+1$ nodes just below. Assign $s, s_1, \ldots, s_k$ to these nodes respectively. If t is of type XI, i.e. t is $p(s_1, \ldots, s_k)$, then p is a subterm of s in some $FP(\lambda p\vec{v}.s)$. In this case there are $k+1$ nodes just below t. Assign $s, s_1, \ldots, s_k$ to these nodes respectively. This completes the description of Y. In general Y will not be wellfounded, as there may be infinitely many iterations of p when t is of type X.

We will assign finite lists of equations and inequations to some of the nodes in Y, and for each list L we also assign a term r. Each list will consist of equations and inequations between terms of the following type : number variables, constants which occur in E, 0, $f_i(v)$, $g_i(v)$, $f_i g_j(v)$, where v is a number variable. The term r will also be of this type. Each list is called a condition. r is called the value of t under the condition L (t is the term assigned to this node of Y). The number variables in a condition L and the value r will be among those which are free in t. A condition L is consistent if there exist numbers such that each equation and inequation in L is satisfied when these numbers are substituted for the variables. Each condition will be consistent, and the conditions assigned to a node will be disjoint, i.e. no list of numbers will satisfy two of them. Suppose L, r are assigned to the node of t, and suppose that the list of numbers $\vec{n}$ satisfy L. Let m be the value of r under this substitution, i.e. m = c of r is a constant c, m = n, $f_i(n)$, $g_i(n)$, $f_i g_j(n)$ if r is v, $f_i(v)$, $g_i(v)$, $f_i g_j(v)$ and n is substituted for v. Then m is the value of t under this substitution.

The conditions and the values will be assigned in steps. Simultaneously we will define a binary relation $\dot{<}$. If X $\dot{<}$ X' then X is a triple (s,L,r), X' is a triple (s',L',r'), s and s' are terms in Y, L and L' are conditions assigned to these terms, with values r, r' respectively. X $\dot{<}$ X' will mean that s is below s' in Y, and the condition L is needed to obtain the condition L'.

Assume $Z = f_Q$ for some fixed number Q.

<u>Step 1</u>: Assign the empty condition together with the value t to each node in Y where t is of type I, II or III. No further conditions will be assigned to these nodes.

<u>Step k(k > 1)</u>: Suppose t is a term at some node in Y. If t is of type IV (i.e. t is $f_j(s)$) then there is one node just below and s is assigned to this node. Suppose a condition L and a value r is assigned to s in an earlier step. Then assign the condition L and the value r' to t, where r' is $f_j(v)$ if r is a number variable v, r' is 0 if r is one of the constants

$0,2,4,6,\ldots,2j$, r' is r if r is another constant, r' is $f_\ell(v)$ if r is $f_i(v)$, where $\ell = \max(i,j)$, r' is $f_j g_i(v)$ if r is $g_i(v)$, r' is $f_\ell g_k(v)$ if r is $f_i g_k(v)$, where $\ell = \max(i,j)$. Let $(s,L,r) < (t,L,r')$; and $X < (t,L,r')$ if $X < (s,L,r)$.

The cases V and VI are similar to case IV.

Case VII: t is $S(\lambda v.s_1)(s_2)$. There are two nodes just below the node of t, one for s_1 and one for s_2. Suppose the contitions $L_1 \ldots L_e$ with values $r_1 \ldots r_e$ are assigned to the node of s_1 in the steps before step k. Let $\vec{u}$ be the list of number variables in $L_1 \ldots L_e$, $r_1 \ldots r_e$ except v.

Lemma 1: There are consistent conditions $M_1 \ldots M_\ell$ in which only the variables $\vec{u}$ occur, and numbers $j_1,\ldots,j_\ell$, all > 0, such that if a list of numbers $\vec{n}$ is substituted for $\vec{u}$ in M_i, $1 \le i \le \ell$, and M_i is satisfied (i.e. each equation and inequation in M_i holds) for $\vec{u} = \vec{n}$, then the function $\lambda v.r$ is f_j, with $j = j_i$. $\lambda v.r$ is the following function: Given a number m. Then there is at most one condition among $L_1 \ldots L_e$ which is satisfied for $v = m$, $\vec{u} = \vec{n}$ (because the conditions $L_1 \ldots L_e$ are disjoint), say L_2. The value of $\lambda v.r$ for $n = m$ is then the number given by r_2. $M_1 \ldots M_\ell$ can be chosen disjoint, and all cases are covered, i.e. if $\lambda v.r$ is some f_i, $i > 0$, for a list of numbers $\vec{n}$, then one of $M_1 \ldots M_\ell$ is satisfied for $\vec{u} = \vec{n}$.

We omit the proof of this lemma and continue the description of step k, case VII. Suppose a condition L and a value q is assigned to the node of s_2 in an earlier step. If L is consistent with a condition M among $M_1 \ldots M_\ell$ (i.e. $L \cup M$, the union of the two lists, is consistent) then assign $L \cup M$ and the value q' to the node of t, where q' is as follows:

Suppose $\lambda v.r$ is f_j, $j > 0$, when M is satisfied. q' is $f_{j-1}(v)$ if q is the variable v. q' is 0 if q is one of the constants $0,2,4,\ldots,2(j-1)$, q' is q if q is another constant. q' is $f_\ell(v)$ if q is $f_i(v)$, where $\ell = \max(i,j-1)$, q' is $f_{j-1}g_i(v)$ if q is $g_i(v)$, q' is $f_\ell g_k(v)$ if q is $f_i g_k(v)$, where $\ell = \max(i,j-1)$.

Let $(s_2,L,q) < (t,L \cup M,q')$; and $X < (t,L \cup M,q')$ if $X < (s_2,L,q)$. If L_i is consistent with M, then $(s_1,L_i,r_i) < (t,L \cup M,q')$, and $X < (t,L \cup M,q')$ if $X < (s_1,L_i,r_i)$, $1 \le i \le \ell$.

Case VIII is similar to case VII.

Case IX: t is $DC(s_1,s_2,\lambda v.\ s_3,\lambda v.\ s_4,s_5)$.

There are five nodes below the node of t, one for each of $s_1 \ldots s_5$. Suppose that conditions L_1, L_2, L_5 and values r_1, r_2, r_5 are assigned to the nodes of s_1, s_2, s_5 respectively at earlier steps. Suppose a condition L_3 and a value r_3 is assigned to the node of s_3 before step k. Let L_3' be the condition obtained from L_3 as follows: Replace v by r_5 in each equation and inequation in L_3. Replace a term $f_i(r_5)$ by 0 if r_5 is among $0,2,4,\ldots,2i$. Replace $f_i(r_5)$ by r_5 if r_5 is another constant. Replace $g_i(r_5)$ by 0 if r_5 is among $0,1,3,5,\ldots,2i+1$, replace it by r_5 if r_5 is another constant. If r_5 is $f_j(u)$ replace $f_i(r_5)$, $g_i(r_5)$, $f_i g_k(r_5)$ by $f_\ell(u)$, $f_j g_i(u)$, $f_\ell g_k(u)$ respectively, where $\ell = \max(i,j)$. If r_5 is $g_j(u)$ replace $g_i(r_5)$ by $g_\ell(u)$, where $\ell = \max(i,j)$. If r_5 is $f_i g_j(u)$ replace $f_k(r_5)$, $g_\ell(r_5)$, $f_k g_\ell(r_5)$ by $f_m g_j(u)$, $f_i g_n(u)$, $f_m g_n(u)$ respectively, where $m = \max(i,k)$, $n = \max(j,\ell)$. Suppose $L_1 \cup L_2 \cup L_3' \cup L_5 \cup \{r_1=r_2\}$ is consistent. Then assign $L_1 \cup L_2 \cup L_3' \cup L_5 \cup \{r_1=r_2\}$ and the value r_3' to the node of t, where r_3' is obtained from r_3 by replacing v by r_5 and then cancelling as above. Suppose a condition L_4 and and a value r_4 is assigned to the node of s_4 before step k. Let L_4', r_4' be as described above. If $L_1 \cup L_2 \cup L_4' \cup L_5 \cup \{r_1 \neq r_2\}$ is consistent, then assign this condition and the value r_4' to the node of t.

Let $X < (t, L_1 \cup L_2 \cup L_3' \cup L_5 \cup \{r_1 = r_2\}, r')$ if X is one of the triples (s_1, L_1, r_1), (s_2, L_2, r_2), (s_3, L_3, r_3), (s_5, L_5, r_5), or X is below (in the ordering $<$) one of these triples. Let $X < (t, L_1 \cup L_2 \cup L_4' \cup L_5 \cup \{r_1 \neq r_2\}, r_4')$ if X is one of the triples (s_1, L_1, r_1), (s_2, L_2, r_2), (s_4, L_4, r_4), (s_5, L_5, r_5), or X is below one of these triples.

Case X: t is $FP(\lambda p\vec{v}.\ s)(s_1 \ldots s_e)$. There are $e+1$ nodes just below the node of t, one for each of $s, s_1 \ldots s_e$. Suppose conditions $L, L_1 \ldots L_e$ and values $r, r_1 \ldots r_e$ are assigned to these nodes respectively at earlier steps. Let L' be the condition obtained from L by replacing $\vec{v}$ by $r_1 \ldots r_e$ in L and then cancelling as in case IX. Let r' be obtained from r in the same way. If $L' \cup L_1 \cup \ldots \cup L_e$ is consistent, assign this condition and the value r' to the node of t. Let $X < (t, L' \cup L_1 \cup \ldots \cup L_e, r')$ if X is one of the triples (s, L, r), $(s_1, L_1, r_1), \ldots, (s_e, L_e, r_e)$, or X is below one of these triples.

Case XI: t is $p(s_1 \ldots s_e)$. Now t is a subterm of s (or t is s), where s occurs as $FP(\lambda p\vec{v}.\ s)$. There are $e+1$ nodes just below the node of t, one for each of $s, s_1 \ldots s_e$. Proceed as in case X.

This completes the description of step k.

Lemma 2: Suppose a condition L and a value r is assigned to the node of a term t. Suppose L is satisfied for a list of numbers $\vec{n}$. Then $t[\vec{n}] = r[\vec{n}]$, where $t[\vec{n}]$ is the number we get by replacing $\vec{n}$ for the free number variables in t, $r[\vec{n}]$ is defined similarly.

Proof: An easy induction on k, where k is the step at which the condition L and the value r was assigned to t. □

Lemma 3: Suppose there is a term t in Y and a list of numbers $\vec{n}$ such that no condition assigned to the node of t is satisfied for $\vec{n}$. Then t is undefined when $\vec{n}$ is substituted for the free number variables in t.

Proof: By regarding cases for t one can prove that there is an infinite descending branch in Y starting at the node of t, and lists $\vec{n}_1, \vec{n}_2, \vec{n}_3, \ldots$ of numbers such that $\vec{n} = \vec{n}_1$, no condition at the k-th node in the branch is satisfied for $\vec{n}_k$, and t_k (the term at the k-th node in the branch) is defined for $\vec{n}_k$ only if t_{k+1} is defined for $\vec{n}_{k+1}$, $k = 1,2,3,\ldots$. Since the branch is infinite, there must be a term $FP(\lambda p\vec{v}.\ s)(\ldots)$ at a node in the branch, and intinitely many terms $p(\ldots)$ at nodes below it. Since $FP(\lambda p\vec{v}.\ s)$ = the least p such that $p = \lambda\vec{v}.\ s$ each $p(\ldots)$ will be undefined when the lists above are substituted for the free variables. Going upwards in the branch from $FP(\lambda p\vec{v}.\ s)(\ldots)$ we see that each term is undefined. Hence t is undefined for $\vec{n}$. □

It remains to construct H'. As in the example we start with expressions $FP(\lambda q\vec{v}.\ t_0)$ such that no fixed point operator is in t_0, and replace it by a finite iteration of q, where the finite iterations $\{q_n\}_{n\in N}$ are defined by: $q_0 = \lambda\vec{v}.\uparrow$ (totally undefined), $q_{n+1} = \lambda\vec{v}.\ t_0$ with the term q_n substituted for q in t_0. The other FP's are removed in the same way. Let H' be the term thus obtained.

Suppose $FP(\lambda q\vec{v}.\ t_0)$ is a term in E such that no fixed point operator FP occurs in t_0. Let Y_0 be a finite part of Y, defined as follows: There is a node in Y such that t_0 is assigned to that node. Let the top node in Y_0 be such a node. If a node P is in Y_0, and the term t is assigned to P we let the nodes just below P be as in Y in the cases I - IX. Case X will not occur because there is no FP in t_0. Case XI: t is $p(s_1 \ldots s_k)$, where p can be q, or p can occur as $FP(\lambda p\vec{u}.\ t_1)$, in which case $FP(\lambda q\vec{v}.\ t_0)$ is a subterm of t_1. There are $k+1$ nodes just below P, one for $s_1 \ldots s_k$, and one for t_0 if p is q, for t_1 if p is not q. We do not add further nodes below the node of t_0 (or t_1).

In the example there is one term $FP(\lambda q\vec{v}.t_0)$ as above, namely $FP[\lambda q\, abcd.\ DC(a,0,Z,\lambda w'''.\ t_4,d)]$. Picture of T_0 in this case:

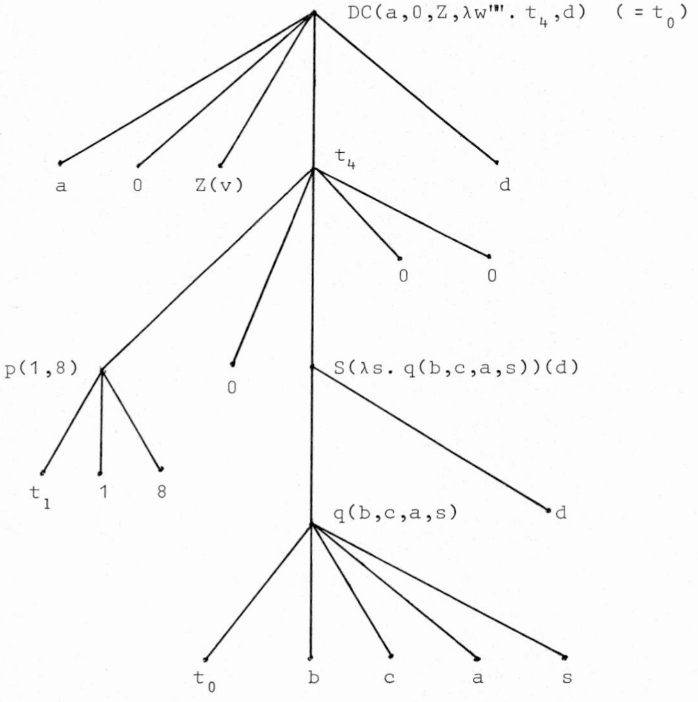

To compute $t_0(a,b,c,d)$ in the case $a \neq 0$, $p(1,8) = 0$, one needs to know $t_0(b,c,a,s)$ for all s.

Back to the general case. As Y_0 is a subset of Y conditions and values are assigned to the nodes of Y_0. The term t_0 is assigned to the top node of Y_0, and maybe also at some of the bottom nodes. In Y there is another copy of Y_0 below each bottom node of t_0, and below each bottom node of t_0 in such a copy there is still another copy of Y_0, etc. So if t_0 is assigned to two different nodes P, Q in Y then Y_1 and Y_2 are identical, where Y_1 and Y_2 are the parts of Y which are below P and Q respectively. Hence if a condition L and a value r is assigned to P in step k, then L and r are also assigned to Q in step k. If L_1, r_1 are assigned to the top node in Y_0, and L_2, r_2 are assigned to a bottom node of t_0 in Y_0, and $(t_0, L_1, r_1) > (t_0, L_2, r_2)$ then L_2, r_2 are assigned before L_1, r_1. Hence L_1 and L_2 are disjoint.

If L_1, r_1 are assigned to the top node in Y_0 there may be conditions and values L_i, r_i, $i = 1,2,3,\ldots,n$, assigned to top nodes of copies of Y_0 in Y (see picture) such that $(t_0, L_1, r_1) > (t_0, L_2, r_2) > \ldots > (t_0, L_n, r_n)$.

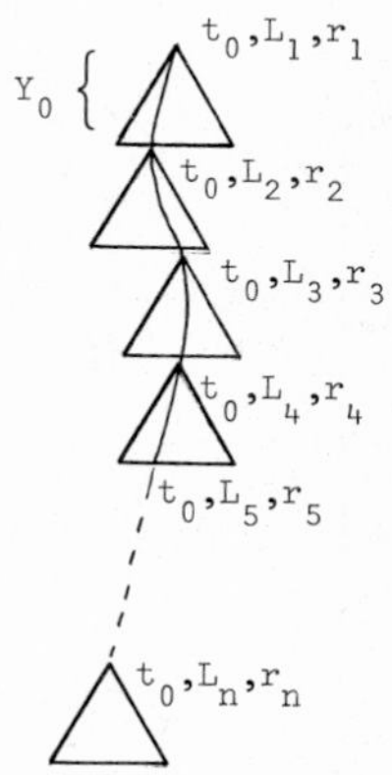

It suffices to prove that there is a number K, independent of Q, x, such that the length of any such chain is at most K (i.e. $n \leq K$). Then we could replace $FP(\lambda q\vec{v}.\ t_0)$ by q_K, the K-th iteration of q.

If S and T do not occur in Y_0 this is easy to prove. For suppose L_1 is satisfied for the numbers $n_1 \ldots n_k$. As $(t_0, L_1, r_1) > (t_0, L_2, r_2)$ there is a list of numbers $m_1 \ldots m_k$ which satisfies

L_2, and each m_i is among $n_1 \ldots n_k$, 0, $\vec{c}$ (the constants in E). L_1 and L_2 are disjoint, so the lists $n_1 \ldots n_k$ and $m_1 \ldots m_k$ are not identical. Given a chain of length n there will be n lists, each of length k, of numbers from $n_1 \ldots n_k$, 0, $\vec{c}$, and none of the lists are identical. Hence n is not greater than the number of ways to choose k elements from $n_1 \ldots n_k$, 0, $\vec{c}$. Let K be this number.

If there are occurrences of S and T in Y_0 the proof is more messy. For in this case there may be variables v in L_2 which are not in L_1. For each such variable v there is a term at a node in Y_0 where v is bounded. These terms can be $S(\lambda v. \ldots)$, or $T(\lambda v. \ldots, \cdot)$, or $T(\cdot, \lambda v. \ldots)$. Suppose $(t_0, L_1, r_1) > (t_0, L_2, r_2)$. From the conditions at the nodes in Y_0 one can find how (L_1, r_1) comes from (L_2, r_2). One can find conditions M and k-tuples of terms $p_1 \ldots p_k$ (in which the variables v just mentioned can occur) such that (t_0, L_1, r_1) is based on the values of t_0 when $v_1 \ldots v_k$ (the free number variables in t_0) are replaced by $p_1 \ldots p_k$, and M is satisfied. M will contain L_2 (with $v_1 \ldots v_k$ replaced by $p_1 \ldots p_k$), and in addition M can contain equations and inequations which stem from other nodes in Y_0. The terms $p_1 \ldots p_k$ are among the following ones: $v_1 \ldots v_k$, 0, $\vec{c}$, $f_i(v_\ell)$, $g_j(v_\ell)$, $f_i g_j(v_\ell)$, v, $f_i(v)$, $g_j(v)$, $f_i g_j(v)$, where $i,j \in N$, $1 \le \ell \le k$, v is a variable as above.

In the example the condition $a \neq 0$, $b \neq 0$, $c = 0$, $p(1,8) = 0$ ($= L_1$) is based on the values of t_0 when a,b,c,d are replaced by b,c,a,s ($= p_1 p_2 p_3 p_4$, v is s), and the condition $a \neq 0$, $b \neq 0$, $c = 0$, $p(1,8) = 0$ ($=$ M) is satisfied. L_2 is the condition $a \neq 0$, $b = 0$, $p(1,8) = 0$.

If v does not occur in M (as in the example) then we can argue as in the case where S and T is not in Y_0. If L is satisfied for $n_1 \ldots n_k$ we can let v be one of these numbers, and hence obtain a list $m_1 \ldots m_k$ from $n_1 \ldots n_k$, 0, $\vec{c}$ such that L_2 is satisfied for $m_1 \ldots m_k$. If v occurs in M this may not be possible. M can for instance contain an equation $f_i(v) = v$, which implies that v is odd, or v is one of the numbers $2(i+1), 2(i+2), \ldots$. Maybe none of these numbers are among $n_1 \ldots n_k$, 0, $\vec{c}$. However, if there is a finite upper bound (independent of Q and x) to the number of f_i's and g_i's in the conditions, this difficulty can be overcome as follows: Given $(t_0, L_1, r_1) > (t_0, L_2, r_2) > \ldots > (t_0, L_n, r_n)$ and L_1 is satisfied for $n_1 \ldots n_k$. Choose lists for $L_2, L_3, \ldots L_n$. If possible choose numbers which already occur in a previous list. If this is not possible choose new numbers. After a while no new f_i, g_i will occur, and it will not be necessary to choose new numbers. An upper bound for n can be found as in the case where S and T are not in Y_0.

So it remains to find an upper bound, independent of Q and x to the number of f_i's and g_i's which can occur in a condition. Which f_i's and g_i's can occur? Of course $f_i \in \vec{f}$, $g_i \in \vec{g}$ and f_Q can occur. Given f_i and the constant $2(i+1)$, f_{i+1} can be defined as follows: $f_{i+1}(s) = 0$ if $s = 2(i+1)$, $= f_i(s)$ if $s \neq 2(i+1)$. But not so many new f_i's can be obtained in this way, as the number of constants is limited. f_{i-1} can be obtained from f_i by: $f_{i-1} = S(f_i)$. So we need an upper bound to the number of new f_i's obtained in this way.

Suppose f_i occurs in M, and a condition on v is expressed by f_i, for instance $f_i(v) = v$. As v occurs as $S(\lambda v, \ldots)$, or $T(\lambda v. \ldots, \cdot)$, or $T(\cdot, \lambda v. \ldots)$ the function $\lambda v. \ldots$ is some f_j or g_j, hence total, hence defined also when $f_i(v) \neq v$, i.e. when $f_i(v) = 0$, v is among $0, 2, 4, \ldots, 2i$. Hence there is a triple X such that $(t_0, L_1, r_1) > X$, and if M' is defined as above (with X in the place of (t_0, L_2, r_2)) then M' implies $f_i(v) = 0$. Let us pass to triples X' such that f_i is in X', and $X > X'$.

After a while we come to the place where f_i is obtained as $S(f_{i+1})$, and then we continue to chase f_{i+1} downwards in $>$. If f_{i+1} is used to put conditions on a v in some M'', we choose triples which implies $f_{i+1}(v) = 0$, because then we can let v be one of the numbers $0,2,4,\ldots,2i$ again (we let v be a number already chosen if possible). In this way we obtain a chain from the top of Y_0 and downwards in Y. It may pass through several copies of Y_0, hence through several nodes of t_0. Picking out these nodes we obtain a chain $(t_0,L_1,r_1) > (t_0,L_2',r_2') > \ldots > (t_0,L_m',r_m')$. Now L_1 is satisfied for $n_1 \ldots n_k$. By the discussion above one can find lists which satisfy $L_2',\ldots,L_m'$, and new numbers occur only in the first lists. As in the case where S and T are not in Y_0 there is an upper bound for the length of the chain, hence for the number of new f_i's

New g_i's can be defined by constants, or be obtained as $T(f_{i+1},g_i)$ ($= g_{i+1}$), in which case f_{i+1} must be obtained first. This proves that there is an upper bound, independent of Q, x, to the number of f_i's and g_i's which can occur in a condition or a value. This also completes the proof of the proposition. □

References

[1] J.E. Fenstad, Recursion theory: An axiomatic approach, Springer Verlag (forthcoming).

[2] J. Moldestad, Computations in higher types, Lecture notes in mathematics no. 574, Springer Verlag 1977.

[3] Y.N. Moschovakis, Abstract first order computability I, Trans. Amer. Math. Soc. 138 (1969) 427-464, and II, 138, 465-504.

[4] Y.N. Moschovakis, Axioms for computation theories - first draft, in: R.O. Gandy and C.E.M. Yates (eds.), Logic Colloquium '69 (North-Holland, Amsterdam 1971) 199-255.

[5] R.A. Platek, Foundations of recursion theory, Ph.D. thesis, Stanford University 1966, not published.

J.E. Fenstad, R.O. Gandy, G.E. Sacks (Eds.)
GENERALIZED RECURSION THEORY II

Set Recursion

Dag Normann

1. Introduction.

In this paper we will define a computation-theory called E-recursion, which will be a theory of partial set-recursive functions, defined on sets and with sets as values. We will use natural numbers as indices.

The original purpose was to develop a theory on the companion of a normal functional of type k+2 such that semirecursion over type k in F and the theory are the same. The motivation for this was that this set-recursion theory might accept priority-arguments, arguments giving results about degrees of functionals. Some results from that program are given in Normann [12] and [13].

The recursion theory we developed for that purpose, happened to be of a more general nature, and not quite unnatural even if one does not have the applications on degrees of functionals in mind.

Moschovakis [10] has constructed essentially the same theory, using inductive schemes and fixpoint operators.

A computation theory on a structure must satisfy certain fundamental properties, composition of recursive functions gives a recursive function, you may diagonalize or compute on indices ($\{e\}(e_1,x) \simeq \{e_1\}(x)$). There will also be some trivial manipulations which are so deeply connected with the structure that they obviously must be computable. In addition there may be some finiteness-properties, search-operators, stage-comparison etc. giving the theory its particular flavour. These properties may either reflect the purpose the theory-maker had with his theory, or they may reflect what the theory-maker thought was a natural notion of computation for the particular structure.

E-recursion will be like α-recursion in the sense that we make a clear distinction between recursion in an object and recursion relative to a relation. This is in contrast with Kleene theory for recursion in higher types. Scheme 8, which is the only place where the nature of higher types is utilized, is a scheme of relativization. It is just natural that a clear distinction here should give a stronger theory.

On the other hand, E-recursion will be like Kleene-recursion in the sense that the computations will be absolute, and not dependent on the domain. In α-recursion we may search through α, while in E-recursion we may essentially just search through ω. E-recursion will thus be weaker than α-recursion.

The domain for E-recursion will be the universe of sets. When we use a set as an object we will treat the set as a finite entity, we may use all information about the set or information uniformly derived from the elements of the set at the same time. If we accept a set R as a relation, we may just ask if a given set x is in R or not, and expect an answer from an oracle for R.

Our scheme <u>iv</u> may be debatable, but that is the one that reflects our distinction between objects and relations.

E-recursion is nothing more than the schemes for the rudimentary functions augmented with a diagonalization scheme. We hereby give the definition:

<u>Definition of E-recursion</u>.

Let $R \subseteq V$ be a relation. We define the partial function recursive relative to R with index e by the following schemes.

<u>i</u> $f(x_1,\ldots,x_n) = x_i$ $\qquad e = \langle 1,n,i\rangle$

<u>ii</u> $f(x_1,\ldots,x_n) = x_i \setminus x_j$ $\qquad e = \langle 2,n,i,j\rangle$

iii $f(x_1,\dots,x_n) = \{x_i,x_j\}$ $e = \langle 3,n,i,j\rangle$

iv $f(x_1,\dots,x_n) \simeq \bigcup_{y\in x_1} h(y,x_2,\dots,x_n)$ $e = \langle 4,n,e'\rangle$ where e' is an index for h

v $f(x_1,\dots,x_n) = h(g_1(x_1,\dots,x_n),\dots,g_m(x_1,\dots,x_n))$

$e = \langle 5,n,m,e',e_1,\dots,e_m\rangle$ where e' is an index for h and $e_1,\dots,e_m$ are indices for $g_1,\dots,g_m$ resp.

vi $f(x_1,\dots,x_n) \simeq x_i \cap R$ $e = \langle 6,n,i\rangle$

vii $f(e_1,x_1,\dots,x_n,y_1,\dots,y_m) \simeq \{e_1\}^R(x_1,\dots,x_n)$

$e = \langle 7,n,m\rangle$

In scheme iv it is understood that the computation terminates only if $h(y,x_2,\dots,x_n)$ terminates for all $y \in x_1$.

The partial functions defined by these schemes are called E-recursive relative to R and they are denoted $\{e\}^R$.

2. Some properties of E-recursion.

All functions that are rudimentary in R will be E-recursive relative to R (E(R)-recursive). Since for each $n \in \omega$ the constant function n is rudimentary, these functions will be E-recursive. Combining schemes i and v we may commute the arguments in the functions.

The schematic definition gives us canonical concepts of

i length of a computation $\| \ \|$

ii subcomputation

iii computation tree

By standard proofs we obtain the recursion theorems and the S^n_m - theorem.

The following lemma will prove that arguments really are 'finite' in the usual sense of generalized recursion theory; and justify the term 'E-recursion'.

Lemma 1.

In E-recursion there is an index e such that for arbitrary $R, x, e_1, \vec{x}$:

$$\{e\}^R(x, e_1, \vec{x}) \simeq \begin{cases} 0 & \text{if } \forall y \in x \quad \{e_1\}^R(y, \vec{x}) \simeq 0 \\ 1 & \text{if } \forall y \in x \quad \{e_1\}^R(y, \vec{x}) \downarrow \text{ and} \\ & \qquad \exists y \in x \quad \{e_1\}^R(y, \vec{x}) \neq 0 \end{cases}$$

where $\downarrow$ means 'has a value'.

Proof. There is a rudimentary function φ such that $\varphi(0) = 0 = \emptyset$ and $\varphi(x) = 1 = \{\emptyset\}$ for all $x \neq 0$. So we may assume that $\{e_1\}$ takes values 0 and 1 only.

Let

$$\{e\}^R(x, e_1, \vec{x}) = \bigcup_{y \in x} \{e_1\}^R(y, \vec{x})$$

As a corollary we will have stage comparison. If a computation σ does not terminate, we write $\sigma\uparrow$ and $\|\sigma\| = \infty$.

Lemma 2.

There is an E-recursive function p such that $p(\sigma_1, \sigma_2)\downarrow$ if and only if $\sigma_1\downarrow$ or $\sigma_2\downarrow$ and then

$$p(\sigma_1, \sigma_2) = \begin{cases} 0 & \text{if } \|\sigma_1\| \leq \|\sigma_2\| \\ 1 & \text{if } \|\sigma_1\| > \|\sigma_2\| \end{cases}$$

Indication of proof. We define p by the recursion-theorem. Essentially there will be 64 cases, one for each pair of schemes used in σ_1 and σ_2. Similar results are well known in the theory of normal functionals, and we regard the methods involved well-known. Moldestad [9] gives an argument similar to our case (iv,v).

As a consequence of stage comparison (Lemma 2) we obtain:

<u>Lemma 3</u>.

In E-recursion there is an index e such that for any $R, e_1, \vec{x}$:

$$\{e\}^R(e_1,\vec{x})\downarrow \iff \exists n \in \omega \quad \{e_1\}^R(n,\vec{x})\downarrow, \text{ and then}$$

$$\{e_1\}^R(\{e\}^R(e_1,\vec{x}),\vec{x})\downarrow$$

This is proved in Grilliot [2], see also Moldestad [9].
This kind of selection operator was first investigated by Gandy [1] and we call it Gandy selection.

<u>Definition</u>.

Let $R \subseteq V$, $\vec{y} \in V^m$. Let φ be a partial map from V^n to V. We say that φ is recursive <u>in</u> y <u>relative to</u> R if there is an index e in E-recursion such that

$$\forall \vec{x} \in V^n \quad (\varphi(\vec{x}) \simeq \{e\}^R(\vec{x},\vec{y}))$$

We then obtain natural definitions of sets <u>recursive</u> and <u>semirecursive</u> in $\vec{y}$ relative to R.

<u>Definition</u>.

Let $A \subseteq V$, $R \subseteq V$. Let the <u>E(R)-recursive closure</u> of A be $\{\{e\}^R(\vec{x})\ ;\ e \in \omega,\ \vec{x} \in A^n,\ n \in \omega\} = M_A(R)$.
If A is E(R)-recursively closed we may split up A as follows:

$\langle M_B(R)\rangle_{B \in {}^fA}$ where ${}^fA$ is the set of finite subsets of A.

If A is E(R)-recursively closed, B a finite subset of A, we say that $C \subseteq A$ is <u>$\Sigma^*_B(R)$-definable</u> if for some Δ_o-formula φ with parameters from $M_B(R)$,

$$x \in C \iff \exists y \in M_{B \cup \{x\}}\ \varphi(x,y)$$

$C \subseteq A$ is Δ^*_B if both C and $A \setminus C$ are Σ^*_B-definable.

Lemma 4.

Let A be $E(R)$-recursively closed and transitive. Then $\langle M_B(R)\rangle_{B \in f_A}$ satisfies Σ^*-collection, i.e.

Let φ be a Δ_o-formula with parameters from $M_B(R)$. Let $u \in M_B$. Assume

$$\forall x \in u\ \exists y \in M_{B \cup \{x\}}(R)\varphi(x,y,R).$$

Then

$$\exists v \in M_B\ \forall x \in u\ \exists y \in v\ \varphi(x,y,R).$$

Proof. Let $\vec{B}$ be a listing of B.
By assumption

$$\forall x \in u\ \exists e \in \omega\ \varphi(x,\{e\}^R(\vec{B},x),R).$$

By Gandy selection we choose one e to each x and use the union scheme to find v.

Lemma 5.

Well-foundedness is Σ^*-definable. (i.e. $\Sigma^*_{\emptyset}$- definable)

Proof. By the recursion-theorem we find an index e such that if y is a well-founded relation on x, then $\{e\}(y,x)\downarrow$ and $\{e\}(y,x)$ is the rank-function of y. So, y is a well-founded relation on $x \iff \exists f \in M_{\{x,y\}}$ (f is a rank function for y).

Theorem 1.

Let A be $E(R)$-recursively closed, B a finite subset of A. $C \subseteq A$ is $E(R)$-semirecursive in $\vec{B}$ if and only if C is $\Sigma^*_B(R)$-definable.

Proof. Assume that C is $E(R)$-semirecursive in $\vec{B}$. Let e be an index such that $x \in C \iff \{e\}^R(x,\vec{B})\downarrow$.
By the recursion theorem we may prove that if $\{e\}^R(x,\vec{B})\downarrow$, then the

computation tree will be in $M_{B \cup \{x\}}(R)$. So

$$x \in C \iff \exists T \in M_{B \cup \{x\}}(R)\ (T \text{ is well-founded and } T \text{ is a computation-tree for } \{e\}(x,\vec{B}))$$

By lemma 5, this is a $\Sigma_B^*(R)$-definition of C.

On the other hand, assume C is $\Sigma_B^*(R)$-definable. As in the proof of lemma 4, we use Gandy selection to find a function that terminates exactly on C.

Definition.

We call a family $\langle M_B \rangle_{B \in {}^fA}$ R-admissible if each M_B is rudimentary closed in R, for each $B, C \in {}^fA$, $M_B \subseteq M_C \iff B \subseteq M_C$, and the family satisfies $\Sigma^*(R)$-collection.

Lemma 6.

Let $\langle M_B \rangle_{B \in {}^fA}$ be R-admissible. Then each M_B is closed under $E(R)$-recursion.

Proof. By induction on the height of a well-founded relation we prove by Σ^*-collection that if y is a well-founded relation on x, then the rank function is in $M_{\{x,y\}}$. So well-foundedness is Σ^*-definable over $\langle M_B \rangle_{B \in {}^fA}$. By the same method we prove that if $\{e\}^R(\vec{B})\downarrow$, then the computation tree is in M_B. The value of a computation is rudimentary in the computation-tree, and lemma 6 follows.

This also shows that the relation $\{e\}^R(\vec{B}) = x$ is $\Sigma^*(R)$-definable over $\langle M_B \rangle_{B \in {}^fA}$.

By lemmas 4 and 6 we see that if A is $E(R)$-recursively closed, then $\langle M_B(R) \rangle_{B \in {}^fA}$ is the finest splitting of A into an R-admissible family.

3. E-recursion and Kleene-recursion.

We are going to prove that E-recursion in a sense generalizes recursion in normal functionals. We will restrict ourselves to a set

I with a canonical pairing operator. Typical examples will be

I = the total functionals of type k (= tp(k)).

I is a transitive set rudimentary closed in R.

We may then identify finite subsets of I with elements of I.

When we are investigating the part of $E(R)$-recursion generated by I, it is natural to seek the least R-admissible family containing $I \cup \{I\}$. It is, however, an advantage to restrict the set of indices to a smaller set. This is covered by the following definition:

Let I be as above, R a relation.

By $\underline{M_a(R;I)}$ we mean $M_{\{a,I\}}(R)$.

By <u>the spectrum of R over I</u> we mean

$$\underline{\mathrm{Spec}(R;I)} = \langle M_a(R;I)\rangle_{a \in I}, \quad M(R;I) = \bigcup_{a \in I} M_a(R;I).$$

$\langle M_a(R;I)\rangle_{a \in I}$ will satisfy $\Sigma^*(R)$-collection over I, and each $M_a(R;I)$ will be rudimentary closed relative to R. Such families are called <u>R-admissible over I</u>. It will follow from our general theory that $\mathrm{Spec}(R;I)$ is the minimal family that is R-admissible over I. A key to this observation is the following definition:

<u>Definition</u>.

<u>a</u> Let $A \subseteq I \times I$ $(= I)$ be a transitive, reflexive relation. Let $a \simeq b$ if $A(a,b)$ and $A(b,a)$. We say that A is a code for a set x if $A/\simeq$ is isomorphic to $\langle TC(\{x\}), \in\rangle$ (TC = transitive closure)

<u>b</u> Let $\langle M_a\rangle_{a \in I}$ be a family over I. We say that $\langle M_a\rangle_{a \in I}$ is locally of type I if for any set x and $a \in I$,

$$x \in M_a \iff x \text{ has a code in } M_a$$

<u>Lemma 7</u>.

$\mathrm{Spec}(R;I)$ is locally of type I.

Proof. By the recursion theorem we define an index e_1 such that if A is a code for x, then $\{e_1\}(A,I) = x$. Again by the recursion theorem one may use e_1 to define an index e_2 such that for any $e \in \omega$, codes $A_1,\ldots,A_n$ for $y_1,\ldots,y_n$, if $\{e\}^R(\dot{y},\ldots,y_n) \simeq x$, then $\{e_2\}^R(e,A_1,\ldots,A_n)$ is a code for x. The definition is by cases according to the schemes, and involves trivial but tedious constructions of codes.

We are now ready to prove

Theorem 2.

Spec(R;I) is the minimal family R-admissible over I.

Proof. We already remarked that Spec(R;I) is R-admissible over I. To prove that Spec(R;I) is included in any family $\langle M_a \rangle_{a \in I}$ R-admissible over I, we prove by induction on the length of the computation that for any $x_1,\ldots,x_n$, if $x_1,\ldots,x_n$ have codes in M_a and $\{e\}^R(x_1,\ldots,x_n)\downarrow$ then both computation-tree and value will be in M_a. Where we in lemma 6 would use Σ^*-collection over a set x, we will here use a code for x and Σ^*-collection over I.

Define the functional $^I E$ by

$$^I E(f) = \begin{cases} 0 & \text{if } \forall a \in I \;\; f(a) = 0 \\ 1 & \text{if } \exists a \in I \;\; f(a) \neq 0 \end{cases}$$

where $f : I \to \mathbb{N}$ is total.

By lemma 1, $^I E$ is E-recursive in I.
Let $I = tp(k)$, $^I E = {}^{k+2}E$. We assume that the reader is acquainted with the basic facts about Kleene-recursion.

Theorem 3.

Let F be a functional of type $k+2$, $C \subseteq tp(k+1)$. Then the following statements are equivalent

i C is Kleene-semirecursive in $^{k+2}E, F$

<u>ii</u> C is E(F)-semirecursive in I

<u>iii</u> C is $\Sigma_I^*(F)$-definable.

<u>Proof</u>. We already proved that <u>ii</u> and <u>iii</u> are equivalent.

<u>i</u> $\Rightarrow$ <u>ii</u> By the recursion theorem for E-recursion we find an index e such that

$$\{e\}^F(e_1,\vec{f}) \simeq \{e_1\}^{\text{Kleene}}(\vec{f},F,{}^{k+2}E)$$

The definition is by cases according to the Kleene-index. For scheme 8, we use schemes <u>iv</u> and <u>vi</u>, the other cases are rudimentary.

<u>ii</u> $\Rightarrow$ <u>i</u> Since C is E(F)-recursive, there is an index e such that $f \in C \iff \{e\}^F(f,I)\downarrow$. The method of proof is to copy the computation $\{e\}^F(f,I)$ as a ${}^{k+2}E,F$-computation on codes. In doing this we need:

In Kleene-recursion there is an index e_1 such that if f and g are characteristic functions for codes for x and y respectively, then

$$\{e_1\}({}^{k+2}E,f,g) \simeq \begin{cases} 0 & \text{if } x = y \\ 1 & \text{if } x \neq y \end{cases}$$

e_1 is found by using the recursion theorem and induction on $\min(\text{rank}(x), \text{rank}(y))$.

We then use e_1 and the recursion theorem to find an index e_2 such that if $f_1,\dots,f_k$ are characteristic functions of codes for $x_1,\dots,x_k$, and $\{e\}^F(x_1,\dots,x_k) \simeq y$, then

$$\lambda a \in I\, \{e_2\}(e,f_1,\dots,f_k,F,{}^{k+2}E,a)$$

is the characteristic function of a code for y. The construction is by induction on the length of the E(F)-computation.

This theorem shows that Kleene-recursion in normal functionals is a special case of E-recursion in relations. We will later see that if we restrict ourselves to regard semi-recursion over I, then we may reduce E(R)-recursion to E(F)-recursion for some F, and thus by theorem 3 to Kleene-recursion in ${}^{k+2}E, F$.

We will need recursive approximations of the spectrum:

<u>Definition</u>.

Let α be an ordinal, A a set. By $M_A^\alpha(R)$ we mean

$$\{\{e\}^R(\vec{B})\ ;\ B \in {}^f A,\ e \in \omega \text{ and the length of the computation is shorter than } \alpha\}.$$

We obtain definitions of $\langle M_B^\alpha(R)\rangle_{B \in {}^f A}$, $\langle M_a^\alpha(R,I)\rangle_{a \in I}$ etc.

From now on, assume that I is a set with a canonical pairing and that $N \subseteq I$.

<u>Definition</u>.

Let $\langle M_a\rangle_{a \in I}$ be a family admissible over I, $C \subseteq M = \bigcup_{a \in I} M_a$. We say that C is <u>weakly Σ^*-definable</u> in a ($w-\Sigma_a^*$) if for some Δ_0-formula φ with parameters from M_a,

$$x \in C \iff \forall b(x \in M_{\langle a,b\rangle} \Rightarrow \exists y \in M_{\langle a,b\rangle}\,\varphi(x,y))$$

The concept is relativized to an arbitrary relation R.

C is $w-\Delta_a^*$ if both C and the complement of C are $w-\Sigma_a^*$.

<u>Lemma 8</u>.

Let R be a relation, $\langle M_a\rangle_{a \in I} = \mathrm{Spec}(R)$.

If $C \subseteq M$ is $\Sigma_a^*(R)$, then C is $w-\Sigma_a^*(R)$.

<u>Proof</u>. Assume $x \in C \iff \exists y \in M_{x,b}\,\varphi(x,y)$.

Let $x \in M_{a,b}$. Then

$$x \in C \iff \exists \alpha \in M_{a,b}(\exists y \in M_{x,b}^\alpha)(\varphi(x,y))$$

It is sufficient to show that the relation $z = M_{x,b}^\alpha$ is $w-\Sigma^*$ and that $\alpha \in M_{a,b}$, $x \in M_{a,b} \Rightarrow M_{x,b}^\alpha \in M_{a,b}$. This is done by regarding the inductive definition of $M_{x,b}^\alpha$.

Lemma 9.

Let R_1 and R_2 be relations.
If R_1 is $w-\Delta^*(R_2)$ then $\mathrm{Spec}(R_1;I) \subseteq \mathrm{Spec}(R_2;I)$, and $\Sigma^*(R_1)$ over $I \subseteq \Sigma^*(R_2)$ over I. (i.e. $\Sigma^*_{\{I\}}(R_1) \cap P(I) \subseteq \Sigma^*_{\{I\}}(R_2) \cap P(I)$)

Proof. Since R_1 is $w-\Delta^*(R_2)$, $\mathrm{Spec}(R_2;I)$ will be R_1-admissible over I. By theorem 2 $\mathrm{Spec}(R_1;I) \subseteq \mathrm{Spec}(R_2;I)$.
Over I, Σ^* and $w-\Sigma^*$ are the same. So it is sufficient to prove that $w-\Sigma^*(R_1) \subseteq w-\Sigma^*(R_2)$. This will again follow if we prove that the $w-\Delta^*(R_2)$-relations are closed under bounded quantification (i.e. $\Delta_o(w\text{-}\Delta^*(R_2)) = w\text{-}\Delta^*(R_2)$). The proof make use of Σ^*-collection. over I and the fact that $\mathrm{Spec}(R_2)$ is locally of type I. The proof is essentially simple.

We are now able to prove that Kleene-recursion in normal functionals of type $k+2$ is essentially the same as E-recursion over I relative to some relation.

Theorem 4.

Let I have a canonical pairing function, $\mathbb{N} \subseteq I$. Let R be a relation. Then there is a total functional F defined on ${}^{\omega}I$ such that

<u>i</u> $\mathrm{Spec}(R;I) = \mathrm{Spec}(F;I)$

<u>ii</u> $\Sigma^*(R)$ over $I = \Sigma^*(F)$ over I.

Proof. Let R be given. By lemma 9 it is sufficient to find F such that F and R are $w-\Delta^*$ in each other.

We define F_α by induction on α. If $F_\alpha(f)$ is undefined for all ordinals α, let $F(f) = 1$.

F_o is the empty function. Let $F_{<\alpha} = \bigcup_{\gamma<\alpha} F_\gamma$.
If f is $E(F_{<\alpha})$-computable by a computation of length $\leq \alpha$, and $F_{<\alpha}(f)$ is undefined, let

$F_\alpha(f) = 0$ if f is a pair $\langle f_1,f_2\rangle$ and f_2 is (the characteristic function of) a code for a set x, rank $x \leq \alpha$ and $x \in R$,

$F_\alpha(f) = 1$ otherwise.

Claim 1. R is $w-\Delta^*(F)$.

Proof. Let $x \in M_a(F)$. x will have a code A in $M_a(F)$. Also $\alpha = \text{rank}\, x \in M_a(F)$.
There will be a set B in $M_a(F)$ that has not been computed before stage α. Let f_1, f_2 be the characteristic functions of B, A and let $f = \langle f_1,f_2\rangle$, f is in $M_a(F)$ and $x \in R \Leftrightarrow F(f) = 0$.
For any a such that $x \in M_a$ we then obtain

$$x \in R \Leftrightarrow \exists f = \langle f_1,f_2\rangle \in M_a \ (f_2 \text{ is a code for } x\,,\ f \notin \langle M_b^{\text{rank}(x)}(F)\rangle_{b \in I} \text{ and } F(f) = 0\,)$$

$$\Leftrightarrow \forall f = \langle f_1,f_2\rangle \in M_a \ (f_2 \text{ is a code for } x \text{ and } f \notin \langle M_b^{\text{rank}(x)}(F)\rangle_{b \in I} \Rightarrow F(f) = 0\,)$$

It follows that $\text{Spec}(R;I) \subseteq \text{Spec}(F;I)$.

Claim 2. The relation

'$\{e_1\}({}^I E,F,\vec{a}) \simeq n$ by a computation shorter than α'

is $E(R)$-recursive.

Proof. This is proved by using the recursion theorem for $E(R)$-recursion on the definition of F.

By claim 2 and Σ^*-collection over I we may prove

Claim 3. If $\{e_1\}({}^I E,F,\vec{a})\downarrow$, then the length of the computation will be in $M_a(R)$.

By claim 2 and 3, $A \in M_a(F) \Rightarrow A \in M_a(R)$, and thus, since both spectra are locally of type I,

$$\mathrm{Spec}(F;I) \subseteq \mathrm{Spec}(R;I)\,.$$

From claim 3 it also follows that if $f \in M_a(R) = M_a(F)$, then for some $\alpha \in M_a(R)$, $F_\alpha(f)$ is defined.

F_α is $E(R)$-recursive in α, I and by lemma 8 $w-\Delta^*$.

For $f \in M_a(R)$ we then have

$$F(f) = 0 \iff \exists \alpha \in M_a\ (F_\alpha(f) = 0)$$
$$\iff \forall \alpha \in M_a\ (F_\alpha(f) \text{ is defined} \Rightarrow F_\alpha(f) = 0)\,.$$

Thus F is $w-\Delta^*(R)$, and theorem 4 is proved.

Remark. If well-foundedness is E-recursive in I (e.g. $I = tp(k)$ for some $k > 0$) the proof of theorem 4 is much simpler. Then just define F by

$$F(f) = \begin{cases} 0 & \text{if } f \text{ is a code for a set } x \text{ in } R \\ 1 & \text{otherwise} \end{cases}$$

4. A hierarchy for the $w-\Sigma^*$-relations in Spec(R).

We will now restrict ourselves to recursion over $I = S \cup {}^{\omega}S$, where S satisfies pairing and contains N. Moldestad [9] developes a notion of recursion in a normal functional over I, and by our results we may as well do $E(R)$-recursion over I for some relation R

Let R be any relation. Let $\mathrm{Spec}(R;I) = \langle M_a \rangle_{a \in I}$, $M = \bigcup_{a \in I} M_a$. We write M_a^α for $M_a^\alpha(R;I)$.

If C is a $\Sigma^*(R)$-subset of M, we obtain recursive approximations C^α by restricting the definition to $\langle M_a^\alpha \rangle_{a \in I}$, i.e.

$$x \in C^\alpha \iff x \in M^\alpha \wedge \exists y \in M_x^\alpha\, \varphi(x,y,R)\,.$$

This approximation is clearly monotone.

If we try to define an approximation of a weak Σ^*-set it is not clear that the approximation will be monotone. If, however, there is a recursive well-ordering of I, or we have some other means of uniformizing recursive sets effectively, we may prove the existence of monotone approximations.

Let $\{\varphi_n\}_{n\in\omega}$ be an effective enumeration of the A_o-formulas. We identify $\omega\times S$ with S. Let $i = \langle n_i, i_1\rangle$ be given, and σ an ordinal.

Define $x\in I^{\sigma}_{i,a}(R)$ if $x\in M^{\omega\cdot\sigma}$ and for all $b\in I$, if $x\in M^{\omega\cdot\sigma}_{i,a,b}$ there is a $y\in M^{\omega\cdot\sigma}_{i,a,b}$ such that $\varphi_{n_i}(x,y,i_1,a)$ holds.

$I_{i,a}(R) = \bigcup_{\sigma\in On} I^{\sigma}_{i,a}(R)$.

Lemma 10.

a $\sigma_1\le\sigma_2 \Rightarrow I^{\sigma_1}_{i,a}(R)\subseteq I^{\sigma_2}_{i,a}(R)$

b $x\in I_{i,a}(R) \Longleftrightarrow \forall b(x\in M_{i,a,b} \Rightarrow \exists y\in M_{i,a,b}\,\varphi_{n_i}(x,y,i_1,a))$.

Proof.

a Let $x\in I^{\sigma_1}_{i,a}$. Assume that $x\in M^{\omega\cdot\sigma_2}_{i,a,b}$ for some given b. Let σ be the least ordinal such that for some c, $x\in M^{\sigma}_{i,a,c}$. Then $\sigma<\omega\cdot\sigma_1$ and $\{c;x\in M^{\sigma}_{i,a,c}\}\in M^{\omega\cdot\sigma_2}_{i,a,b}$. Then this set contains an element c_1 recursive in i,a,b, the computation takes place in $M^{\omega\cdot\sigma_2}_{i,a,b}$. But $\exists y\in M^{\omega\cdot\sigma_1}_{i,a,c_1}\,\varphi_{n_i}(x,y,i_1,a)$. Now $M^{\omega\cdot\sigma_2}_{i,a,c_1}\subseteq M^{\omega\cdot\sigma_2}_{i,a,b}$, so $\exists y\in M^{\omega\cdot\sigma_2}_{i,a,b}\,\varphi_{n_i}(x,y,i_1,a)$.

In this argument we need several places that $\omega\cdot\sigma_2$ is a limit ordinal, in order to carry through some manipulations of finite length inside $M^{\omega\cdot\sigma_2}_{i,a,b}$.

b $\Rightarrow$ Assume $x\in I_{i,a}(R)$. Then after some σ, $x\in I^{\sigma}_{i,a}(R)$. By the argument of a we see that if $x\in M_{i,a,b}$ for some b, then there is a $y\in M_{i,a,b}$ such that $\varphi_{n_i}(x,y,i_1,a)$.

$\Leftarrow$ Assume $\forall b(x\in M_{i,a,b}\Rightarrow \exists y\in M_{i,a,b}\,\varphi_{n_i}(x,y,i_1,a))$.

Let σ_o be such that $x\in M^{\sigma_o}$. By induction, let σ_{i+1} be such that $x \in M^{\omega\cdot\sigma_i}_{i,a,b} \Rightarrow \exists y \in M^{\omega\cdot\sigma_{i+1}}_{i,a,b}\ \varphi_{n_i}(x,y,i_1,a)$.

σ_{i+1} will exist by Σ^*-collection.

Let $\sigma = \sup \sigma_i$. Then $x \in I^{\sigma}_{i,a}$.

5. Further results and historical remarks.

The approach of Moschovakis [10] is somewhat different from the one given here. He introduces the recursive functions and functionals via fix-point operations. His theory is also generalized to arbitrary partial set functions and consistent functionals. It may be regarded as a weakness of E-recursion theory that it is relativized to relations and not to functions. Lemma 7 and Theorem 4 will be false if we relativize to arbitrary functions, but correct in some interesting cases.

It is possible to derive suitable versions of the Grilliot-selection theorem (Harrington-MacQueen [6] or Moldestad [9]). Moschovakis has done so for his equivalent theory. When we apply E-recursion on problems about degrees, we use theorems 3 and 4 and Grilliot-selection for Kleene-recursion. In these applications (Normann [12]) we also use the results from section 4 of this paper.

It may be proved that given a positive Σ^*-inductive definition of the form

$$x \in \Gamma(R) \Longleftrightarrow \exists y \in M_x \forall z \in y\, \varphi(y,z,x\,R)$$

where φ is an open formula, the fix-point will be Σ^*-definable. The details of that proof are worked out in Gurrik [3].

Defining a natural concept of normal computation theory we may characterize those normal theories that are equivalent to $E(R)$-recursion over I for some R. This was originally proved for $I = \omega$ by Harrington, Kechris, Simpson [7] and for $I = tp(k)$, $k > 0$ by Kechris [8] and Normann [11].

In Normann [14] we regard set-recursion with a natural jump operator, and obtain a connection to strong recursion for the superjump as defined by Harrington in [4] and [5]. In that paper we also give a precise formulation of the characterization above, and prove a similar result for set-recursion with jump.

The companion theory for recursion in higher types has been steadily developing. Harrington [5] gave the first version of the Spector-Gandy theorem (our Theorem 1). D.B. MacQueen [15] proved the original version of theorem 2. Further developments are found in Kechris [8] and Normann [11]. Kolaitis [16] has proved similar results in the context of inductive operators.

References.

1. R.O. Gandy, General recursive functionals of finite type and hiearchies of functionals, in: Ann. Fac. Sci. Univ. Clermont-Ferrand No 35 (1967) 5-24.
2. T.J. Grilliot, Selection functions for recursive functionals. Notre Dame Jour. Formal Log. (1969) 225-234.
3. Gurrik, Cand.real. thesis, in Norwegian.
4. L.A. Harrington, The superjump and the first recursively Mahlo ordinal, in J.E. Fenstad and P.G. Hinman (eds.) Generalized Recursion Theory (North-Holland, Amsterdam 1974) 43-52.
5. L.A. Harrington, Contributions to recursion theory on higher types, Ph.D. Thesis M.I.T. 1973.
6. L.A. Harrington and D.B. MacQueen, Selection in abstract recursion theory, Journ. Symb. Log. 41 (1976) 153-158.
7. L.A. Harrington, A.S. Kechris and S.G. Simpson, 1-envelopes of type 2 objects, American Math. Soc. Notices 20 (1973) A-587.
8. A.S. Kechris, The structure of envelopes: A survey of recursion in higher types, M.I.T. Logic Seminar Notes, December 1973.
9. J. Moldestad, Computations in Higher Types, Lecture Notes in Mathematics 574, Springer-Verlag 1977.

10. Y.N. Moschovakis, Recursion in the Universe of Sets, Mimeographed notes, 1976.

11. D. Normann, Imbedding of Higher Type Theories, Preprint Series in Mathematics No 16, Oslo 1974.

12. D. Normann, Degrees of functionals, Preprint Series in Mathematics No 22, Oslo 1975.

13. D. Normann, Recursion in ^3E and a splitting theorem, to appear in the proceedings from the Fourth Scandinavian Conference of Logic. Jyväskylä 1976.

14. D. Normann, A Jump Operator in Set Recursion, to appear.

15. D.B. MacQueen, Post's problem for recursion in higher types, Ph.D. Thesis, M.I.T. 1972.

16. Ph.G. Kolaitis, Recursion in a Quantifier VS. Elementary induction, to appear.

J.E. Fenstad, R.O. Gandy, G.E. Sacks (Eds.)
GENERALIZED RECURSION THEORY II

A NON-ADEQUATE ADMISSIBLE SET WITH A GOOD DEGREE-STRUCTURE

by

Dag Normann
Oslo

Viggo Stoltenberg-Hansen
Oslo

Moschovakis [2] defined a set of axioms for generalized computation theories. He conjectured that the so called Friedberg-theories would be suitable for priority arguments. This conjecture was disproved by Simpson [3], when he, using AD, defined a Friedberg-theory with a negative solution to Post's problem. Simpson's example also serves as a counterexample to the regular set theorem.

Stoltenberg-Hansen [4] defined a Friedberg-theory Θ to be adequate if there is a semi-recursive prewellordering $<$ whose initial segments are uniformly Θ-finite, having the following property:

There is an initial segment $<^*$ (not necessarily proper) of $<$ such that

i There is a recursive function π mapping each element of the domain of Θ to a $<^*$-bounded subset of field($<^*$) such that $a \neq b \Rightarrow \pi(a) \cap \pi(b) = \emptyset$.

ii Any semirecursive $<^*$-bounded subset of field($<^*$) is Θ-finite.

iii The ordertype of $<^*$ is a limit ordinal.

Many results about r.e. degrees using a method of finite priority are provable for adequate Friedberg-theories, e.g. the Friedberg-Muchnic solution to Post's problem. Furthermore, in adequate theories every r.e. degree is regular.

Some similar results have been obtained by Simpson for his thin admissible sets.

A typical example of a Friedberg-theory is the recursion theory obtained from an admissible, resolvable set with urelements. In this paper we will prove that $L(\omega_1)_{V^\omega(\mathbb{Q})}$ is not adequate, where $\omega_1 = \omega_1^{ck}$ and $V^\omega(Q)$ is a countable-dimensional vector space over $\mathbb{Q}$. We also prove that if $\mathfrak{M}$ is a structure having a representative in $L(\alpha)$ where α is admissible, then $L(\alpha)_{\mathfrak{M}}$ will be admissible and we may imbed the $L(\alpha)$-degrees into the $L(\alpha)_{\mathfrak{M}}$-degrees. If the representative is natural, the imbedding will be an isomorphism on the Σ_n-degrees for all n. Thus we solve all degree-theoretic problems (for definable degrees) for $L(\omega_1)_{V^\omega(\mathbb{Q})}$ by reducing them to $L(\omega_1)$. In addition it follows from theorem 2 that every r.e. degree is regular for this structure, thus showing that adequacy is not necessary for the regular set theorem.

In solving problems for the non-adequate structure $L(\omega_1)_{V^\omega(\mathbb{Q})}$ we used the adequate structure $L(\omega_1)$, so this is not a way of getting around adequateness, but of reducing to adequateness.

We will use notation from Barwise [1]. The results left without proof or with just a trace of proof will all be trivial for readers with background from Barwise [1] or a similar exposition.

In the sequel we let $\mathfrak{M} = \langle M ; R_1, \ldots, R_l \rangle$ be a structure for the language L and we let $L(\alpha)_{\mathfrak{M}}$ be admissible relative to the language $L^* = L(\in)$.

Definition 1

i $\mathfrak{M}$ is _imbeddable_ into $L(\alpha)$ if there is $\mathfrak{M}' = \langle M', R_1', \ldots, R_l' \rangle \in L(\alpha)$ isomorphic to $\mathfrak{M}$. $\mathfrak{M}'$ is called a _representative_ for $\mathfrak{M}$.

ii A representative $\mathfrak{M}'$ for $\mathfrak{M}$ is a _natural representative_ if the set of finite $\tau : \mathfrak{M} \to \mathfrak{M}'$ which can be extended to an isomorphism from $\mathfrak{M}$ to $\mathfrak{M}'$ is $L(\alpha)_{\mathfrak{M}}$-recursive.

Assume $\mathfrak{M}$ is imbeddable into $L(\alpha)$ and fix a representative $\mathfrak{M}'$

for $\mathfrak{M}$. Let σ be an isomorphism between $\mathfrak{M}$ and $\mathfrak{M}'$. We are going to define an interpretation $I = I_\sigma$ of $L(\alpha)_{\mathfrak{M}}$ into $L(\alpha)$ such that each element $x \in L(\alpha)_{\mathfrak{M}}$ is interpreted by a unique element $I(x) \in L(\alpha)$. Define

$$I_\sigma(p) = \langle 0, \sigma(p) \rangle$$
$$I_\sigma(a) = \langle 1, \{I_\sigma(x); x \in a\} \rangle .$$

Let $\mathcal{F}_1, \ldots, \mathcal{F}_N$, $\mathcal{J}$ and $\mathcal{D}$ be the functions used by Bawise [1] to generate $L(\alpha)_{\mathfrak{M}}$. It is a tedious exercise to define $L(\alpha)$-recursive functions $\mathcal{F}_1^I, \ldots, \mathcal{F}_N^I$, $\mathcal{J}^I$ and $\mathcal{D}^I$ dependent only on $\mathfrak{M}'$ and not on the particular isomorphism σ such that if say $\mathcal{F}_i(x,y) = z$ then $\mathcal{F}_i^I(I(x), I(y)) = I(z)$.

We may now use these functions to define $L^I(M,\beta)$, the interpretation of $L(M,\beta)$, for each $\beta < \alpha$. Let $L^I(M,\alpha) = \bigcup_{\beta<\alpha} (L^I(M,\beta))_2$. Then $I: L(M,\alpha) \to L^I(M,\alpha)$ is a bijection. $L^I(M,\alpha)$ is $L(\alpha)$-recursive.

Suppose Ψ is a formula of the language L^* without parameters. Then Ψ^I is obtained in the usual way by replacing each symbol of L^* appearing in Ψ by a symbol for a suitable $L(\alpha)$-recursive relation, and restricting quantifiers to $L^I(M,\alpha)$. Thus for example $\in$ is replaced by E where E is the $L(\alpha)$-recursive relation

$$xEy \iff y = \langle 1, (y)_2 \rangle \ \& \ x \in (y)_2 .$$

Finally we use the notation that if σ is an automorphism on $\mathfrak{M}$, then $\bar{\sigma}$ is the extension of σ onto $L(\alpha)_{\mathfrak{M}}$ defined by

$$\bar{\sigma}(p) = \sigma(p)$$
$$\bar{\sigma}(a) = \{\bar{\sigma}(x); x \in a\} .$$

Lemma 1

i There is a total $L(\alpha)_{\mathfrak{M}}$-recursive function ρ such that for each $x, \rho(x)$ is a set of good Σ_1-definitions for x with parameters from $M \cup \{L(M,\beta); \beta < \alpha\}$.

ii Let σ be an automorphism on $\mathfrak{M}$. Then for each formula Ψ of L^*

$$L(\alpha)_{\mathfrak{M}} \models \Psi(x_1,\ldots,x_n) \iff L(\alpha)_{\mathfrak{M}} \models \Psi(\bar{\sigma}(x_1),\ldots,\bar{\sigma}(x_n)).$$

Furthermore $\bar{\sigma}(L(M,\beta)) = L(M,\beta)$ for each $\beta < \alpha$ so for each definable set $A \subseteq L(\alpha)_{\mathfrak{M}}$ there are $p_1,\ldots,p_k \in M$ such that $\bar{\sigma}''A$ is determined by the values of σ on $p_1,\ldots,p_k$.

iii Let σ be an isomorphism from $\mathfrak{M}$ onto $\mathfrak{M}'$. Then for each formula Ψ of L^*

$$L(\alpha)_{\mathfrak{M}} \models \Psi(x_1,\ldots,x_n) \iff L(\alpha) \models \Psi^{I_\sigma}(I_\sigma(x_1),\ldots,I_\sigma(x_n))$$

Furthermore Ψ^{I_σ} and $I_\sigma(L(M,\beta)) = L^{I_\sigma}(M,\beta)$ are dependent on $\mathfrak{M}'$ but not on the particular σ, so for each definable set $A \subseteq L(\alpha)_{\mathfrak{M}}$ there are $p_1,\ldots,p_k \in M$ such that $I_\sigma''A$ is determined by the values of σ on $p_1,\ldots,p_k$.

For $A,B \subseteq L(\alpha)_{\mathfrak{M}}$ we let $A \leq B$ mean that A is recursive in B, where "recursive in" is the $L(\alpha)_{\mathfrak{M}}$ analogue of "α-recursive in".

<u>Theorem 1</u>

Assume $\mathfrak{M}$ is imbeddable into $L(\alpha)$. If $A,B \subseteq L(\alpha)$ then $A \leq B$ in $L(\alpha)$ if and only if $A \leq B$ in $L(\alpha)_{\mathfrak{M}}$.

<u>Proof</u>: We first show that if $W \subseteq L(\alpha)$ is $L(\alpha)_{\mathfrak{M}}$- r.e. then W is in fact $L(\alpha)$-r.e. So suppose $W \subseteq L(\alpha)$ is definable over $L(\alpha)_{\mathfrak{M}}$ by a Σ_1-formula Ψ using parameters $y_1,\ldots,y_k \in L(\alpha)_{\mathfrak{M}}$. Then

$$x \in W \iff L(\alpha)_{\mathfrak{M}} \models \Psi(x,y_1,\ldots,y_k) \iff L(\alpha) \models \Psi^I(I(x),I(y_1),\ldots,I(y_k)).$$

But the last relation is Σ_1 over $L(\alpha)$ since Ψ is Σ_1 and $I \restriction L(\alpha)$ is $L(\alpha)$-recursive.

Let $Pu(a) = \{x \in a; sp(x) = \emptyset\}$ where sp is the support function. Suppose $A \leq B$ in $L(\alpha)$ via a reduction procedure W, i.e.

for each $a_1, a_2 \in L(\alpha)$,

$$a_1 \subseteq A \ \& \ a_2 \cap A = \emptyset \iff \exists b_1, b_2(\langle a_1, a_2, b_1, b_2\rangle \in W \ \& \ b_1 \subseteq B \ \& \ b_2 \cap B = \emptyset)$$

where W is $L(\alpha)$-r.e. Then $A \leq B$ in $L(\alpha)_{\mathfrak{m}}$ via the reduction procedure $W_1 = \{\langle a_1, a_2, b_1, b_2\rangle ; \langle a_1, Pu(a_2), b_1, b_2\rangle \in W\}$.

Now suppose $A \leq B$ in $L(\alpha)_{\mathfrak{m}}$ via the reduction procedure V. Let $V_1 = \{\langle Pu(a_1), Pu(a_2), Pu(b_1), Pu(b_2)\rangle ; Pu(b_1) = b_1 \ \& \ \langle a_1, a_2, b_1, b_2\rangle \in V\}$. Then V_1 is $L(\alpha)_{\mathfrak{m}}$-r.e. and a subset of $L(\alpha)$, so V_1 is $L(\alpha)$-r.e. Clearly $A \leq B$ in $L(\alpha)$ via V_1.

Corollary. There is an imbedding i of the $L(\alpha)$-degrees into the $L(\alpha)_{\mathfrak{m}}$-degrees given by $i(L(\alpha)\text{-deg}(A)) = L(\alpha)_{\mathfrak{m}}\text{-deg}(A)$ where $A \subseteq L(\alpha)$.

Using the regular set theorem for $L(\alpha)$ and the deficiency set for a regular $L(\alpha)_{\mathfrak{m}}$-r.e. set we have

Theorem 2

The imbedding i is an isomorphism on the r.e. degrees if and only if every $L(\alpha)_{\mathfrak{m}}$-r.e. degree contains a regular $L(\alpha)_{\mathfrak{m}}$-r.e. set.

In particular we have that i is an isomorphism on the r.e. degrees if $L(\alpha)_{\mathfrak{m}}$ is adequate.

Theorem 3

Assume that $\mathfrak{m}'$ is a natural representative of $\mathfrak{m}$. Let σ_o be a fixed isomorphism of $\mathfrak{m}$ onto $\mathfrak{m}'$, $I = I_{\sigma_o}$ the derived imbedding of $L(\alpha)_{\mathfrak{m}}$ into $L(\alpha)$.

If $A \subseteq L(\alpha)_{\mathfrak{m}}$ is 1st order definable over $L(\alpha)_{\mathfrak{m}}$, then A and $I''A$ have the same $L(\alpha)_{\mathfrak{m}}$-degree.

Corollary. Let α, $\mathfrak{m}$ and $\mathfrak{m}'$ be as in theorem 3. Then there is an isomorphism between the Σ_n-degrees of $L(\alpha)$ and the Σ_n-degrees of $L(\alpha)_{\mathfrak{m}}$.

Proof of corollary: It follows from theorem 3 that the imbedding of degrees described in theorem 1 will be onto the definable $L(\alpha)_{\mathfrak{M}}$-degrees.

Proof of theorem 3: Let A be defined by a formula using parameters $p_1,\ldots,p_k,L(M,\alpha_1),\ldots,L(M,\alpha_n)$ only. Let $\tau = \sigma_o \restriction \{p_1,\ldots,p_k\}$. Then for any isomorphism $\sigma: \mathfrak{M} \to \mathfrak{M}'$, if $\sigma \restriction \{p_1,\ldots,p_k\} = \tau$, then $I''A = I_\sigma''A$.

The following relation will be recursive:

$R(x,y) \iff$ There is a $\tau' \supseteq \tau$ such that τ' is defined on the set of parameters from M used in one of the definitions in $\rho(x)$ (where ρ is as in lemma 1 (i)), τ' may be extended to an isomorphism $\sigma: \mathfrak{M} \to \mathfrak{M}'$ and $I_\sigma(x) = y$. ($I_\sigma(x)$ depends only on τ').

Let $R_1(x,z) \iff \exists y(R(x,y)\ \&\ z = \{w;wEy\})$.

Then

$$x \subseteq A \iff \exists z(R_1(x,z)\ \&\ z \subseteq I''A) \quad \text{and}$$

$$x \cap A = \emptyset \iff \exists z(R_1(x,z)\ \&\ z \cap I''A = \emptyset).$$

This reduces A to $I''A$. To obtain the other reduction, define $R_2(u,v) \iff u = \{x;\exists y \in v\, R(x,y)\}$.

R_2 will be recursive. Moreover, given v, the set of x such that for some $y \in v, R(x,y)$ holds, will be of bounded constructible rank in $L(\alpha)_{\mathfrak{M}}$. Thus there will be a set u in $L(\alpha)_{\mathfrak{M}}$ such that $R_2(u,v)$ will hold.

But then

$$v \subseteq I''A \iff \exists u(R_2(u,v)\ \&\ u \subseteq A)$$

$$v \cap I''A = \emptyset \iff \exists u(R_2(u,v)\ \&\ u \cap A = \emptyset).$$

This ends the proof of theorem 3.

Theorem 4.

Let $\mathfrak{M} = V^{\omega}(\mathbb{Q})$ be a countably-dimensional vectorspace over the field $\mathbb{Q}$. Then $L(\omega_1)_{\mathfrak{M}}$ will be admissible, resolvable but not adequate.

Proof: Both admissibility and resolvability are trivial. Let $<$ be a Σ_1-prewellordering of $L(\omega_1)_{\mathfrak{M}}$. Let $<^*$ be an initial segment of $<$ such that $L(\omega_1)_{\mathfrak{M}}$ is recursively projectible into field$(<^*)$, and $\|<^*\|$ is a limit ordinal. Let π be a projection. Let $<$ and π be definable in parameters $p_1,\ldots,p_k \in M$. There will be a finite-dimensional subspace $\mathfrak{M}_o$ of $\mathfrak{M}$ such that $p_1,\ldots,p_k \in M_o$. For any two elements $r_1, r_2 \in M \setminus M_o$ there will be an automorphism τ on $\mathfrak{M}$ such that $\tau(r_1) = r_2$ and τ is the identity on M_o. It follows that $\pi(r_1)$ and $\pi(r_2)$ will be at the same $<^*$-level. Now, let $r \in M \setminus M_o$. Let π_1 be a recursive projection of ω_1 into ω. Define $\pi_2(\alpha) = \pi(1+\pi_1(\alpha)\cdot r)$ (where $\cdot$ is the scalar multiplication of the vector space). The image of this projection will be semi-recursive but not $L(\omega_1)_{\mathfrak{M}}$-finite. It will, however be bounded in $<^*$. This shows that $L(\omega_1)_{\mathfrak{M}}$ cannot be adequate.

Remark. The only property about $<$ used was its definability. Moreover, the same argument shows that $L(\alpha)_{\mathfrak{M}}$ is not thin in the sense of Simpson [3].

We end this paper by listing some results obtained by similar methods. We will only give indications of proofs.

Definition 2.

Let $\mathfrak{M} = \langle M; R_1,\ldots,R_l\rangle$ be a structure. We call $\mathfrak{M}$ semirigid if there is a finite subset A of M such that the identity on M is the only $\mathfrak{M}$-automorphism extending the identity on A.

Theorem 5.

i If $L(\alpha)_{\mathfrak{M}}$ contains a well-ordering of M then $\mathfrak{M}$ is semirigid.

ii If α is admissible, $\mathfrak{M}$ semirigid and $\mathfrak{M}$ has a natural representative $\mathfrak{M}' \in L_\alpha$, then there is a wellordering of M in $L(\alpha)_{\mathfrak{M}}$.

Indication of proofs.

i If $<$ is a relation and σ is an automorphism that is the identity on the parameters defining $<$, $\bar{\sigma}$ will be a $<$-automorphism. If $<$ is a well-ordering, then the identity is the only $<$-automorphism.

ii An isomorphism $\sigma ; \mathfrak{M} \to \mathfrak{M}'$ will be uniquely decided by its values on a finite set $A \subseteq M$, and thus $L(\alpha)_{\mathfrak{M}}$-recursive. M' is well-orderable in $L(\alpha)_{\mathfrak{M}}$, so M will be well-orderable.

Some examples of structures that are not semirigid but have natural representatives in $L(\omega_1)$ will be:

$\mathfrak{M}_1 = V^\omega(\mathbb{Q})$, $\mathfrak{M}_2 = \langle M_2 \rangle$ where M_2 is countable,
$\mathfrak{M}_3 = \langle \mathbb{Q}, < \rangle$ and $\mathfrak{M}_4$ = The field of algebraic numbers.

Among those, only $L(\omega_1)_{\mathfrak{M}_1}$ is not adequate. In fact, $L(\alpha)_{\mathfrak{M}}$ is adequate for all admissible α if $\mathfrak{M}$ is the trivial structure or a dense linear ordering. (To prove this we must carefully define a prewell-ordering of $HF(M)$. For uncountable $\mathfrak{M}$ we use in addition an absoluteness argument.)

Our last result will be a criterion for adequateness:

Theorem 6.

Let α be admissible, α^* the Σ_1-projectum of α. Let $\mathfrak{M}$ be a structure and assume that $\mathfrak{M}$ has a representative in $L(\alpha^*)$. Then $L(\alpha)_{\mathfrak{M}}$ is admissible and adequate.

Indication of proof: We prove two statements:

i If $\beta < \alpha^*$ and $A \subseteq L(M,\beta)$ is Σ_1 over $L(\alpha)_{\mathfrak{M}}$ then $A \in L(\alpha)_{\mathfrak{M}}$.

Proof: By the imbedding A will be mapped onto $I''A$, a Σ_1-set bounded in $L(\alpha^*)$. It follows that $I''A \in L(\alpha)$, and hence we may use a bounded existential quantifier to define A. Thus $A \in L(\alpha)_{\mathfrak{M}}$.

ii $L(\alpha)_{\mathfrak{M}}$ may be projected into $L(M,\alpha^*)$.

Proof: We project $L(\alpha)_{\mathfrak{M}}$ into $\alpha^* \times HF(M)$ by using the projection of α to α^* on the parameters defining the elements of $L(\alpha)_{\mathfrak{M}}$.

Corollary.

Let $\mathfrak{M} = V^{\omega}(Q)$, $\alpha > \omega$ admissible. Then $L(\alpha)_{\mathfrak{M}}$ is adequate if and only if $\alpha^* > \omega$.

References

1. J. Barwise, Admissible sets and structures, Springer Verlag, 1975.

2. Y.N. Moschovakis, Axioms for computation theories, first draft, in R.O. Gandy and C.M.E. Yates (eds.), Logic Colloquim '69. North-Holland 1971, pp. 199-255.

3. S.G. Simpson, Post's problem for admissible sets, in J.E. Fenstad and P.G. Hinman (eds.), Generalized Recursion Theory, North-Holland 1974, pp. 437-441.

4. V. Stoltenberg-Hansen, Finite injury arguments in infinite computation theories, Oslo Preprint Series 1977.

J.E. Fenstad, R.O. Gandy, G.E. Sacks (Eds.)
GENERALIZED RECURSION THEORY II

On the $\forall\exists$-Sentences of α-Recursion Theory

Richard A. Shore*

Department of Mathematics

Cornell University, Ithaca, New York and

University of Illinois, Chicago Circle, Chicago, Illinois

This paper is at least in some ways a survey of three major areas of α-recursion theory: the lattices of α-r.e. sets, α-degrees and α-r.e. degrees. Its major emphasis is on the last of these areas. We have, in addition, limited ourselves by adopting a single viewpoint from which we will survey the work to date in these areas. In each case we focus on the progress made towards deciding the appropriate $\forall\exists$ theory (i.e. all sentences with only one alternation of quantifiers). An added perspective will be provided by continuously pointing out both the analogues and differences with ordinary recursion theory (henceforth ORT). In particular, the last section contains a construction furnishing the first known difference between the theories of the α-r.e. degrees (for some α) and the ω-r.e. degrees. We will also indicate how certain open problems in the theory of the ω-r.e. degrees can be settled for these α in such a way as to greatly simplify the $\forall\exists$-theory of the α-r.e. degrees. In addition, a few other conjectures and more general suggestions for future investigations in both the three areas specifically discussed and some related ones are scattered throughout the paper. Basic definitions and background information as well as a view of some other areas in α-recursion theory can be found in S. Simpson's article in this volume. For a semi-historical survey and introduction to the sub-

*The author would like to thank the University of Illinois, Chicago Circle for defraying some of the costs of attending the symposium. The preparation of this paper was partially supported by NSF Grant MCS77-04013.

ject as well as a rather complete annotated bibliography we refer the reader to [26].

1. α-R.E. SETS

We begin with a brief look at the lattices $\mathcal{E}(\alpha)$ of α-r.e. sets. (For a much more thorough survey the reader should consult M. Lerman's article in this volume.) As in ordinary recursion theory the underlying plan of investigation of these lattices has been to analyze their structural properties with an eye towards decision procedures for their elementary theories (or perhaps towards proving them undecidable). Along these lines, as in ORT, attention has focused on quotient lattices, $\mathcal{E}^*(\alpha)$, of $\mathcal{E}(\alpha)$ by the ideal of "finite" sets. Of course by Lerman [12] the elementary theories of $\mathcal{E}(\alpha)$ and $\mathcal{E}^*(\alpha)$ are equidecidable. Note that in this setting the appropriate definition of "finite" is what Lerman has called α^*-finite: the set and every one of its α-r.e. subsets is α-finite. With this definition the "finite" sets do in fact form an ideal (even a definable one) in $\mathcal{E}(\alpha)$. Indeed Lerman [12] has shown that in a strong sense this is the only reasonable definition of "finite" in the setting of these lattices.

At the first level of quantifier complexity it is fairly straightforward to show that all the $\mathcal{E}^*(\alpha)$ have the same decidable $\exists$-theory [11]. (The language used here is the strong one of Lachlan [5] in which quantifiers range over $\mathcal{E}^*(\alpha)$ but the operation of complementation is allowed in addition to union and intersection. A predicate saying when such a Boolean combination is an element of $\mathcal{E}^*(\alpha)$ is also included.) It is at the next ($\exists\forall$) level that the standard first non-trivial question about the r.e. sets appears: Is there a maximal element in $\mathcal{E}^*(\alpha)$?

Here Lerman's work [13] supplies an answer which is the best structural result to date on this whole class of lattices in that it gives a complete answer for every α by exactly pinning down the combinatorial properties of those ordinals with maximal sets. As the required property is a fairly simple counting of α (to be precise one needs an α-recursive function f such that $\lim_{\sigma\to\alpha}\lim_{\tau\to\alpha} f(x,\sigma,\tau)$ gives a one-one map of α into ω), it also points to a unique role for ω and true finiteness even in generalized recursion theory (GRT) where one tries to replace them with other domains and notions. Other differences between various $\mathcal{E}^*(\alpha)$ and $\mathcal{E}^*(\omega)$ of a basically set theoretic nature also appear in the study of maximal sets. Thus for example Leggett [8] shows how the order types of the complements of α-r.e. sets play a key role even in the structure of maximal α-r.e. sets. Other examples appear also in [2]. As one might expect however such properties play lesser roles in the more purely recursion theoretic settings of degree theory.

Now the existence of maximal elements in $\mathcal{E}^*(\omega)$ is not only historically the first important question at the second level of quantifier complexity but is also a key ingredient in Lachlan's decision procedure for the $\forall\exists$ theory of $\mathcal{E}^*(\omega)$ [5]. The other main ingredient is a strengthened version of the major subset construction. (A is major in B if for every r.e. W, $B \cup W =^* \omega$ implies that $A \cup W =^* \omega$, where $*$ indicates modulo finite sets.) Here too Lerman [11] has shown that when α is countable and sufficiently nice ($\alpha^* = \alpha = \sigma 2p(\alpha) = t\sigma 2p(\alpha)$) one can carry out such a construction. Indeed the $\forall\exists$ theory of $\mathcal{E}^*(\alpha)$ is exactly the same as that of $\mathcal{E}^*(\omega)$. On the other hand when α is sufficiently like $\aleph_1^L$ one can also do the strengthened major subset argument. In this case too Lerman uses it to give a necessarily different decision procedure for the $\forall\exists$ theory.

Although one needs somewhat more than the major subset

construction in these arguments one does not as yet have even the usual version for every α. The best result seems to be that if the Σ_2 cofinality of α ($\sigma 2\,\mathrm{cf}(\alpha)$) is greater than or equal to the Σ_2 projection of α ($\sigma 2\mathrm{p}(\alpha)$) then every non-$\alpha$-recursive α-r.e. set has a major α-r.e. subset [10]. Now if $\sigma 2\,\mathrm{cf}(\alpha) < \sigma 2\mathrm{p}(\alpha)$ there are a few nice structural results on hyperhypersimple sets [2] but the major subset question remains open and in general the $\forall\exists$ theory of $\mathcal{E}^*(\alpha)$ may prove difficult to analyze though it should be quite interesting.

2. α-DEGREES

The theory of the α-degrees (under α-reducibility) is the least developed of the three areas we are considering. While it is certainly true that less effort has been devoted to this area, there may be other reasons as well for its lagging development. In particular the basic motivations for definitions and constructions in α-recursion theory as well as other generalized recursion theories have come mainly from the study of recursiveness and recursive enumerability. Thus they may not be as well suited to general degree theory which concerns itself with quite arbitrary sets. Indeed most of the results and constructions that do exist in this area have a much more set-theoretic character than those in the other ones. In fact the newest of them [3] use techniques from the recent combinatorial solutions to certain cases of the singular cardinal problem. Nonetheless the general notion of relative recursion on arbitrary sets seems to be an important one for GRT and so we hope for continued progress in this area of α-recursion theory.

The first questions one asks about $\mathcal{D}_\alpha$, the ordering of the α-degrees, are the simple embedding problems in the style of Kleene and Post [4]. Now it does not seem to be too difficult to show directly that various partial orderings can be embedded in the α-degrees by combining the Kleene-Post type of arguments with some tricks peculiar to α-recursion theory. The only proofs of such results that have appeared in the literature, however, are ones giving the embeddings into the α-r.e. degrees [14] which require a priority argument as well. In any case as every α-finite partial ordering is so embeddable we can conclude that the $\exists$-theory of $(\mathcal{D}_\alpha, \leq)$ is the same decidable theory for every α.

In ORT the easiest questions at the next ($\forall\exists$) quantifier level are again of the Kleene-Post type but in various relativized forms. The simplest cases for example being the theorems that (a) $(\forall A)(\exists B,C)[B|C \;\&\; A\leq B,C]$ and (b) $(\forall A)(\exists B)(A|B)$. The constructions needed for these theorems in ORT differ very little from the unrelativized embeddings (although the second does require one important additional idea). Ordinarily one thinks instead of the minimal degree construction as the next level of complexity. Now as it turns out it is not only the next question but in its most general form it's the only other question at this level. To be precise the minimal degree type constructions and the Kleene-Post ones together suffice to decide the full $\forall\exists$ theory of $\mathcal{D}_\omega$. (We made this observation in response to a question of C. Jockusch. It was also made independently by M. Lerman while R.Soare suggested the use of theorem 2.1 to replace the ad hoc construction we first used to prove the result.) One needs only the following two theorems:

THEOREM 2.1 [4]. For any given sets $A_1,\ldots,A_n$ there are sets $B_1,\ldots,B_m$ such that no B_i is recursive in the join of $A_1,\ldots,A_n$ and all the other B_j and, for each i, $A_{j_1},\ldots,A_{j_k}$, $A_i\leq A_{j_1}\vee\ldots\vee A_{j_k}\vee B_1\vee\ldots\vee B_m$ only if $A_i\leq A_{j_1}\vee\ldots\vee A_{j_k}$.

THEOREM 2.2 [15]. Every finite lattice can be embedded as an initial segment of $\mathcal{D}_\omega$.

To see that these theorems suffice note that any $\forall\exists$ sentence is equivalent to a disjunction of ones of the form $\forall\vec{x}\ \exists\vec{y}[D(\vec{x}) \to \bigvee D_i(\vec{x},\vec{y})]$ where the $D(\vec{x})$ and $D_i(\vec{x},\vec{y})$ are complete atomic diagrams in the displayed variables. We claim that a sentence of this form will be true in $\mathcal{D}_\omega$ just in case $D(\vec{x})$ is inconsistent or some $D_i(\vec{x},\vec{y})$ is consistent with the x's forming an initial segment of $\mathcal{D}_\omega$. The necessity of this condition follows immediately from theorem 2.2. Its sufficiency is an easy consequence of theorem 2.1. (Let the x_j's be the $A_1,\ldots,A_n$ of the theorem and define the required y_k's by using the $B_1,\ldots,B_m$ it gives so that

$$y_k = \bigvee\{x_j \mid D_i(\vec{x},\vec{y}) \to x_j \leq y_k\} \vee$$
$$\bigvee\{y_\ell \mid D_i(\vec{x},\vec{y}) \to y_\ell \leq y_k\} \vee B_k\ .)$$

Now in α-recursion theory some progress has been made on these questions but genuine results quickly came to a halt. Indeed it was only as we were writing this paper that we learned of any real results on the Kleene-Post type questions. Current techniques can fairly easily handle the simple versions giving (a) above if $\bar{\bar{\alpha}}$ is regular and (b) if in addition $V = L$. Moreover the techniques can be combined to show that $\mathcal{D}_\alpha$ satisfies the full theorem 2.1 if $\bar{\bar{\alpha}}$ is regular and $V = L$. On the other hand S. Friedman [3] has just shown that it fails very dramatically for many ordinals. A typical case is $\alpha = \aleph_{\omega_1}$ for which assuming $V = L$ he shows that the degrees above $\emptyset'$ are well-ordered by $\leq_\alpha$. His techniques include ones from β-recursion theory and from recent set-theoretic work on the singular cardinal problem. In fact assuming only GCH below $\bar{\bar{\alpha}}$ he shows that if $\bar{\bar{\alpha}}$ is singular of uncountable cofinality then there is a set A above which any two sets are comparable at least with respect to $\leq_{w\alpha}$. As in the singular cardinal problem however the cofinality ω case seems much more difficult. We would propose $\alpha = \aleph_\omega$ as the prime case to consider and the existence of α-incomparable α-degrees above $\emptyset'$ as a test question (with say $V = L$).

On the other hand, although considerable effort has been devoted to the minimal degree problem the results are only partial as well and no counterexamples at all have come to light. The actual state of the known results is methodologically a bit strange. By using a fairly effective construction (and a priority argument) one can produce minimal degrees if α is sufficiently admissible ([27] and [18]). Alternatively one can use a totally non-effective counting of α to construct one [20]. This is an anomaly that we would very much like to see cleared up. Again we have always viewed $\aleph_\omega^L$ (assuming it's not countable) as the key test case but perhaps in view of S. Friedman's work one should consider $\aleph_{\omega_1}^L$ as well. Some constraints (at least below $\emptyset'$) are given in [1].

Thus we see that the main open questions about the α-degrees are at the $\forall\exists$ quantifier level and center around theorems 2.1 and 2.2. The general plan is to decide the $\forall\exists$ theory of the $\mathcal{D}_\alpha$'s. We might hope for example that if $V = L$ and $\bar{\bar{\alpha}}$ is regular it is the same as for $\mathcal{D}_\omega$. If in addition $\bar{\bar{\alpha}}$ is countable or Σ_2 admissible this result seems only somewhat beyond the reach of current techniques.

We would also like to at least mention two other areas of general α-degree theory. The first is that of the jump operator. A fairly reasonable definition seems at hand and some nice results along the lines of the Friedberg completeness theorem of ORT have been proved by Simpson [34]. Much remains to be done however. The second is the theory of degrees below $\emptyset'$ which is currently developing and flowering in ORT but has been almost entirely neglected in α-recursion theory. It should prove especially interesting when α is not Σ_2 admissible.

3. α-R.E. DEGREES

The notions of recursive enumerability and relative recursiveness are, of course, central to generalizations of recursion theory. In addition to these considerations however there is another major methodological motivation for the effort that has gone into the study of the α-r.e. degrees. It is in this setting in ORT that the specific technology of recursion theory (i.e., priority arguments) is most fully developed. (Soare [35] contains a good introduction to these methods in ORT.) Thus an analysis of the special tools and techniques of recursion theory calls for an extended investigation of these methods and so of the r.e. degrees in GRT. One hopes to discover not only the analogies with ORT but also the essences and limits of these analogies and the various possible divergent behavior of r.e. degrees in GRT beyond these limits. Indeed along these lines we can trace a two-fold progression in the work to date in α-recursion theory. One path of development pursues the elucidation of the structure of the α-r.e. degrees (under relative α-recursiveness). The other follows the development of the different types of priority arguments needed to establish the structural results.

The first step along both paths was a positive solution of Post's problem for every admissible α by Sacks and Simpson [25]. In terms of the structure of the α-r.e. degrees this is obviously the first step: there are incomparable (and hence incomplete) α-r.e. degrees. Methodologically it supplied a general procedure for handling the basic finite injury priority arguments. Their solution was especially interesting from the viewpoint of GRT in that the techniques had a strong model theoretic flavor that is entirely absent in ORT. Thus they not only use the recursion theoretic notion of a Σ_1 projection of α for indexing requirements but in key cases they exploit the sequence of Σ_1 submodels of L_α to calculate bounds on injury sets. Of course such reflection phenomena are by now standard topics in all areas of GRT.

A more purely recursion theoretic approach to the finite injury argument was then developed by Lerman [14]. It exploits a direct approximation to a slightly more complicated projection (tame Σ_2) to index requirements. This effectively replaces the use of both the Σ_1 projection and the sequence of Σ_1 submodels in [25]. As a sample application of this new method Lerman showed that every α-finite partial ordering can be embedded in the α-r.e. degrees for every α. (Of course, the methods of [25] would also have sufficed to carry out this finite injury type construction.) This then determined the $\exists$-theory of the α-r.e. degrees under $\leq_\alpha$. Indeed as Lerman points out the same construction gives the corresponding result for the language having a join operation as well as $\leq_\alpha$.

The next level of complexity of priority arguments involves ones for which the preservations and injuries are still bounded but not in the uniform way possible in the above constructions. The typical example in ORT is Sacks' splitting theorem [22, §5]:

$$(\forall C > 0)(\exists A, B)(A \vee B \equiv C \;\&\; A, B < C).$$

We proved this theorem for every admissible α [28] by introducing a technique for treating α-finite sets of requirements as single requirements. Combined with the ideas of Σ_1 projections and α-recursive approximations to the $\sigma 2cf(\alpha)$ this method seems to suffice for arguments of this type. Indeed it works even in somewhat weaker settings than α-recursion theory. Thus for example if an admissible set or structure for a generalized recursion theory A has what Simpson [33] calls a thin pre-wellordering (i.e., a many-one A-recursive projection into an ordinal that induces a pre-wellordering with A-finite levels) then one can use this method to carry out such finite injury priority arguments on A. Now it is not clear what sort of axioms (such as having a thin pre-well-ordering) must be added on to the usual ones for admissibility to allow such constructions to succeed. Something is definitely needed, however, as Harrington has recently constructed an admissible set A in which every A-r.e. set is A-recursive or A-complete. That a pre-wellordering alone is probably not sufficient is indicated by some results of Simpson using AD [33].

Moving up the line of priority arguments one must next confront the infinite injury methods. Note that in terms of quantifier complexity we remain at the $\forall\exists$ level even in the language with just $\leq$ as the basic examples of these methods in ORT are the density theorem [23],

$$(\forall A < B)(\exists C)(A < C < B),$$

and the minimal pair constructions of [6] and [37],

$$(\exists A, B > 0)(\forall C)(C \leq A \ \&\ C \leq B \rightarrow C \equiv 0).$$

Along these lines we have a positive solution to the density question for every α [29] which exploits the techniques of the splitting theorem together with α-recursive approximations to even more complex projections and cofinality functions. (One must relativize them to the set A given in the hypotheses of the theorem.) The minimal pair problem has proven a bit more stubborn. Work to date by Lerman and Sacks [17], Shore [30] and Maass [18] has produced positive results for most admissible ordinals but not all. The main source of the difficulty in the remaining cases is that $\sigma 2cf(\alpha) < \sigma 2p(\alpha)$, although even some such α have been handled.

Our analysis is thus also blocked at the $\forall\exists$ level for the α-r.e. degrees. However even in ORT the structure of the r.e. degrees at this level of complexity is quite unsettled. For example there are many finite non-distributive lattices for which it is not known if they can be embedded in the ω-r.e. degrees. These are all $\exists\forall$ questions in the language with just $\leq$ as are some other important open problems in the theory of the ω-r.e. degrees. We believe however that we can in ORT decide [32] the $\forall\exists$ sentences of the form $\forall\vec{x}\ \exists\vec{y}\ (D(\vec{x}) \rightarrow D(\vec{x},\vec{y}))$ where $D(\vec{x})$ and $D(\vec{x},\vec{y})$ are complete atomic diagrams in the indicated variables. (What's lacking from the full $\forall\exists$ theory is just the possibility of taking disjuncts of diagrams, $\bigvee D_i(\vec{x},\vec{y})$, on the right.)

The key ingredients of this result are somewhat strengthened versions of the density theorem along the lines of [21] and the embeddings of the distributive lattices in the ω-r.e. degrees [36]. The former type of argument seems accessible to the methods of α-recursion theory used for the density theorem while the latter is essentially just the minimal pair construction. Thus from the viewpoint of structural analysis as well as methodology the minimal pair problem seems to be the next step for the α-r.e. degrees. Its solution would either bring us roughly up to par with ORT or establish a difference at the $\forall\exists$ quantifier level presumably based on the interplay between the Σ_2 projection and cofinality of α when $\sigma 2cf(\alpha) < \sigma 2p(\alpha)$.

The strong failure of Σ_2 admissibility indicated by $\sigma 2cf(\alpha) < \sigma 2p(\alpha)$ is however a two-edged sword. On the one side it has blocked the generalizations to all α of several key results in ORT. On the other side it is the tool needed for some α to cut through obstacles that are either much more difficult or impossible to handle in ORT. As a first example consider the problem of extending the decision procedure discussed above to all $\forall\exists$ sentences (i.e., allowing more than one disjunct on the right of the matrix). One is immediately confronted with a new type of problem epitomized in ORT by Lachlan's non-diamond theorem [6]:

$$(\forall A, B > 0)(\exists C)[0 < C < A, B \vee A, B < C < 1].$$

Thus no pair of elements joining to $\emptyset'$ can be a minimal pair. If one is to carry on the decision procedure one must also answer the question left open in [6] of whether such a pair can have any infimum at all. (The relevant sentence is $(\forall A, B, D)(\exists C)$ $[D \leq A, B \rightarrow D < C \leq A, B \vee A, B \leq C \leq 1]$.) Now in ORT all that [6] tells us (by relativization) is that this is true if A or B is low (i.e., $A' \equiv \emptyset'$ or $B' \equiv \emptyset'$). For many α with $\sigma 2cf(\alpha) < \sigma 2p(\alpha)$ such as $\aleph_\omega^L$ this is really all that we need, for by [34] every incomplete $\aleph_\omega^L$-r.e. degree is low. Now Lerman [16] has shown that the non-diamond theorem holds for every α and we are confident that the relativization to low r.e. degrees needed for the general result presents no serious problems. This should then greatly simplify the $\forall\exists$ theories of the α-r.e. degrees for many α and bring us that much closer to a decision procedure for them. Indeed we suspect (or better hope) that the problem of embedding the non-distributive lattices in the α-r.e. degrees will then be the only serious one left.

In a slightly different direction we have been able to exploit this same property of $\aleph_\omega^L$ and related ordinals with $\sigma 2cf(\alpha) < \sigma 2p(\alpha)$ to establish an actual difference between the theories of the r.e. degrees for ω and α. The differences arise at the $\forall\exists$ level if $\vee$ is allowed in the language and at the $\forall\exists\forall$ level if only $\leq$ is used. The question involved is whether one can combine the splitting and density theorems to show that

$$(\forall A < B)(\exists C, D)[A < C, D < B \ \& \ C \vee D = B].$$

Lachlan has shown by a quite difficult construction [7] that this is not true for the ω-r.e. degrees. On the other hand we will show in the next section that this sentence is true for many α. Again the reason we can carry out the construction for these α is that A must be low. In ORT the analogous result is due to Robinson [21] and it says, of course, that the sentence is true if A is low.

In yet a third direction one sees the condition that $\sigma 2cf(\alpha) < \sigma 2p(\alpha)$ entering directly into results on the jump operator as the sole determining characteristic in Maass's work in this volume. He shows that there is a high α-r.e. set A (i.e., $A' \equiv \emptyset''$) if and only if $\sigma 2p(\alpha) \geq \sigma 2cf(\alpha)$ thus correcting an error in [31]. His work also brings out another important view of this situation. If one considers the structure $\langle L_\alpha, A\rangle$ where A is a complete α-r.e. set and $\sigma 2cf(\alpha) < \sigma 2p(\alpha)$ then the situation looks very much like that in β-recursion theory when $\sigma 1p(\beta) > \sigma 1cf(\beta)$. Admissibility fails in a very strong way. We had thus expected an interaction with β-recursion theory for such ordinals in questions about degrees above $0'$. These expectations have in fact just been fulfilled. As we were writing this paper we received the announcement [3] of results by S. Friedman mentioned in section 2. Another corollary of his methods in β-recursion theory was that for many admissible α of uncountable cofinality and $\sigma 2cf(\alpha) < \sigma 2p(\alpha)$ such as $\aleph_{\omega_1}^L$ there are no incomplete degrees α-r.e. in $0'$ not α-recursive in $0'$. Many cases remain open however. In particular our favorite paradigm $\aleph_\omega^L$ is not touched by these methods as yet. We still view it as the prime target for future work. More information on β-recursion theory can be found in S. Friedman's article in this volume. Of special interest to α-recursion theory when $\sigma 2cf(\alpha) < \sigma 2p(\alpha)$ is Maass's work [19] on the admissible collapse which gives a nice framework for working with such structures. We should also note that even when $\langle L_\alpha, A\rangle$ is admissible it presents some interesting problems not found in the unrelativized case.

We would also like to mention one last area of ORT connected with the jump operator and the r.e. degrees that is relatively undeveloped in α-recursion theory: highness and lowness. Although we have here used the definitions of these ideas in terms of the jump operator there are other possibilities and it is not even clear what the best choices are. One version of highness has arisen in work on minimal pairs [30]. Another view comes up in trying to connect lattice properties in $\mathcal{E}^*(\alpha)$ with degree theoretic ones. Work here has been done on maximal sets by Leggett [9].

4. COMBINING THE SPLITTING AND DENSITY THEOREMS

In this section we will establish the difference discussed above between the first order theories of the r.e. degrees for ω and certain other α's. As indicated it basically stems from our earlier result [31] that if there is precisely one non-hyperregular α-r.e. degree then (and only then) every incomplete α-r.e. set is low. The prime examples of course are the $\aleph_\lambda^L$ for $\lambda < \aleph_\lambda^L$ such as $\aleph_\omega^L$ and $\aleph_{\omega_1}^L$. We first prove the analog of Robinson's theorem [21] in ORT for every admissible ordinal α:

THEOREM 4.1. For any α-r.e. $A < B$ with $A' \equiv 0'$ there are α-r.e. C and D such that $A < C < B$, $A < D < B$ and $C \vee D \equiv B$.

It then follows that if there is precisely one non-hyperregular α-r.e. degree we can always combine the splitting and density theorem in contrast to Lachlan's result [7] for ORT.

The proof of this theorem in α-recursion theory bears much the same relation to our proof of the splitting theorem there as Robinson's proof does to Sacks' splitting theorem in ORT. Thus it is basically a finite injury argument of the unbounded type rather than an infinite injury argument. This is possible since the lowness of A allows us to employ an α-recursive approximation to A' that tells us which preservations are actually needed in that they arise from computations that are correct on A. (Of course, we use the recursion theorem to ask the right questions about the construction while it is actually in progress.) In the proof of the density theorem, on the other hand, one has no control over the preservations associated with computations that are incorrect on A. These preservations are in fact the source of the infinite injury nature of that construction.

As our construction here is basically that of the splitting theorem it would be very helpful if the reader had some familiarity with this argument as presented in [28] or better [26] where we give a more heuristic view than here. For unexplained notation and especially for information on our approximation methods we refer to [29, §1,3]. We now suppose that we are given α-r.e. sets $A < B$ which we may assume to be regular by [24]. A is also assumed to be low and A^σ, B^σ denote the elements enumerated in A or B before stage σ. Our proof begins with a lemma on low sets.

LEMMA 4.2. If A is α-r.e. and low then

a) A is hyperregular and

b) there is an α-recursive $\{0,1\}$ - valued function $g(\sigma,x)$ such that for every x $A'(x) = \lim_{\sigma\to\alpha} g(\sigma,x)$.

Proof. a) The proof of theorem 2.3 of [31] actually shows that if A is not hyperregular then $\emptyset'' \leq_{w\alpha} A'$. Thus as A is low $A' \leq_\alpha O'$ and the non-hyperregularity of A would give $O'' \leq_{w\alpha} O'$ for a contradiction.

b) Let K be a regular α-r.e. set of α-degree $\emptyset'$. As A is low we can compute A' from K via some reduction procedure with index δ. We use the approximation to this reduction to define $g : g(\sigma,x) = \{\delta\}_\sigma^{K^\sigma}(x)$. As is shown in [28] this approximation is correct in the limit. □

We now let ρ be the $\Sigma_1(A)$ projectum of α and γ be its Σ_2 cofinality. We then have the associated functions $f:\alpha \xrightarrow{1-1} \rho$ and $h : \gamma \xrightarrow{\text{cofinal}} \alpha$ which are $\Sigma_1(A)$ and Σ_2 respectively. As in [29] they have natural α-recursive approximations $f^\sigma(x)$ and $h^\sigma(x)$. <u>Warning</u>: We are thinking of our reduction procedures as being indexed below ρ by f. To abbreviate notation a bit we will omit f^{-1} in the corresponding notations. Thus $[\delta]^A$ is short for $[f^{-1}\delta]^A$ and $\{\delta\}^{A^\sigma \vee C_i^\sigma}(x)$ is "short" for $\{f^{-1}\delta\}_\sigma^{A^\sigma \vee C_i^\sigma}(x)$ as used in [29].

Our construction will split the given set B into two disjoint α-r.e. sets C_0 and C_1 so that $C_0 \vee C_1 = B$. The requirements are then $[\delta]^{A \vee C_i} \neq B$. We assign them priorities (in the limit) in blocks given by $fh(\epsilon)$ for $\epsilon < \gamma$. It then suffices to show that these requirements are satisfied since we can then let $A \vee C_i$ be the sets required in the theorem. We now simultaneously describe the construction and define some key notions.

CONSTRUCTION. Stage σ. For each $\epsilon < \delta$ and $i = 0,1$ find the least x such that there is no preservation requirement of index $\langle \epsilon, i, \delta, x\rangle$ for any $\langle \epsilon, i\rangle$-active δ. See if there is an $\langle \epsilon, i\rangle$-active δ of priority ϵ (i.e., $\delta < f^\sigma h^\sigma(\epsilon)$) with $\{\delta\}_\sigma^{A^\sigma \vee C_i^\sigma}(x) = B^\sigma(x)$. If so consider the one with the least computation giving this result. Now continue the enumeration of A and calculation of $g(\gamma, k(\sigma, \epsilon, i, x))$ until one finds a τ such that either A^τ contains an element assumed to be out of A for this computation or out of the one associated with some preservation of index $\langle \epsilon, i, \eta, y\rangle$ for $y < x$ and some $\langle \epsilon, i\rangle$-active η or one calculates that $g(\tau, k(\sigma, \epsilon, i, x)) = 1$.

(In general we say that a computation is shown to be A-incorrect by the element enumerated at stage τ if that element is assumed to be out of A in the computation. We say that a preservation of index $\langle \varepsilon, i, \delta, y\rangle$ is shown to be A-incorrect if either its associated computation or the computation associated with any preservation of index $\langle \varepsilon, i, \beta, y'\rangle$ with $y' < y$ and β $\langle \varepsilon, i\rangle$-active is shown to be A-incorrect. A computation or preservation is A-incorrect if it is shown to be so at some stage. Otherwise it is A-correct.)

In the first case above we have no desire to preserve the computation. In the latter our approximation to A' is saying that we will get an A-correct computation and so we try to preserve the ones we have. To be precise if there is no δ as required or we find ourselves in the first case above we do nothing for ε and i and go on to the next step. Otherwise we create a preservation requirement with index $\langle \varepsilon, i, \delta, x\rangle$ consisting of all the elements assumed to be out of C_i in this computation.

We now put the element z of B enumerated at σ into whichever of C_0, C_1 will cause the least damage. Again to be precise let ε_i be the least ε for which z is in a preservation requirement of index $\langle \varepsilon, i, \eta, y\rangle$ for any η and y. If $\varepsilon_0 \leq \varepsilon_1$ we put z into C_1. Otherwise we put it into C_0. We then cancel all preservation requirements for $i = 1$ or 0 respectively which contain z. We next cancel all preservation requirements which are shown to be A-incorrect by the element of A enumerated at stage σ.

Note that we say a reduction procedure δ is $\langle \varepsilon, i\rangle$-active at σ unless we have a preservation requirement of index $\langle \varepsilon, i, \delta, y\rangle$ created at a stage $\tau < \sigma$ such that $\{\delta\}_{\tau}^{A^{\tau} \vee C_i^{\tau}}(y) \neq B^{\sigma}(y)$. Thus δ being $\langle \varepsilon, i\rangle$-inactive means that we had previously preserved a computation of equality but y has entered B and the computation from $A \vee C$ via δ has remained valid.

Of course the function g is the one given in Lemma 4.2. The definition of k is rather more complicated. We first set

$\ell(\sigma,\epsilon,i) = \cup\{\tau<\sigma \mid (\exists\, \eta,y)$ (some A-correct preservation requirement of index $\langle \epsilon,i,\eta,y\rangle$ is cancelled at stage $\tau)\}$. We then let $k(\sigma,\epsilon,i,x)$ be an index for the following $\Sigma_1(A)$ sentence:

$(\exists\, \eta \geq \ell(\sigma,\epsilon,i))$[At stage η we have for each $y<x$ a preservation requirement of index $\langle \epsilon,i,\beta,y\rangle$ for some $\langle\epsilon,i\rangle$-active β which is A-correct and we are considering creating one with index $\langle \epsilon,i,\delta,x\rangle$ whose associated computation is also A-correct].

Now this sentence itself requires an index for the construction which in turn depends on k. This self-reference is justified, of course, by an appeal to the recursion theorem. That k gives an index for the above sentence means that it is true if and only if $k(\sigma,\epsilon,i,x) \in A'$. Finally the search for the desired τ at each step σ must terminate. If no τ exists such that A^τ shows that one of the computations involved is A-incorrect then they are all A-correct. In this case σ itself is the witness η needed to make the sentence indexed by $k(\sigma,\epsilon,i,x)$ true. We then have that $k(\sigma,\epsilon,i,x) \in A'$ and so $\lim_{\tau\to\alpha} g(\tau,k(\sigma,\epsilon,i,x)) = 1$. Thus we eventually find a τ such that $g(\tau,k(\sigma,\epsilon,i,x)) = 1$.

We must now verify that the construction succeeds. As it is α-recursive the C_i are α-r.e. Moreover they are regular by the regularity of B. As in the splitting theorem $C_i \leq B$: To find $C_i \cap \beta$ just find a stage σ such that $B^\sigma \cap \beta = B \cap \beta$. We then have $C_i \cap \beta = C_i^\sigma \cap \beta$. It is also obvious that $C_0 \cup C_1 = B$ so $B \leq C_0 \vee C_1$. We thus know that $A \leq A \vee C_i \leq B$ and $(A \vee C_0) \vee (A \vee C_1) \equiv B$. To conclude the proof of the theorem it suffices to show that $B \not\leq A \vee C_i$, $i = 0,1$. The key here is of course the priority lemma.

LEMMA 4.3. For each $\epsilon < \gamma$ and $i = 0,1$ there is a stage $\sigma_{\epsilon,i}$ after which no preservation of index $\langle \epsilon,i,\delta,x\rangle$ is created or cancelled for any δ and x.

Proof: We proceed by induction on $\langle \epsilon, i \rangle$. Let σ_ϵ be the bound given by induction for $\langle \epsilon, 1 \rangle$. We wish to consider $\langle \epsilon+1, 0 \rangle$. We may of course assume that $f^\sigma h^\sigma(\epsilon) = fh(\epsilon)$ for $\sigma > \sigma_\epsilon$. Now any A-correct preservation of index $\langle \epsilon, 0, \delta, y \rangle$ can be cancelled at some stage $\tau > \sigma_\epsilon$ only if we enumerate a z in B at stage τ which is in some preservation of index $\langle \epsilon', 1, \delta', y' \rangle$ with $\epsilon' < \epsilon$. By assumption all such preservations are created before stage σ_ϵ and they only contain elements less than σ_ϵ. As B is regular there is a bound τ_0 on the stages at which such elements are enumerated in B. Thus after τ_0 no A-correct preservation of index $\langle \epsilon, 0, \delta, y \rangle$ is ever cancelled.

Next consider the set $W = \{\delta < fh(\epsilon) \mid (\exists \tau > \tau_0)$ (δ is $\langle \epsilon, 0 \rangle$-inactive at stage τ and the associated preservation requirement is A-correct)$\}$. W is $\Sigma_1(A)$ and a subset of $fh(\epsilon) < \rho$. Thus it is α-finite. As A is hyperregular the map taking $\delta \in W$ to its witness $\tau > \tau_0$ is bounded by τ_1. After τ_1 no $\delta < fh(\epsilon)$ can become $\langle \epsilon, 0 \rangle$ - inactive. So any preservation of index $\langle \epsilon, 0, \delta, y \rangle$ created after stage τ_1 (or existing at τ_1 with δ $\langle \epsilon, 0 \rangle$-active) must have an associated computation giving the correct value of $B(y)$. If there were such preservations for every y then we could compute B α-recursively in A (as checking for A-correctness is α-recursive in A) for a contradiction. Let x be the least number for which there is no such preservation requirement and let $\tau_2 \geq \tau_1$ be a stage by which we have such preservation requirements for all $y < x$.

We next consider $g(\sigma, k(\tau, \epsilon, 0, x))$ for $\sigma \geq \tau > \tau_2$. By our choice of $\tau_0 \leq \tau_2$ $\ell(\tau, \epsilon, 0)$ has stabilized, i.e. $\ell(\tau, \epsilon, 0) = \ell(\tau_2, \epsilon, 0)$ for every $\tau > \tau_2$. Thus k has also stabilized say at $k(\epsilon, 0, x)$, i.e. $k(\tau, \epsilon, 0, x) = k(\epsilon, 0, x)$ $\forall \tau > \tau_2$. By our choice of x we see that the sentence coded by $k(\epsilon, 0, x)$ is false and so $g(k(\epsilon, 0, x)) = 0$.

We can then choose a $\tau_3 \geq \tau_2$ so that $g(\sigma, k(\tau, \varepsilon, 0, x)) = g(\sigma, k(\varepsilon, 0, x)) = 0$ for every $\sigma, \tau > \tau_3$.

Finally we claim that no preservation requirement of index $\langle \varepsilon, 0, \delta, y \rangle$ is created at any stage $\tau > \tau_3$. This can happen for any y only if it happens first for x. Suppose we are considering some computation $\{\delta\}_\tau^{A^\tau \vee C_0^\tau}(x) = B^\tau(x)$ at stage $\tau > \tau_3$. By our choice of τ_3 we can never find a $\sigma > \tau$ with $g(\sigma, k(\tau, \varepsilon, 0, x)) = 1$. Thus we must find that this computation is A-incorrect when we do the search at stage τ. Of course we then do not create any preservation and go on to the next step.

This concludes the argument for $\langle \varepsilon+1, 0 \rangle$ and τ_3 is the required $\sigma_{\varepsilon+1,0}$. The argument for $\langle \varepsilon+1, 1 \rangle$ is exactly the same but we begin at τ_3. All that's left is to verify that for a limit ordinal $\lambda < \gamma$ the $\sigma_{\varepsilon,i}$ for $\varepsilon < \lambda$ are bounded. (The argument for $\langle \lambda, 0 \rangle$ is then a repeat of the above but beginning after this bound.) All that's involved here is to note that the map $\langle \varepsilon, i \rangle \mapsto \sigma_{\varepsilon,i}$ is Σ_2. As $\lambda < \gamma$ the Σ_2 cofinality of α, there is a bound on the range of this map as required. □

We can now conclude the proof of theorem 4.1 by establishing the following:

LEMMA 4.4. $B \not\leq A \vee C_i$ for $i = 0, 1$.

Proof: If not we can compute B from $A \vee C_i$ via some reduction procedure $\delta < fh(\varepsilon)$ for some $\varepsilon < \gamma$. Let x be as in the proof of the above lemma (i.e. the least element for which we never create a preservation requirement after stage $\sigma_{\varepsilon,i}$). Now if δ were $\langle \varepsilon, i \rangle$-inactive at any stage $\tau > \sigma_{\varepsilon,i}$ then the associated computation would be correct on both A and C_i. (Otherwise the preservation would be cancelled at some stage after $\sigma_{\varepsilon,i}$ contradicting the definition of $\sigma_{\varepsilon,i}$.) We would then have a correct computation from $A \vee C_i$ via δ giving incorrect information about B contrary to our choice of δ. Thus δ is never $\langle \varepsilon, i\rangle$-inactive at a stage $\tau > \sigma_{\varepsilon,i}$. As A and C are regular there is a $\tau > \sigma_{\varepsilon,i}$ such that $\{\delta\}_\tau^{A^\tau \vee C_i^\tau}(x) = [\delta]^{A \vee C_i}(x) = B(x) = B^\tau(x)$ and the least such computation is in fact the correct one via δ. We would then try to create a preservation requirement of index $\langle \varepsilon, i, \delta, x\rangle$. As the computation is A-correct and $\tau > \sigma_{\varepsilon,i}$ the first possibility in our search can never happen. We must then create such a requirement again contradicting the choice of $\sigma_{\varepsilon,i}$.

$\square$

References

[1] C.T. Chong, Generic sets and minimal α-degrees, to appear.

[2] __________ and M. Lerman, Hypersimple α-r.e. sets, Ann. Math. Logic 9(1975), 1-48.

[3] S. Friedman, A counterexample to Post's Problem, circulated manuscript.

[4] S.C. Kleene and E.L. Post, The upper semi-lattice of degrees of recursive unsolvability, Ann. Math. Ser. 2, 59(1954) 397-407.

[5] A.H. Lachlan, The elementary theory of recursively enumerable sets, Duke Math. J. 35(1968) 123-146.

[6] __________, Lower bounds for pairs of recursively enumerable degrees, Proc. Lon. Math. Soc. (3) 16(1966) 537-569.

[7] __________, A recursively enumerable degree which will not split over all lesser ones, Ann. Math. Logic 9(1975) 307-365.

[8] A. Leggett, Maximal α-r.e. sets and their complements, Ann. Math. Logic 6(1974) 293-357.

[9] __________, α-Degrees of maximal α-r.e. sets, to appear.

[10] __________ and R.A. Shore, Types of simple α-recursively enumerable sets, J. Symb. Logic 41(1976) 681-694.

[11] M. Lerman, On the elementary theories of some lattices of α-recursively enumerable sets, to appear.

[12] __________, Ideals of generalized finite sets in lattices of α-recursively enumerable sets, Z. f. math. Logik und Grund. d. Math. 22(1976) 347-352.

[13] __________, Maximal α-r.e. sets, Trans. Ann. Math. Soc. 188 (1974) 341-386.

[14] __________, On suborderings of the α-recursively enumerable α-degrees, Ann. Math. Logic 4(1972) 369-392.

[15] __________, Initial segments of the degrees of unsolvability, Ann. Math. 93(1971) 365-389.

[16] __________, Least upper bounds for minimal pairs of α-r.e. degrees, J. Symb. Logic 39(1974) 49-56.

[17] __________ and G.E. Sacks, Some minimal pairs of α-recursively enumerable degrees, Ann. Math. Logic 4(1972) 415-442.

[18] W. Maass, On minimal pairs and minimal degrees in higher recursion theory, to appear.

[19] __________, Inadmissibility, tame r.e. sets and the admissible collapse, to appear.

[20] J. MacIntyre, Minimal α-recursion theoretic degrees, J. Symb. Logic 38(1973) 18-28.

[21] R.W. Robinson, Interpolation and embedding in the recursively enumerable degrees, Ann. Math. 93(1971) 285-314.

[22] G.E. Sacks, Degrees of unsolvability, Annals of Mathematical Studies no.55, Princeton University Press, Princeton, New Jersey, 1963.

[23] __________, The recursively enumerable degrees are dense, Ann. Math. 80(1964) 300-312.

[24] __________, Post's problem, admissible ordinals and regularity, Trans. Am. Math. Soc. 124(1966) 1-23.

[25] __________ and S.G. Simpson, The α-finite injury method, Ann. Math. Logic 4(1972) 323-367.

[26] R.A. Shore, α-Recursion theory, in Handbook of Mathematical Logic, J. Barwise ed., North-Holland, Amsterdam, 1977.

[27] __________, Minimal α-degrees, Ann. Math. Logic 4(1972) 393-414.

[28] __________, Splitting an α-recursively enumerable set, Trans. Am. Math. Soc. 204(1975) 65-77.

[29] __________, The recursively enumerable α-degrees are dense, Ann. Math. Logic 9(1976) 123-155.

[30] __________, Some more minimal pairs of α-recursively enumerable degrees, Z. f. Math. Logik und Grund. der Math., to appear.

[31] __________, On the jump of an α-recursively enumerable set, Trans. Am. Math. Soc. 217(1976) 351-363.

[32] __________, On the elementary theory of the recursively enumerable degrees, in preparation.

[33] S.G. Simpson, Post's problem for admissible sets, in Generalized Recursion Theory, Proceedings of the 1972 Oslo Symposium, J.E. Fenstad and P.G. Hinman eds., North Holland, Amsterdam 1974, 437-441.

[34] __________, Degree theory on admissible ordinals, ibid., 165-194.

[35] R.I. Soare, The infinite injury priority method, J. Symb. Logic, 41(1976) 513-530.

[36] S.K. Thomason, Sublattices of the recursively enumerable degrees, Z. f. Math. Logik und Grund. d. Math. 17(1971) 273-280.

[37] C.E.M. Yates, A minimal pair of r.e. degrees, J. Symb. Logic 31(1966) 159-168.

J.E. Fenstad, R.O. Gandy, G.E. Sacks (Eds.)
GENERALIZED RECURSION THEORY II

SHORT COURSE ON ADMISSIBLE RECURSION THEORY[1]

Stephen G. Simpson
Department of Mathematics
The Pennsylvania State University
University Park, Pennsylvania 16802

CONTENTS

§ 1. Introduction 355

§ 2. Primitive recursively closed ordinals 356

§ 3. The $\sum_n$ selection theorem 361

§ 4. Admissible ordinals: some examples 367

§ 5. Degree theory on admissible ordinals 371

§ 6. Reflecting ordinals 374

§ 7. The S_n hierarchy 377

§ 8. Oracles, fans and theories 380

§ 9. The basis problem: applications to logic 383

§10. Conclusion 387

References 387

§1. INTRODUCTION.

This paper consists of slightly expanded notes from my course of three lectures at the Second Symposium on Generalized Recursion Theory, held at the University of Oslo, June 13-17, 1977. I would like to thank the organizers and sponsors of the Symposium for making it possible for me to give these lectures.

Although the lectures dealt only with admissible ordinals, the discussion was far from complete. My aim was to give, not a definitive treatment or even an overview, but rather a quick introduction to the subject as a whole and to some topics of current interest. I am preparing a book-length treatment which is to be published by Springer-Verlag in their series, Perspectives in Mathematical Logic.

[1]Preparation of this paper was partially supported by NSF Grant MPS 76-05993.

§2. PRIMITIVE RECURSIVELY CLOSED ORDINALS.

In this section we define the notion of a primitive recursively (p.r.) closed ordinal. We then show how to generalize some theorems from ordinary recursion theory (on the natural numbers) to an arbitrary p.r. closed ordinal.

Let OR be the class of all ordinals. An n-ary function $F : OR^n \to OR$ is said to be primitive recursive if it is generable from the initial functions

$$F(x_1,\ldots,x_n) = x_i, \qquad 1 \le i \le n;$$

$$F(x) = x+1;$$

$$F(x) = 0;$$

$$F(x,y,u,v) = \begin{cases} x & \text{if } u < v, \\ y & \text{otherwise;} \end{cases}$$

by applications of the schemata of substitution

$$F(\vec{x}) = G(H_1(\vec{x}),\ldots,H_m(\vec{x}))$$

and primitive recursion

$$F(y,\vec{x}) = G(\sup_{z<y} F(z,\vec{x}),y,\vec{x}).$$

Familiar examples of primitive recursive functions are the usual operations of ordinal arithmetic

$$x+y = \begin{cases} x & \text{if } y = 0, \\ \sup_{z<y} (x+z)+1 & \text{otherwise;} \end{cases}$$

$$x \cdot y = \sup_{z<y} x \cdot z+1;$$

$$x^y = \begin{cases} 1 & \text{if } y = 0, \\ \sup_{z<y} x^z \cdot x & \text{otherwise.} \end{cases}$$

There is also a primitive recursive pairing function

$$J : OR^2 \xrightarrow[\text{onto}]{1\text{-}1} OR$$

such that $x,y \leq J(x,y)$ for all $x,y \in OR$. Namely

$$J(x,y) = T(x+y)+x$$

where

$$T(z) = \sum_{w<z} w+1$$

$$= \sup_{w<z} T(w)+w+1.$$

An n-ary relation $R \subseteq OR^n$ is said to be _primitive recursive_ if its characteristic function

$$\chi_R(x_1,\ldots,x_n) = \begin{cases} 1 & \text{if } R(x_1,\ldots,x_n), \\ 0 & \text{otherwise} \end{cases}$$

is primitive recursive. The class of primitive recursive relations is closed under Boolean operations and the bounded quantification schema

$$R(y,\vec{x}) \leftrightarrow (\exists z < y)S(z,y,\vec{x}).$$

Primitive recursive functions on the ordinals are both a generalization and an extension of primitive recursive functions on ω. On the one hand, it is easy to see that for primitive recursive F, if $x_1,\ldots,x_n < \omega$ then $F(x_1,\ldots,x_n) < \omega$, and the restriction of F to ω is primitive recursive in the sense of Kleene [18]. On the other hand, primitive recursive functions on OR are different in spirit from those on ω, because the computation trees can be infinite. For instance, the computation of $\omega^\omega = F(\omega,\omega)$ where $F(x,y) = x^y$ as defined above involves the evaluation of ω^n for $n = 0,1,2,\ldots$ and so is ineluctably infinite.

A nonzero ordinal α is said to be _primitive recursively closed_ if $F(x_1,\ldots,x_n) < \alpha$ whenever $x_1,\ldots,x_n < \alpha$ and F is primitive recursive. For instance, as remarked above, ω is p.r. closed. Since there are only countably many primitive recursive functions, a simple argument shows that there exists a proper class of p.r. closed ordinals.

FROM NOW ON, α IS A FIXED P.R. CLOSED ORDINAL. We proceed to set up a generalized recursion theory on the ordinals less than α.

Definition 2.1. A set $X \subseteq \alpha$ is said to be α-recursively enumerable (abbreviated α-RE) if there exist a primitive recursive relation R and a parameter $p < \alpha$ such that

$$X = \{x < \alpha : (\exists y < \alpha) R(x,y,p)\}.$$

In ordinary recursion theory, the two most basic results on recursively enumerable sets are the Enumeration Theorem and the Selection Theorem. We now generalize these results to α-recursion theory.

Enumeration Theorem 2.2. There exists an α-RE relation $W(e,x)$ such that for each α-RE set X there exists $e < \alpha$ such that

$$X = W_e = \{x < \alpha : W(e,x)\}.$$

We omit the tedious but straightforward proof of the Enumeration Theorem.

Definition 2.3. A partial function $f : \alpha \overset{p}{\to} \alpha$ (p for "partial") is said to be α-recursive if the relation

$$\text{graph}(f) = \{(x,y) : f(x) \simeq y\}$$

is α-RE. A subset of α is said to be α-recursive if its characteristic function is α-recursive.

Selection Theorem 2.4. Given an α-RE relation $R(x,y)$, we can find a partial α-recursive function f such that f is a selector for R, i.e. for all x, $f(x)$ is defined if and only if $\exists y R(x,y)$ in which case $R(x,f(x))$.

Proof. Let

$$R(x,y) \leftrightarrow \exists z P(x,y,z,p)$$

where $p < \alpha$, P is primitive recursive, and $x,y,z < \alpha$. Put $f(x) \simeq y$ if and only if

$$\exists z(P(x,y,z,p) \ \& \ \forall y',z'[J(y',z') < J(y,z) \rightarrow \neg P(x,y',z',p)])$$

where J is the primitive recursive pairing function. It is not hard to verify that f is a partial α-recursive selector for R, Q.E.D.

The Enumeration and Selection Theorems team up to yield several basic results about α-RE sets. For instance

Theorem 2.5. There exist α-RE sets A and B such that $A \cap B = \emptyset$ but A and B are α-recursively inseparable, i.e. there is no α-recursive set X such that $A \subseteq X$ and $X \cap B = \emptyset$.

Proof. Let f be a partial α-recursive selector for the α-RE relation

$$R(x,y) \leftrightarrow J(x,y) \in W_x.$$

It is easy to verify that

$$A = \{x : f(x) \simeq 0\}$$

and

$$B = \{x : f(x) \simeq 1\}$$

have the desired properties, Q.E.D.

From the truism that finite sets play an important role in ordinary recursion theory, it follows that no generalization of recursion theory would be complete without a notion of generalized finiteness. Accordingly, we make the following definition.[2]

[2]At this point in the text, an important definition has been omitted. Please turn to page 390.

Proposition 2.7. There is a primitive recursive relation $D(x,y)$ such that for any α-finite set K there exists $x < \alpha$ such that

$$K = D_x = \{y < x : D(x,y)\}.$$

Proof. The proof of the Enumeration Theorem 2.2 shows that the relation W can be written in the form

$$W(x,y) \leftrightarrow \exists z T(x,y,z)$$

where T is primitive recursive. We put

$$D(J(w,x),y) \leftrightarrow (\exists z < w) T(x,y,z).$$

An important link between α-recursion theory and axiomatic set theory can be established as follows. Define a mapping

$$\ell : \mathrm{OR} \to \{\mathrm{sets}\}$$

by

$$\ell(x) = \{\ell(y) : y \in D_x\}.$$

Then, for all p.r. closed ordinals α, we have:

Proposition 2.8

(i) $\ell[\alpha] = \{\ell(x) : x < \alpha\}$ is equal to L_α, the αth level of the constructible hierarchy (called M_α by Gödel [7]).

(ii) $X \subseteq \alpha$ is α-RE if and only if it is $\sum_1(L_\alpha)$ in the sense of Jensen [13].

(iii) $X \subseteq \alpha$ is α-finite if and only if $X \in L_\alpha$.

It is well known that L_α is a model of a substantial fragment of Zermelo-Fraenkel set theory. In particular, L_α is closed under the rudimentary set-theoretical operations $F_1,\ldots,F_8$ of Gödel [8]. This means that intuition acquired from the study of models of set theory can be applied profitably in α-recursion theory.

More details on primitive recursive functions, p.r. closed ordinals, and the constructible hierarchy can be found in Jensen-Karp [14].

§3. THE $\sum_n$ SELECTION THEOREM.

As in the previous section, α is a fixed but arbitrary p.r. closed ordinal. Having discussed α-analogs of recursively enumerable and finite sets, we now turn to an α-analog of the arithmetical hierarchy.

Definition 3.1 A set $X \subseteq \alpha$ is called $\sum_n$ if there exist a primitive recursive relation R and a parameter $p < \alpha$ such that

$$X = \{x : \exists y_1 \forall y_2 \ldots y_n R(x,y_1,y_2,\ldots,y_n,p)\}$$

where all variables range over ordinals $< \alpha$.

Thus X is $\sum_1$ if and only if it is α-RE. It can be shown that X is $\sum_n$ if and only if it is $\sum_n(L_\alpha)$ in the sense of Jensen [13].

In ordinary recursion theory, it is well known that many theorems about recursively enumerable sets can be lifted to the context of $\sum_n^0$ sets, $n = 2,3,\ldots$. In this section we shall consider the $\sum_n(L_\alpha)$ versions of the Enumeration and Selection Theorems.

$\sum_n(L_\alpha)$ Enumeration Theorem 3.2 Let $n \geq 1$. There exists a $\sum_n$ relation $S \subseteq \alpha \times \alpha$ such that for each $\sum_n$ set X there exists $e < \alpha$ such

$$X = \{x : S(e,x)\}.$$

Proof. This is an easy consequence of the special case $n = 1$ (Theorem 2.2).

$\sum_n(L_\alpha)$ Selection Theorem 3.3 (Jensen) Let $n \geq 1$. For each $\sum_n$ relation $R(x,y)$ there exists a $\sum_n$ selector, i.e. a partial function $f : \alpha \overset{p}{\rightarrow} \alpha$ such that the graph of f is $\sum_n$ and f is a selector for R.

Without exaggeration it may be said that the proof of the $\sum_n(L_\alpha)$ Enumeration Theorem is a straightforward generalization of the corresponding proof in orindary recursion theory. Surprisingly, the same is not true of the $\sum_n(L\ )$ Selection Theorem. All known proofs of the latter use a Löwenheim-Skolem argument which is embodied in Lemma 3.5 below. This argument has no counterpart in ordinary recursion theory.

The proof of the $\sum_n(L_\alpha)$ Selection Theorem which we present below is due to Sy Friedman and is a simplification of Jensen's proof [13] which is itself a simplification of the original proof in Jensen [12].

Let X be a set of ordinals. We write π_X for the <u>principal function</u> of X, i.e. the function which enumerates the elements of X in increasing order. The domain of π_X is an ordinal type(X), the order type of X.

A nonempty set of ordinals is said to be p.r. closed if it is closed under the primitive recursive functions.

<u>Lemma</u> 3.4 Let X be a p.r. closed set of ordinals and let $\pi : \beta \xrightarrow[\text{onto}]{1-1} X$ where $\pi = \pi_X$, $\beta = \text{type}(X)$ Then

(i) β is p.r. closed;

(ii) $\pi F(x_1,\ldots,x_n) = F(\pi x_1,\ldots,\pi x_n)$

for all $x_1,\ldots,x_n < \beta$ and primitive recursive F.

We omit the proof of this lemma which is essentially a straightforward induction on the schemata for primitive recursive functions.

It is interesting to note that Lemma 3.4 implies easily that $L = \{\ell(x) : x \in OR\}$ is a model of the Generalized Continuum Hypothesis.

<u>Definition</u> The $\sum_n$ <u>projectum</u> of α (denoted ρ_n or ρ_n^α) is the least ρ such that not every $\sum_n(L_\alpha)$ subset of ρ is α-finite.

<u>Lemma</u> 3.5 (Jensen) Let $n \geq 1$ and assume the $\sum_n$ selection thoerem. Then ρ_n^α can be characterized as the least ρ such that there exists a $\sum_n$ monomorphism $f : \alpha \xrightarrow{1-1} \rho$.

Proof. We first show that the existence of such a monomorphism implies $\rho \geq \rho_n$. To see this, put

$$E = \{fx : fx \notin D_x\}.$$

Then E is a Σ_n subset of ρ. We claim that E is not α-finite. If it were, we would have $E = D_x$ for some $x < \alpha$. Then

$$fx \in D_x \leftrightarrow fx \in E$$
$$\leftrightarrow fx \notin D_x$$

a contradiction. Hence $\rho \geq \rho_n$.

It remains to construct a Σ_n monomorphism $f_n : \alpha \xrightarrow{1-1} \rho_n$. We shall do this by a Skolem hull argument.

Let $A \subseteq \rho_n$ be $\Sigma_n(L_\alpha)$ and not α-finite. Let $p < \alpha$ be such that A is Σ_n in the parameter p. By the Σ_n Enumeration and Selection theorems, let $q < \alpha$ be such that every relation which is Σ_n in a parameter r has a selector which is Σ_n in parameters q and r.

Let X be the "Σ_n Skolem hull" of

$$X_0 = \{x : x < \rho_n \text{ or } x = p \text{ or } x = q\},$$

i.e. X is the set of all y such that $\{y\}$ is Σ_n in a finite set of parameters from X_0. An easy argument with the pairing function shows that there exists a partial epimorphism

$$g : \rho_n \xrightarrow[\text{onto}]{p} X$$

which is Σ_n in parameters p and q.

Let $\beta = \text{type }(X)$ and $\pi = \pi_X: \beta \xrightarrow[\text{onto}]{1-1} X$. Lemma 3.4 tells us that π is an elementary embedding with respect to primitive recursive relations. Since X is a Σ_n Skolem hull, it follows that π is an elementary embedding with respect to Σ_n relations. Thus from g we get a partial epimorphism

$$\bar{g} : \rho_n \xrightarrow[\text{onto}]{p} \beta$$

which is Σ_n in parameters $\bar{p} = \pi^{-1}(p)$ and $\bar{q} = \pi^{-1}(q)$.

We claim that $\beta = \alpha$. To see this, note that $\pi x = x$ for all $x < \rho_n$ since $\rho_n \subseteq X_0 \subseteq X$. Hence A is $\Sigma_n(L_\beta)$ in parameter $\bar{p}$. Thus, if β were less than α, A would be α-finite, a contradiction.

We now have a Σ_n partial epimorphism $\bar{g} : \rho_n \xrightarrow[\text{onto}]{p} \alpha$. Let f_n be a Σ_n selector for the relation $\bar{g}^{-1}$. Then $f_n : \alpha \xrightarrow{1-1} \rho_n$ and Lemma 3.5 is proved.

From now on let $f_n : \alpha \xrightarrow{1-1} \rho_n$ be a fixed Σ_n monomorphism. We define a well-ordering $<_n$ of α by

$$x <_n y \leftrightarrow f_n x < f_n y.$$

(For convenience we put $\rho_0 = \alpha$, $f_0 =$ identity map, and $x <_0 y$ if and only if $x < y$.)

The next lemma is a weak generalization of the fact that the class of primitive recursive relations is closed under bounded quantification.

<u>Lemma</u> 3.6 (S. Friedman) Let $Q(x,y,z)$ be a Σ_n relation and assume the Σ_k selection theorem for all k, $1 \leq k \leq n$. Then the relation

$$S(x,z) \leftrightarrow (\forall y <_n z) Q(x,y,z)$$

is Σ_{n+1}.

Proof. For $n = 0$ this is trivial so assume $n \geq 1$. We have

$$\begin{aligned} S(x,z) &\leftrightarrow (\forall y <_n z) Q(x,y,z) \\ &\leftrightarrow (\forall y <_n z) \exists u P(x,y,z,u) \end{aligned}$$

where the negation of P is Σ_{n-1}. By induction on n we may assume that our lemma holds with n replaced by $n-1$. Hence the negation of

$(\exists u <_{n-1} v)P(x,y,u,v)$ is $\sum_n$. This observation will play a key role in both of the following cases.

Case I: the $\sum_n$ cofinality of ρ_{n-1} is less than ρ_n, i.e. there exists a $\sum_n$ function $g : \kappa \to \alpha$, $\kappa < \rho_n$ such that the range of $f_{n-1}g$ is unbounded in ρ_{n-1}. It follows by the definition of ρ_n that $S(x,z)$ implies the existence of an α-finite function $h : f_n z \to \kappa$ such that $u <_{n-1} ghf_n y$ for all $y <_n z$. Thus

$$S(x,z) \leftrightarrow \exists h(\forall y <_n z)(\exists u <_{n-1} ghf_n y)P(x,y,z,u)$$

$$\leftrightarrow \exists h \forall y[f_n y < f_n z \to (\exists u <_{n-1} ghf_n y)P(x,y,z,u)]$$

which is easily seen to be $\sum_{n+1}$.

Case II: the $\sum_n$ cofinality of ρ_{n-1} is $\geq \rho_n$. It follows by $\sum_n$ selection that $S(x,z)$ implies the existence of v such that $u <_{n-1} v$ for all $y <_n z$. Thus we have

$$S(x,z) \leftrightarrow \exists v(\forall y <_n z)(\exists u <_{n-1} v)P(x,y,z,u)$$

$$\leftrightarrow \exists v \forall y[f_n y < f_n z \to (\exists u <_{n-1} v)P(x,y,z,u)]$$

which is easily seen to be $\sum_{n+1}$. This completes the proof.

<u>Proof of the $\sum_{n+1}$ (L_α) Selection Theorem</u>. Suppose that $R(x,y)$ is $\sum_{n+1}$. We may write

$$R(x,y) \leftrightarrow \exists w Q(x,y,w)$$

where the negation of Q is $\sum_n$. Define

$$Q^*(x,y,w) \leftrightarrow$$
$$Q(x,y,w) \;\&\; \forall y',w'[J(y',w') <_n J(y,w) \to \neg Q(x,y',w')].$$

By induction on k we may assume the $\sum_k$ selection theorem, $1 \le k \le n$. Hence by Lemma 3.6, Q^* is $\sum_{n+1}$. It follows that

$$R^*(x,y) \leftrightarrow \exists w Q^*(x,y,w)$$

is a $\sum_{n+1}$ selector for R. This completes the proof.

Because of its great importance, we restate the following result from the proof of the $\sum_n$ Selection Theorem.

Theorem 3.7 (Jensen) For $n \ge 1$, the $\sum_n$ projectum of α can be characterized as the least ρ such that there exists a $\sum_n$ monomorphism $f : \alpha \xrightarrow{1-1} \rho$.

Proof. Immediate from 3.3 and 3.5.

For later use (in §6) we introduce the following notion which is a variant of the $\sum_n$ projectum.

Definition 3.8

(i) A set $X \subseteq \alpha$ is said to be Δ_n (or $\Delta_n(L_\alpha)$) if both X and $\alpha - X$ are $\sum_n(L_\alpha)$.

(ii) The Δ_n projectum of α (denoted η_n or η_n^α) is the least η such that not every Δ_n subset of η is α-finite.

Theorem 3.9 (Jensen) The Δ_n projectum of α can be characterized as the least η such that there exists a $\sum_n$ epimorphism $h : \eta \xrightarrow{\text{onto}} \alpha$.

Proof. We construct a $\sum_n$ epimorphism $h : \eta_n \xrightarrow{\text{onto}} \alpha$; the other half of the theorem is an easy diagonalization as in the proof of Lemma 3.5.

Clearly $\rho_n \le \eta_n \le \rho_{n-1}$. Our first claim is that the $\sum_n$ cofinality of ρ_{n-1} is at most η_n (cf. proof of Lemma 3.6). To see this, let $B \subseteq \eta_n$ be Δ_n and not α-finite. We have

$$x \in B \leftrightarrow \exists y P(x,y)$$

and

$$x \notin B \leftrightarrow \exists y Q(x,y)$$

where the negations of P and Q are $\sum_{n-1}$. Define $g : \eta_n \to \alpha$ by letting gx be the $<_{n+1}$-least y such that $P(x,y)$ or $Q(x,y)$ holds. By Lemma 3.6 g is $\sum_n$. Furthermore, if the range of $f_{n-1}g$ were bounded below ρ_{n-1}, there would exist z such that

$$x \in B \leftrightarrow (\exists y <_{n-1} z)P(x,y)$$

whence B would be α-finite by definition of ρ_{n-1} (unless $\eta_n = \rho_{n-1}$ in which case our claim is trivially true anyway). Thus the range of $f_{n-1}g$ is unbounded in ρ_{n-1} and the first claim is proved.

We now write

$$f_n y = z \leftrightarrow \exists w R(y,z,w)$$

where the negation of R is $\sum_{n-1}$. Define $h : \eta_n \to \alpha$ by

$$h(J(x,z)) = y \leftrightarrow$$
$$\exists w R(y,z,w) \vee (y = 0 \ \& \ (\forall w <_{n-1} gx)\neg R(y,z,w)).$$

Then h is $\sum_n$ by 3.6 and it is easy to check that h is epimorphic. This completes the proof.

A corollary of the proof is that $\eta_n = \max\{\rho_n, \kappa_n\}$ where κ_n is the $\sum_n$ cofinality of ρ_{n-1}.

§4. ADMISSIBLE ORDINALS : SOME EXAMPLES.

In §§2 and 3 we saw that some of the simplest results of ordinary recursion theory (those whose proofs use only explicit diagonalization or minimization) can be generalized to the context of an arbitrary p.r. closed ordinal α. In order to generalize the deeper theorems (proved by wait-and-see or priority techniques), it has often been found convenient to impose strong additional hypotheses on α. The most popular of these hypotheses is admissibility.

Definition 4.1 A nonzero ordinal α is said to be $\sum_n$ admissible if for every α-finite set $K \subseteq \alpha$ and $\sum_n$ function $f : \alpha \to \alpha$, the set

$$f[K] = \{f(x) : x \in K\}$$

is α-finite.

Thus $\sum_n$ admissibility is equivalent to saying that L_α (regarded as a model of a fragment of set theory) satisfies the axiom of replacement for $\sum_n$ formulae.

The special case $n = 1$ is of particular importance.

Definition 4.2 α is admissible if it is $\sum_1$ admissible, i.e. the image of an α-finite set under an α-recursive function is α-finite.

It is not hard to see that every admissible ordinal is p.r. closed. The main reason for singling out $\sum_1$ admissibility is that that hypothesis is satisfied in a variety of concrete situations which are of independent interest. Among the best known examples of admissible ordinals are the following.

Example 4.3 The first admissible ordinal is of course ω.

Example 4.4 (Kreisel-Sacks [22]) The second admissible ordinal is ω_1^{CK} = Church-Kleene ω_1 = the least nonrecursive ordinal. Moreover, a subset of ω is ω_1^{CK}-finite if and only if it is hyperarithmetical, and ω_1^{CK}-RE if and only if it is Π_1^1.

Let α and β be p.r. closed ordinals with $\alpha < \beta$. We say that α is β-stable if L_α is a $\sum_1$ elementary submodel of L, i.e. if for every primitive recursive relation R and parameter $p < \alpha$,

$$(\exists y < \beta)R(p,y) \to (\exists y < \alpha)R(p,y).$$

It is easy to see that if α is β-stable then α is admissible.

Example 4.5 (Takeuti [53]). Let M be a transitive model of Zermelo-Fraenkel set theory, and let α be an uncountable cardinal in the sense of M. Then α is β-stable where $\beta = OR \cap M$. Hence α is admissible.

(Note: If α is a regular cardinal in the sense of M, then α is $\sum_n$ admissible for all n. However, if $\alpha = \aleph_\omega^M$ and M satisfies V = L, then α is not even $\sum_2$ admissible since the function $n \to \aleph_n^M$ is $\sum_2(L_\alpha)$.)

Example 4.6 (Takeuti [54]) Let δ_2^1 be the least non-Δ_2^1 ordinal. Then δ_2^1 is admissible, and a subset of ω is δ_2^1-finite (respectively δ_2^1-RE) if and only if it is Δ_2^1 (respectively $\sum_2^1$). Moreover δ_2^1 can be characterized as the first OR-stable ordinal.

Some further examples of admissible ordinals arise from Kleene's theory of recursion in higher types (Kleene [19], [20]). For any functional $F = {}^nF$ of type n, we denote by ω_1^F the least ordinal not recursive in F.

Example 4.7 (Richter [35]). Let E_1 be the type 2 Tugué functional, i.e. $E_1(f) = 1$ if $\{(m,n) : f(2^m.3^n) = 1\}$ is a well-ordering, 0 otherwise. Then $\omega_1^{E_1}$ is admissible and a subset of ω is $\omega_1^{E_1}$-finite (respectively $\omega_1^{E_1}$-RE) if and only if it is recursive in E_1 (respectively the domain of a partial function which is partial recursive in E_1). Moreover $\omega_1^{E_1}$ can be characterized as the first recursively inaccessible ordinal, i.e. the first admissible ordinal which is a limit of admissible ordinals.

Example 4.8 Similar results hold for certain other normal type 2 functionals such as the superjump of E_1, etc. etc.

Example 4.9 (Harrington [10]). Let $S = {}^3S$ be the type 3 superjump, i.e. $S(F)(n) = 1$ if $\{n\}^F(n) = 1$, 0 otherwise, where $F = {}^2F$. Then ω_1^S is admissible, and a subset of ω is ω_1^S-finite (respectively ω_1^S-RE) if and only if it is recursive in S (respectively semirecursive in S*, the "partial normalization" of S). Moreover ω_1^S can be characterized as the first recursively Mahlo ordinal, i.e. the first admissible ordinal α such that every α-recursive normal $f : \alpha \to \alpha$ has an admissible fixed point.

Example 4.10 (Harrington [11]) Let $CL = {}^3CL$ be the type 3 inductive closure operator, i.e. $CL(\Gamma) = \Gamma_\infty$ where $\Gamma_\sigma = \bigcup\{\Gamma(\Gamma_\xi) : \xi < \sigma\}$ for all $\Gamma : P(\omega) \to P(\omega)$. Then ω_1^{CL} is admissible, and a subset of ω is ω_1^{CL}-finite (respectively ω_1^{CL}-RE) if and only if it is recursive in CL (respectively semi-recrusive in CL*, the "partial normalization" of CL). Moreover ω_1^{CL} can be characterized as the first α-stable ordinal where α is the first <u>nonprojectible</u>, i.e. the first admissible α such that $\alpha = \rho_1^\alpha > \omega$.

<u>Example</u> 4.11 (Harrington) Similar results hold for the type n superjump, the type n inductive closure operator, and various other objects of type $n \geq 3$ such as the Kolmogorov R-operator.

Yet more examples are provided by the theory of nonmonotonic inductive definability (Richter-Aczel [36]). Given an operator $\Gamma : P(\omega) \to P(\omega)$ we put $\Gamma_\sigma = \cup\{\Gamma(\Gamma_\xi) : \xi < \sigma\}$ and define the closure ordinal $|\Gamma|$ to be the least σ such that $\Gamma_\sigma = \Gamma_{\sigma+1} = \Gamma_\infty$. A subset of ω is said to be C-<u>inductively definable</u> if it is many-one reducible to Γ_∞ for some Γ such that

$$\{(n,X) : n \in \Gamma(X)\} \in C.$$

We denote by $|C|$ the supremum of the ordinals $|\Gamma|$ where Γ is as above.

<u>Example</u> 4.12 (Richter-Aczel [36]) For $n \geq 1$, the ordinal $|\Sigma_n^0|$ is admissible and in fact can be characterized on the first Σ_{n+1} reflecting ordinal (see Definition 6.1 and Theorem 6.2). Moreover a subset of ω is Σ_n^0 inductively definable if and only if it is $|\Sigma_n^0|$-RE.

<u>Example</u> 4.13 (Richter-Aczel [36]) The ordinal $|\Pi_1^1|$ is admissible and can be characterized as the first Π_1^1 reflecting ordinal (see Theorem 6.3). Moreover a subset of ω is Π_1^1 inductively definable if and only if it is $|\Pi_1^1|$-RE.

<u>Example</u> 4.14 (Richter-Aczel [36]) The ordinal $|\Sigma_1^1|$ is admissible and can be characterized as the first Σ_1^1 reflecting ordinal (see Theorems 6.4, 6.5 and 6.6). Moreover, a subset of ω is Σ_1^1 inductively definable if and only if it is $|\Sigma_1^1|$-RE.

The power of admissible recursion theory as applied to concrete examples is well illustrated by the following result.

<u>Theorem</u> 4.15 (Kreisel-Sacks [22]) Let α be an admissible ordinal whose Σ_1 projectum is ω. Then there exists an α-RE set $M \subseteq \omega$ which is <u>maximal</u>, i.e. $\omega - M$ is infinite but for any α-RE set $A \subseteq \omega$ either $(\omega-M) \cap A$ or $(\omega-M) \cap (\omega-A)$ is finite.

Note that the special hypothesis $\rho_1^\alpha = \omega$ is satisfied for a wide class of admissible ordinals α, in particular all of the admissible ordinals mentioned in Examples 4.3, 4,4, and 4.6-4.14. In each of these examples, the α-RE subsets of ω are of independent interest. For instance, Theorem 4.15 applied to Example 4.4 gives us a maximal Π_1^1 subset of ω. This is a transparent "structural" (in fact lattice-theoretic) property of the Π_1^1 subsets of ω. It could not have been discovered by "classical" methods since the proof involves a priority argument.

§5. DEGREE THEORY ON ADMISSIBLE ORDINALS.

There are many basic facts of ordinary recursion theory which generalize straightforwardly to an arbitrary admissible ordinal. Often these generalizations would present difficulties if α were assumed merely to be p.r. closed. FROM NOW ON α IS A FIXED ADMISSIBLE ORDINAL. Then we have:

5.1. A set is α-finite if and only if it is α-recursive and bounded.

5.2. An α-RE set which is not α-finite is the range of an α-recursive monomorphism $f : \alpha \xrightarrow{1-1} \alpha$.

5.3 (Generalization[3] of Rice's theorem [34]). A collection C of α-RE sets is <u>completely</u> α-RE (i.e. $\{x : W_x \in C\}$ is α-RE) if and only if there exists an α-RE set R such that

$$C = \{A : A \text{ is } \alpha\text{-RE} \ \& \ \exists x(x \in R \ \& \ D_x \subseteq A)\}.$$

5.4 (Corollary of 5.3). An α-RE set A is α-finite if and only if $\{B : B \text{ is } \alpha\text{-RE} \ \& \ A \subseteq B\}$ is completely α-RE.

The proofs of 5.1-5.4 are left to the reader. (Note: In 5.3 and throughout this paper, we tacitly assume that the enumeration W_x $(x < \alpha)$ is <u>principal</u> in the sense that given any other α-recursive enumeration W'_x $(x < \alpha)$ of the α-RE sets, there exists an α-recursive function f such that $W'_x = W_{f(x)}$ for all x. The existence of a principal α-recursive enumeration of the α-RE sets is easily

[3]This result was conjectured at the Symposium by Professor Ershov.

established.)

In addition to the simple facts such as 5.1-5.4, many of the deeper, more difficult theorems concerning degrees of unsolvability have been generalized to α-recursion theory. The proofs of these generalizations are usually <u>not</u> straightforward. The basic degree-theoretic definitions are as follows:

<u>Definition</u>. For $A,B \subseteq \alpha$ we say that A is α-<u>recursive in</u> B (abbreviated $A \leq_\alpha B$) if there exists an α-RE set W such that for all x and y,

$$D_x \subseteq A \;\&\; D_y \cap A = \emptyset \;\leftrightarrow$$
$$\exists u,v[J(J(x,y),J(u,v)) \in W \;\&\; D_u \subseteq B \;\&\; D_v \cap B = \emptyset].$$

<u>Definition</u>. The α-<u>jump</u> of A (denoted A^*) is defined by

$$z \in A^* \;\leftrightarrow$$
$$\exists x,y[J(x,y) \in W_z \;\&\; D_x \subseteq A \;\&\; D_y \cap A = \emptyset].$$

The following facts are easily verified:

5.5. $A \leq_\alpha A$.

5.6. If $A \leq_\alpha B$ and $B \leq_\alpha C$ then $A \leq_\alpha C$.

5.7. If $A \leq_\alpha B$ then $A^* \leq_\alpha B^*$.

5.8. $A \leq_\alpha A^*$ and not $A^* \leq_\alpha A$.

5.9. $\emptyset^*$ is an α-RE set.

5.10. $A \leq_\alpha \emptyset^*$ for all α-RE sets A.

An α-<u>degree</u> is an equivalence class under the equivalence relation $A \leq_\alpha B$ and $B \leq_\alpha A$. It is customary to use the same notation for α-degrees as for ordinary degrees of unsolvability (the special case $\alpha = \omega$). In particular an α-degree $\underset{\sim}{a}$ is said to be α-RE if it contains an α-RE set. By 5.9 and 5.10, $\underset{\sim}{0}'$ is the largest α-RE degree.

The following theorems concerning α-RE degrees are full generalizations of well-known results for $\alpha = \omega$.

Theorem 5.11 (Sacks-Simpson [41] generalizing the Friedberg-Muchnik theorem). There exist α-RE degrees $\underset{\sim}{a}$ and $\underset{\sim}{b}$ which are incomparable, i.e. neither $\underset{\sim}{a} \le \underset{\sim}{b}$ nor $\underset{\sim}{b} \le \underset{\sim}{a}$.

Theorem 5.12 (Simpson [49] generalizing a result of Sacks [37]). There exists an α-RE degree a which is low, i.e. $\underset{\sim}{a} > \underset{\sim}{0}$ and $\underset{\sim}{a}' = \underset{\sim}{0}'$.

Theorem 5.13 (Shore [43] generalizing the Splitting Theorem of Sacks [37]). If $\underset{\sim}{a}$ is a nonzero α-RE degree, then there exist incomparable α-RE degrees $\underset{\sim}{b}$ and $\underset{\sim}{c}$ such that $\underset{\sim}{b} \cup \underset{\sim}{c} = a$.

Theorem 5.14 (Shore [44] generalizing the Density Theorem of Sacks [38]). If $\underset{\sim}{a}$ and $\underset{\sim}{b}$ are α-RE degrees with $\underset{\sim}{a} < \underset{\sim}{b}$ then there exists an α-RE degree $\underset{\sim}{c}$ such that $\underset{\sim}{a} < \underset{\sim}{c} < \underset{\sim}{b}$.

Theorem 5.15 (Lerman [26] generalizing a result of Lachlan [24]). If $\underset{\sim}{a}$ and $\underset{\sim}{b}$ are nonzero α-RE degrees with $\underset{\sim}{a} \cap \underset{\sim}{b} = \underset{\sim}{0}$ then $\underset{\sim}{a} \cup \underset{\sim}{b} < \underset{\sim}{0}'$.

If in addition α is Σ_2 admissible, the following results have been obtained:

Theorem 5.16 (Shore [45] generalizing a result of Sacks [37]). There exists an α-RE degree $\underset{\sim}{a}$ which is high, i.e. $\underset{\sim}{a} < \underset{\sim}{0}'$ and $\underset{\sim}{a}' = \underset{\sim}{0}''$.

Theorem 5.17 (Shore [42] generalizing a result of Sacks [37]). There exists an α-degree $\underset{\sim}{m} < \underset{\sim}{0}'$ which is minimal, i.e. $\underset{\sim}{0} < \underset{\sim}{m}$ and there is no α-degree $\underset{\sim}{a}$ such that $\underset{\sim}{0} < \underset{\sim}{a} < \underset{\sim}{m}$.

Theorem 5.18 (Lerman-Sacks [29] and Shore [47] generalizing a result of Lachlan [24] and Yates [55]). There exist nonzero α-RE degrees $\underset{\sim}{a}$ and $\underset{\sim}{b}$ such that $\underset{\sim}{a} \cap \underset{\sim}{b} = \underset{\sim}{0}$.

It is conjectured that Theorems 5.17 and 5.18 hold for arbitrary admissible ordinals α. On the other hand, the following result shows that no such generalization of Theorem 5.16 is possible.

Theorem 5.19 (Shore [45]). Let α be the admissible ordinal $\aleph_\omega^M$ where M is a transitive model of ZF+V = L. Then every α-degree $\underset{\sim}{a}$ such that $\underset{\sim}{0} < \underset{\sim}{a} < \underset{\sim}{0}'$ is low. Hence no α-RE degree is high.

To complete the picture presented by Theorems 5.16 and 5.19, Maass [31] has announced that for an admissible ordinal α, high α-RE degrees exist if and only if the Σ_2 cofinality of α is greater than or equal to the Σ_2 projectum of α.

We end this section with a word about the proofs of Theorems 5.11-5.18. These proofs are for the most part straightforward generalizations of the proofs for $\alpha = \omega$, provided one assumes plenty of admissibility, e.g. Σ_2 and Σ_3 admissibility. However, if α is assumed merely to be Σ_1 or Σ_2 admissible, grave difficulties present themselves. Namely, the natural priority indexing in terms of the ordinals less than α may cause requirements to interfere with each other to the extent that some of them can never be satisfied. The key to the solution of these difficulties usually lies in a reindexing of the requirements in terms of the ordinals less than δ, where δ is a suitably chosen projectum of α. The choice of δ is dictated by the nature of the conflicts between the requirements. Usually δ is not one of the Jensen projecta ρ_n^α or η_n^α (discussed in §3), but rather a recursion-theoretically natural projectum such as the tame S_2 or S_3 projectum (these particular projecta were introduced by Lerman [25], [27]). However, it frequently happens that Jensen-style characterizations of δ (analogous to Theorems 3.7 and 3.9) play a role in the proof that all requirements are eventually satisfied.

The reader who wishes to learn more about α-degrees for admissible ordinals α may consult the survey papers by Simpson [49] and Shore [46], [48]. Recently there has been much progress on the problem of generalizing the Friedberg-Muchnik theorem and other results to the context of p.r. closed ordinals which are not admissible. This work has been discussed at the Oslo Symposium by Friedman [5] and Maass [31].

Another large area which we shall not discuss here is the work that has been done on the lattice of α-RE sets. A fine survey paper on this subject is Lerman [28].

§6. REFLECTING ORDINALS.

Some very interesting classes of admissible ordinals were discovered in the late 1960's by P. Aczel and W. Richter. The basic definition is as follows.

Definition 6.1 (Richter-Aczel [36]) A nonzero ordinal α is said to be $\sum_n$ reflecting if for every primitive recursive relation R and parameter $p < \alpha$, if

$$(\exists x_1 < \alpha)(\forall x_2 < \alpha)\ldots x_n < \alpha)R(x_1,\ldots,x_n,p)$$

then there exists $\beta < \alpha$ such that

$$(\exists x_1 < \beta)(\forall x_2 < \beta)\ldots x_n < \beta)R(x_1,\ldots,x_n,p).$$

Equivalently, for every $\sum_n$ formula φ and parameter $p < \alpha$,

$$L_\alpha \models \varphi[p] \quad \text{implies} \quad (\exists\beta < \alpha)L_\beta \models \varphi[p].$$

The latter form of the definition bears more than a little resemblance to the well-known Lévy-Montague reflection principle of Zermelo-Fraenkel set theory.

The relative sizes of the $\sum_n$ reflecting ordinals $(n \geq 2)$ and of the Π^1_1 and Σ^1_1 reflecting ordinals (defined similarly) have been studied and the following results have been obtained.

Theorem 6.2 (Richter-Aczel [36]) Let α be an ordinal.

(i) α is $\sum_2$ reflecting if and only if it is a limit ordinal.

(ii) α is $\sum_3$ reflecting if and only if it is greater than ω and admissible.

(iii) If α is $\sum_4$ reflecting then it is recursively inaccessible, recursively Mahlo, etc. and is in fact the recursion-theoretic analog of a Π^1_1 indescribable cardinal.

(iv) If α is $\sum_{n+1}$ reflecting $(n \geq 2)$ then it is a limit of $\sum_n$ reflecting ordinals, etc. etc.

Theorem 6.3 (Richter-Aczel [36]) A countable ordinal α is Π^1_1-reflecting if and only if it is α^+ stable. (Here α^+ denotes the first admissible ordinal greater than α.)

<u>Theorem</u> 6.4 (Richter-Aczel [36]) If α is countable and α^{++} stable, then it is Σ^1_1 reflecting.

<u>Theorem</u> 6.5 (Aanderaa [1]) If α is countable and Σ^1_1 reflecting, then it is a limit of Π^1_1 reflecting ordinals.

We now present a simple proof of Aanderaa's theorem. This proof does not seem previously to have appeared in print (or even in multilith), although it is widely known.

<u>Proof of Aanderaa's theorem 6.5</u>. Let α be countable and Σ^1_1 reflecting.

Case I: α is Π^1_1 reflecting. It is easy to see that Π^1_1 reflection is expressible by a Σ^1_1 sentence ψ. Thus, in this case, $L_\alpha \models \psi$. Hence by Σ^1_1 reflection there are unboundedly many $\beta < \alpha$ such that $L_\beta \models \psi$, i.e. β is Π^1_1 reflecting.

Case II: α is not Π^1_1 reflecting. Then by 6.3 there exists $y < \alpha^+$ such that for some primitive recursive relation R and parameter $p < \alpha$, $R(p,y)$ holds but $\neg(\exists z < \alpha)R(p,z)$. Such a y will be called a failure of stability for α, and R and p will be called witnesses for y.

Let y_0 be the first failure of stability for α. Let R_0 and p_0 be witnesses for y_0. It is not difficult to devise a Σ^1_1 formula $\varphi(x)$ such that for all infinite $\beta > p_0$, $L_\beta \models \varphi[p_0]$ if and only if there exists a model $\langle M,E\rangle$ of $KP+V = L$ which is an end extension of $L_{\beta+1}$ such that β is not M-stable (i.e. L_β is not a Σ_1 elementary submodel of $\langle M,E\rangle$) and R_0 and p_0 are witnesses for the first failure of stability of β within $\langle M,E\rangle$.

Clearly $L_\alpha \models \varphi[p_0]$ since we may take $\langle M,E\rangle = \langle L_{\alpha^+},\in\rangle$. On the other hand, α is Σ^1_1 reflecting. Hence there are unboundedly many $\beta < \alpha$ such that $L_\beta \models \varphi[p_0]$.

We claim that any such β is Π^1_1 reflecting. To see this, let $\langle M,E\rangle$ be an end extension of $L_{\beta+1}$ as stipulated. By Ville's lemma (Barwise [3, p. 243]), $\langle M,E\rangle$ is an end extension of L_{β^+}. Let z_0 be the first failure of stability for β within $\langle M,E\rangle$. Then R_0 and p_0 are witnesses for z_0 so in particular $R_0(p_0,z_0)$ holds. Hence z_0 is not less than α. But $\beta^+ \leq \alpha$ since α is admissible. Hence z_0 is not less than β^+. Hence β is β^+ stable. Hence by

6.3 β is Π_1^1 reflecting.

This completes the proof.

The next theorem was first stated in some unpublished notes of P. Aczel. The proof can be found in Abramson-Sacks [2].

Theorem 6.6 (H. Friedman and R. Gostanian [9]) Let α be countable, admissible, and locally countable (i.e. every ordinal less than α is countable in the sense of L_α). Then α is Σ_1^1 reflecting if and only if it is not Gandy. (An admissible ordinal α is said to be Gandy if each ordinal less than α^+ is the order type of an α-recursive well-ordering.)

We end this section by presenting a new theorem which gives a sufficient (but not necessary) condition for a p.r. closed ordinal to be Σ_{n+2} reflecting. In §7 we shall see how this theorem can be used to construct counterexamples in α-recursion theory.

Theorem 6.7 (Simpson [50]) Let α be a p.r. closed ordinal which is greater than ω and equal to its own Δ_n projectum (Definition 3.8). Then α is Σ_{n+2} reflecting.

Proof (sketch). We use a Skolem hull argument similar to the one in the proof of Lemma 3.5. Let $\varphi(x)$ be a Σ_{n+2} formula and p a parameter such that $L_\alpha \models \varphi[p]$. Using the Σ_n selection theorem we can construct a certain Skolem hull $X \subseteq \alpha$ such that X is the range of a Σ_n function $g : \delta \to \alpha$, $p < \delta < \alpha$. Let $\beta = \text{type}(X)$. As in the proof of Lemma 3.5, we can argue that $L_\beta \models \varphi[p]$. Furthermore there is a Σ_n epimorphism $\bar{g} : \delta \xrightarrow{\text{onto}} \beta$. Hence by Theorem 3.9 $\beta < \alpha$. This completes the proof.

Corollary 6.8. Suppose that L_α is a model of Zermelo set theory. Then L_α is Σ_n reflecting for all n.

Corollary 6.9 (Jensen [13]) A p.r. closed ordinal α is admissible if and only if it is equal to its own Δ_1 projectum.

§7. THE S_n HIERARCHY.

One of the most basic ideas of ordinary recursion theory is the approximation of nonrecursive functions by recursive ones. (Such approximations figure prominently in all priority arguments.) Generalizing this idea to an admissible ordinal α, we obtain the S_n hierarchy.

<u>Definition</u> 7.1 (Lerman [27]) A function $f : \alpha \to \alpha$ is said to be S_{n+1} if there exists an α-recursive function g such that for all x,

$$f(x) = \lim_{y_1} \lim_{y_2} \ldots \lim_{y_n} g(x, y_1, y_2, \ldots, y_n).$$

Here the limits are taken in the discrete topology, i.e. $\lim_y h(y) = z$ if and only if

$$\exists w \forall y (y > w \to h(y) = z).$$

It is easy to show that $f : \alpha \to \alpha$ is S_1 if and only if it is α-recursive, and S_2 if and only if it is $\sum_2$. More generally, if α is $\sum_n$ admissible, then f is S_{n+1} if and only if it is $\sum_{n+1}$.

<u>Definition</u> 7.2 (Lerman [27]) (i) A subset of α is said to be S_n if and only if its characteristic function is S_n.

(ii) The S_n <u>projectum</u> of α is the least β such that not every S_n subset of β is α-finite.

We have the following Jensen-style characterization of the S_3 projectum:

<u>Theorem</u> 7.3. The S_3 projectum of α is the least β such that there exists an S_3 epimorphism $h : \beta \xrightarrow{\text{onto}} \alpha$.

An interesting application of the S_n hierarchy was made by Lerman in his solution of the maximal set problem. An α-RE set M is said to be <u>maximal</u> if $\alpha - M$ is not α-finite but for every α-RE set R either $(\alpha - M) \cap R$ or $(\alpha - M) \cap (\alpha - R)$ is α-finite. It is a classical theorem of Friedberg that maximal sets exist for $\alpha = \omega$. On the other hand, Sacks [39] observed that maximal sets do not exist when $\alpha = \aleph_1^M$ where M is a transitive model of $ZF + V = L$. Thus the question arose: for which admissible ordinals α do maximal α-RE sets exist? This question was answered by Lerman as follows:

Theorem 7.4 (Lerman [27]) There exists a maximal α-RE set if and only if the S_3 projectum of α is ω.

In his paper [27] Lerman left open a number of questions concerning the precise relationship between the S_n and $\sum_n$ hierarchies for $n \geq 3$. In particular, he asked whether the S_3 projectum and the Δ_3 projectum always coincide, and whether a $\sum_3$ function $f : \alpha \to \alpha$ is necessarily S_n for some n. We now produce an example to answer these questions in the negative.

Example 7.5 (Simpson [50]) We exhibit an admissible ordinal $\alpha > \omega$ which is equal to its own S_n projectum for all n, but whose Δ_3 projectum is ω.

Let α be the first admissible ordinal such that L_α is a model of the axiom of infinity, the power set axiom, and the $\sum_2$ comprehension axiom. (The last condition means that α is equal to its own $\sum_2$ projectum.) It is not hard to show that these conditions are expressible as a $\sum_5$ sentence ψ such that $L_\alpha \models \psi$. Thus α is not $\sum_5$ reflecting. Hence, by Theorem 6.7, α is greater than its Δ_3 projectum. (By going back to the proof of 6.7, we can actually show that the Δ_3 projectum of α is ω.)

It remains to show that α is its own S_n projectum, for all n. To see this, note first that

$$\alpha = \sup_{k<\omega} \aleph_k^{L_\alpha} ,$$

i.e. α is the supremum of the first ω infinite ordinals which are cardinals in the sense of L_α. It is not hard to see that the function $k \to \aleph_k^{L_\alpha}$ is $\sum_2(L_\alpha)$. Thus the $\sum_2$ cofinality of α is ω.

Now suppose that $B \subseteq \beta < \alpha$ is S_{n+1}. We have

$$\begin{aligned} \chi_B(x) &= \lim_{y_1} \dots \lim_{y_n} g(x,y_1,\dots,y_n) \\ &= \lim_{k_1} \dots \lim_{k_n} g(x,\aleph_{k_1}^{L_\alpha},\dots,\aleph_{k_n}^{L_\alpha}) \end{aligned}$$

where g is α-recursive. Clearly the predicate

$$g(x,\aleph_{k_1}^{L_\alpha},\ldots,\aleph_{k_n}^{L_\alpha}) = 1$$

is Δ_2. Hence B is α-finite since $\eta_2^\alpha = \alpha$. Thus α is its own S_{n+1} projectum, Q.E.D.

Remark 7.6. For any admissible ordinal α, the following inequalities hold:

$$\eta_3 \leq S_3 \text{ projectum} \leq \rho_2 .$$

By Example 7.5 the first inequality can be strict. A similar example shows that the second inequality can be strict while the S_3 projectum is ω. Thus Lerman's criterion for the existence of maximal α-RE sets could not have been stated conveniently in terms of the Jensen projecta η_n^α and ρ_n^α. In other words, Lerman's Theorem 7.4 is a genuine application of the S_n hierarchy.

§8. ORACLES, FANS AND THEORIES.

We are going to discuss a certain recursion-theoretic topic which has applications to logic on a countable admissible ordinal. The study of these logics was pioneered by Barwise in the 1960's. It is reasonable to look for applications of α-recursion theory to α-logic since, after all, ordinary recursion theory originated in the study of logical systems (Gödel [6]).

The basic recursion theoretic notion which we shall require is the notion of a "fan". Roughly, an α-fan is a nonempty subset of the powerset of α whose complement is α-recursively enumerable. In order to make sense out of this, we must first answer the question: what do we mean by an α-recursively enumerable collection of subsets of α?

Actually, this question is rather subtle and has at least three answers corresponding to three different intuitively natural notions of α-recursive oracle computation. Each of the three is useful in certain contexts. We shall simply list the definitions and refer to them uncritically by number (1,2,3).

FROM NOW ON, α IS A FIXED BUT ARBITRARY COUNTABLE ADMISSIBLE ORDINAL. Note that the hypothesis of countability is satisfied in most of the examples of §4. We identify a subset A of α with its characteristic function χ_A so that 2^α, the set of all functions from α into $\{0,1\}$, is just the power set of α.

<u>Definition</u> 8.1 A set $S \subseteq 2^\alpha$ is said to be α-RE <u>in sense 1</u> if there exists an α-RE set $W \subseteq \alpha$ such that

$$X \in S \leftrightarrow$$
$$\exists u,v[J(u,v) \in W \ \& \ D_u \subseteq X \ \& \ D_v \cap X = \emptyset].$$

<u>Definition</u> 8.2 A set $S \subseteq 2^\alpha$ is said to be α-RE <u>in sense 2</u> if there exist a primitive recursive relation $R \subseteq 2^\alpha \times \alpha \times \alpha$ and a parameter p such that

$$X \in S \leftrightarrow (\exists y < \alpha)R(X,y,p).$$

Here we are using the notion of a primitive recursive functional $F : \alpha^m \times (2^\alpha)^n \to \alpha$ which is defined in the obvious way using the initial functional

$$F(x,A) = \chi_A(x).$$

An equivalent formulation is that there exist a Σ_1 formula $\varphi(x)$ and a parameter p such that

$$X \in S \quad \leftrightarrow <L_\alpha[X],\in,X> \models \varphi[p]$$

where $L_0[X] = \emptyset$, $L_{\xi+1}[X] = \text{fodo}(<L_\xi[X],\in,X>)$, and $L_\lambda[X] = \bigcup\{L_\xi[X] : \xi < \lambda\}$ for limit ordinals λ (the relative constructible hierarchy).

<u>Definition</u> 8.3 A set $S \subseteq 2^\alpha$ is said to be α-RE <u>in sense 3</u> if there exists a finite set $E \cup \{e\}$ of Kripke [23] equations such that

$$X \in S \leftrightarrow E \cup \Delta_X \vdash e$$

where Δ_X is the diagram of X, i.e.

$$\Delta_X = \{f(x) = 1 : x \in X\} \cup \{f(x) = 0 : x \in \alpha - X\}.$$

Equivalently, there exist a $\sum_1$ formula $\varphi(x)$ and a parameter p such that

$$X \in S \leftrightarrow \langle L_{\alpha^X}[X], \in, X\rangle \models \varphi[p]$$

where α^X is the least $\beta \geq \alpha$ such that $\langle L_\beta[X], \in, X\rangle$ is an admissible structive.

Remark 8.4 To each of the three oracle notions 1, 2, and 3 there are corresponding notions of relative α-recursiveness ($A \leq_\alpha B$ for $A, B \in 2^\alpha$) and the α-jump operator (A^* for $A \in 2^\alpha$). The reader may convince himself that the degree theoretic concepts discussed in §5 are the ones which correspond to Definition 8.1. The parallel concepts which correspond to Definition 8.3 are usually called α-calculability degrees and α-calculability jump (Sacks-Simpson [41]) because of the connection with the Kripke equation calculus [23].

Remark 8.5 It is easy to see that α-RE in sense 1 implies α-RE in sense 2, and α-RE in sense 2 implies α-RE in sense 3, but no other implications hold in general. However, 1 and 2 coincide for sets $X \in 2^\alpha$ which are regular, i.e. $X \cap \beta$ is α-finite for all $\beta < \alpha$. Also, 2 and 3 coincide for sets X which are subgeneric, i.e. $\alpha^X = \alpha$. Thus all three notions agree if we restrict attention to sets which are both regular and subgeneric (Sacks [39]). It turns out that these doubly well-behaved sets are plentiful, e.g. they are comeager in the topology of α-finite neighborhood conditions (Lowenthal [30]).

At last we are ready for the main definition of this section.

Definition 8.6 An α-fan of type 1, 2 or 3 is a nonempty set $P \subseteq 2^\alpha$ such that $2^\alpha - P$ is α-RE in sense 1, 2 or 3 respectively.

A link between fans and logic is established by the following result.

Theorem 8.7 Let T be a consistent, α-RE set of sentences in the logic

$$\mathcal{L}_\alpha = \mathcal{L} \cap L_\alpha$$

of Barwise [3, p. 243]. Then the set of all complete, consistent extensions of T is an α-fan of type 3.

Actually, we shall be interested not in the full Barwise logic $\mathcal{L}_\alpha$ so much as its propositional part. The propositional fragment will suffice to illustrate most of the points we want to make. Thus we have:

Definition 8.8 The propositional α-logic on atoms a_x $(x < \alpha)$ with α-finite formulae φ_x $(x < \alpha)$ is defined as follows:

$$\varphi_{3x} = a_x \; ;$$
$$\varphi_{3x+1} = \neg\varphi_x \quad \text{(negation)} \; ;$$
$$\varphi_{3x+2} = \bigvee_{y \in D_x} \varphi_y \quad (\alpha\text{-finite disjunction}).$$

A model of a set T of formulae is a set $A \subseteq \alpha$ such that every formula of T is true under the truth assignment

$$a_x \to \begin{cases} \text{true} & \text{if } x \in A, \\ \text{false} & \text{otherwise.} \end{cases}$$

A theory is a set of formulae which is consistent (i.e. has a model) and is closed under logical consequence.

Theorem 8.9 Let P be a subset of 2^α. The following are equivalent.

(i) P is an α-fan of type 2;

(ii) P is the set of all models of some α-RE theory.

In the next section we shall discuss certain recursion-theoretic results which are conveniently stated in terms of α-fans. We shall then draw corollaries concerning theories in propositional α-logic.

§9. THE BASIS PROBLEM: APPLICATIONS TO LOGIC.

As in the previous section, α is a countable admissible ordinal. The basis problem for α is the problem of "effectively" choosing an element from a given α-fan. This problem is of great interest because of the possibility of applications to α-logic. In the special case $\alpha = \omega$, the basis problem has been studied from this viewpoint by Kreisel [21] and others. Many of the results in this section are generalizations of known results for $\alpha = \omega$.

The earliest and best known basis result for $\alpha = \omega$ is a "negative" one which was pointed out by Kleene [17]. Its generalization to arbitrary α is as follows:

Theorem 9.1 There exists an α-fan P (of type 1) such that no element of P is α-recursive.

Proof. Let A and B be disjoint α-recursively inseparable α-RE sets (Corollary 2.5). Put $P = \{X : X$ separates A and $B\}$.

Corollary 9.2 There exists an α-RE theory T which is essentially undecidable, i.e. not included in any α-recursive theory.

Proof. Let T be the theory corresponding to the α-fan P via Theorem 8.9. It is not hard to see that a theory is essentially undecidable if (and only if) it has no α-recursive model (equivalently, no α-recursive complete extension). Thus T is essentially undecidable.

A strengthening of Theorem 9.1 is the following:

Theorem 9.3 (Simpson [51]) There exists an α-fan P (of type 1) such that:

(i) P has positive measure (in the usual product measure on 2^{α}, cf. Lowenthal [30]);

(ii) if $X \in P$ then X is biimmune, i.e. any α-RE set which is included in X or in $\alpha - X$ is α-finite.

Proof. For $\alpha = \omega$ this is proved by Jockusch [15] so assume $\alpha > \omega$. Then we may take P to be the set of $X \in 2^{\alpha}$ such that both X and $\alpha - X$ meet every infinite, α-finite set. It is easy to check that P is a fan and has measure 1.

Corollary 9.4 There exists an essentially undecidable α-RE theory T such that:

(i) no completion of T is a Boolean combination of α-RE sets;

(ii) T is α-recursively separable from $\{\varphi : \neg\varphi \in T\}$.

Here again T is the theory corresponding to P under 8.9. Statement 9.4(ii) is proved by a measure theoretic argument. For $\alpha = \omega$, 9.4(i) and 9.4(ii) are due to Specker [52] and Ehrenfeucht [4] respectively.

In order to obtain "positive" results on the basis problem, the following recursion theoretic rendering of Barwise's theorem [3] is a useful tool.

Theorem 9.5 ("Barwise completeness") Let $S \subseteq 2^{\alpha} \times \alpha$ be a relation which is α-RE (in sense 3). Then $\{y : \forall X S(X,y)\}$ is an α-RE subset of α.

In logical terms, Theorem 9.5 says that the set of consequences of an α-recursive set of formulae is α-RE. The following corollary says that an α-recursive set of formulae is consistent if each α-finite subset is consistent.

Corollary 9.6 ("Barwise compactness") If S is as in Theorem 9.5 then

$$\forall X \exists y\, S(X,y) \rightarrow \exists z \forall X (\exists y < z) S(X,y).$$

Proof. Suppose not. Let $W = W_e = \{w : \exists x T(e,w,x)\}$ be an α-RE set which is not α-recursive. Define W^* to be the set of w such that

$$\exists X \forall y (S(X,y) \rightarrow (\exists x < y) T(e,w,x)).$$

Then $W \subseteq W^*$. Moreover the complement of W^* is α-RE by 9.5. Hence W is a proper subset of W^*. Hence $\exists X \forall y \neg S(X,y)$ by definition of W^*.

Corollary 9.7 A set $X \in 2^{\alpha}$ is α-recursive if and only if $\{X\}$ is an α-fan (of type 3).

Theorem 9.5 is used in the proof of the following two "positive" basis theorem which for $\alpha = \omega$ are due to Jockusch and Soare [16].

Theorem 9.8 (Simpson [51]) Let P be an α-fan (of type 3) and assume that no element of P is α-recursive. Then P has $2^{\aleph_0}$ elements X such that X is almost α-recursive, i.e. for all $g : \alpha \rightarrow \alpha$ such that g is α-recursive in X (in sense 3) there exists an α-recursive function $f : \alpha \rightarrow \alpha$ such that $g(y) < f(y)$ for all y. Furthermore, given any countable collection C of sets which are not α-recursive, we can arrange that no member of C is α-recursive in

X (in sense 3).

Note that an almost α-recursive set is automatically subgeneric. Thus 9.8 gives a new proof of Sacks' theorem [40] that if $\alpha > \omega$ then $\alpha = \omega_1^X$ for some $X \subseteq \omega$. (Just apply the theorem to the α-fan $\{X \subseteq \omega : \omega_1^X \geq \alpha\}$.) An interesting model theoretic consequence of 9.8 is that there exists an almost α-recursive model of ZF+V = L which is an end extension of L_α.

Theorem 9.8 and the following one are proved by constructing a descending sequence $P = P_0 \supseteq P_1 \supseteq \ldots$ such that X is the unique member of $\cap\{P_n : n < \omega\}$. We omit further details.

<u>Theorem</u> 9.9 (Simpson [51]) Assume that the Δ_2 projectum of α is ω. Let P be an α-fan (of type 3). Then P has an element which is subgeneric and whose α-calculability degree $\underset{\sim}{a}$ has $\underset{\sim}{a}^{\mathcal{L}} = \underset{\sim}{0}^{\mathcal{L}}$ (cf. Remarks 8.4 and 8.5).

We finish with a "negative" result on the basis problem. The fan P in the following theorem is obtained by generalizing a priority construction of Jockusch and Soare [16] for $\alpha = \omega$.

<u>Theorem</u> 9.10 (Simpson [51]) There exists an α-fan P (of type 1) with no α-recursive elements such that if Q is any proper subfan of P then P-Q is an α-fan (of type 1). Furthermore, if X and Y are distinct elements of P then:

(i) X is regular and subgeneric (cf. Remark 8.5);

(ii) the α-degree $\underset{\sim}{a}$ of X realizes least possible jump, i.e. $\underset{\sim}{a}^{\mathcal{L}} = \underset{\sim}{a} \cup \underset{\sim}{0}^{\mathcal{L}}$;

(iii) the α-degrees of X and Y are incomparable;

(iv) no non α-recursive set is α-recursively truth table reducible to both X and Y (in sense 3).

If we let T be the theory corresponding to P under 8.9, we obtain the following result which for $\alpha = \omega$ is due to Martin and Pour-El [33].

<u>Corollary</u> 9.11 (Simpson [51]) There exists an essentially undecidable α-RE theory T such that any α-RE theory which includes T is α-finitely axiomatizable over T.

Another interesting property, which follows from 9.10(iv), is that every almost α-recursive element of P is of minimal α-degree. This observation plus 9.8 gives a new proof of MacIntyre's theorem [32] that there exists a minimal α-degree.

§10. CONCLUSION.

In these lectures I hope to have convinced the student that admissible recursion theory is an important, perhaps even central, branch of generalized recursion theory. This is so because of the wide variety of methods employed (priority, forcing, Skolem hull, reflection, compactness) and the numerous applications to other branches (higher types, inductive definability, descriptive set theory, admissible logics). The reader who wishes to see the subject developed at a more leisurely pace, with all the gory details (especially priority arguments), is referred to the papers listed below and to my forthcoming book.

REFERENCES

[1] S. Aanderaa, Inductive definitions and their closure ordinals, Generalized Recursion Theory, North-Holland, 1974, pp. 207-220.

[2] F. Abramson and G.E. Sacks, Uncountable Gandy ordinals, J. London Math. Soc. (2) 14 (1976), 387-392.

[3] K.J. Barwise, Infinitary logic and admissible sets, J. Symb. Logic 34(1969), 226-252.

[4] A. Ehrenfeucht, Separable theories, Bull. Acad. Polon. Sci. Ser. Math. Astron. Phys. 9(1961), 17-19.

[5] S. Friedman, this symposium.

[6] K. Gödel, Über formal unentscheidbare Sätze der Principa Mathematica und verwandter Systeme I, Monatshefte fur Math. und Phys. 38(1931), 173-198.

[7] K. Gödel, Consistency proof for the generalized continuum hypothesis, Proc. Nat. Acad. Sci. 25(1939), 220-224.

[8] K. Gödel, The Consistency of the Axiom of Choice and the Generalized Continuum Hypothesis with the Axioms of Set Theory, second printing, Annals of Math. Studies No. 3, Princeton University Press, 1951, 74pp.

[9] R. Gostanian, The next admissible ordinal, Ph.D. Dissertation, New York University, 1971, 68pp.

[10] L. Harrington, The superjump and the first recursively Mahlo ordinal, Generalized Recursion Theory, North-Holland, 1974, pp. 43-52.

[11] L. Harrington, this symposium.

[12] R.B. Jensen, Stufen der konstruktiblen Hierarchie, Habilitationsschrift, University of Bonn, 1967, 105pp.

[13] R.B. Jensen, The fine structure of the constructible hierarchy, Annals of Math. Logic 4(1972), 229-308.

[14] R.B. Jensen and C. Karp, Primitive recursive set functions, Proc. Symp. Pure Math. XIII part 1, Amer. Math. Soc., 1971, pp. 143-176.

[15] C.G. Jockusch, Jr., Π_1^0 classes and Boolean combinations of recursively enumerable sets, J. Symb. Logic 39(1974), 95-96.

[16] C.G. Jockusch, Jr. and R.I. Soare, Π_1^0 classes and degrees of theories, Trans. Amer. Math. Soc. 173(1972), 33-56.

[17] S.C. Kleene, Recursive functions and intuitionistic mathematics, Proceedings of the International Congress of Mathematicians (Cambridge, Mass., 1950), vol. 1, 1952, pp. 679-685.

[18] S.C. Kleene, Introduction to Metamathematics, Van Nostrand, 1952, 550 pp.

[19] S.C. Kleene, Recursive functionals and quantifiers of finite type I, Trans. Amer. Math. Soc. 91(1959), 1-52.

[20] S.C. Kleene, Recursive functionals and quantifiers of finite type II, Trans. Amer. Math. Soc. 108(1963), 106-142.

[21] G. Kreisel, A variant to Hilbert's theory on the foundations of arithmetic, Brit. J. Phil. Sci. 4(1953), 107-129.

[22] G. Kreisel and G.E. Sacks, Metarecursive sets, J. Symb. Logic 28(1963), 304-305; J. Symb. Logic 30(1965), 318-338.

[23] S. Kripke, Transfinite recursion on admissible ordinals I, II (abstracts), J. Symb. Logic 29(1964), 161-162.

[24] A.H. Lachlan, Lower bounds for pairs of recursively enumerable degrees, Proc. London Math. Soc. 16(1966), 537-569.

[25] M. Lerman, On suborderings of the α-recursively enumerable α-degrees, Annals of Math. Logic 4(1972), 369-392.

[26] M. Lerman, Least upper bounds for minimal pairs of α-RE α-degrees, J. Symb. Logic 39(1974), 49-52.

[27] M. Lerman, Maximal α-RE sets, Trans. Amer. Math. Soc. 188(1974), 341-386.

[28] M. Lerman, Lattices of α-recursively enumerable sets, this volume.

[29] M. Lerman and G.E. Sacks, Some minimal pairs of α-recursively enumerable degrees, Annals of Math. Logic 4(1972), 415-442.

[30] F. Lowenthal, Measure and categoricity in α-recursion, Set Theory and Hierarchy Theory, Lecture Notes in Mathematics No. 537, Springer-Verlag, 1976, pp. 185-201.

[31] W. Maass, this symposium.

[32] J.M. MacIntyre, Minimal α-recursion theoretic degrees, J. Symb. Logic 38(1973), 18-28.

[33] D.A. Martin and M.B. Pour-El, Axiomatizable theories with few axiomatizable extensions, J. Symb. Logic 35(1970), 205-209.

[34] H.G. Rice, On completely recursively enumerable classes and their key arrays, J. Symb. Logic 21(1956), 304-308.

[35] W. Richter, Constructively accessible ordinal numbers, J. Symb. Logic 33(1968), 43-55.

[36] W. Richter and P. Aczel, Inductive definitions and reflecting properties of admissible ordinals, Generalized Recursion Theory, North-Holland, 1974, pp. 301-381.

[37] G.E. Sacks, Degrees of Unsolvability, second edition, Annals of Math. Studies No. 55, Princeton University Press, 1966, 175 pp.

[38] G.E. Sacks, The recursively enumerable degrees are dense, Annals of Math. 80(1964), 300-312.

[39] G.E. Sacks, Post's problem, admissible ordinals, and regularity, Trans. Amer. Math. Soc. 124(1966), 1-24.

[40] G.E. Sacks, Countable admissible ordinals and hyperdegrees, Advances in Math. 20(1976), 213-262.

[41] G.E. Sacks and S.G. Simpson, The α-finite injury method, Ananls of Math. Logic 4(1972), 343-367.

[42] R.A. Shore, Minimal α-degrees, Annals of Math. Logic 4(1972), 393-414.

[43] R.A. Shore, Splitting an α-recursively enumerable set, Trans. Amer. Math. Soc. 204(1975), 65-78.

[44] R.A. Shore, The recursively enumerable α-degrees are dense, Annals of Math. Logic 9(1976), 123-155.

[45] R.A. Shore, On the jump of an α-recursively enumerable set, Trans. Amer. Math. Soc. 217(1976), 351-363.

[46] R.A. Shore, α-Recursion theory, Handbook of Mathematical Logic, North-Holland, 1977, pp. 653-680.

[47] R.A. Shore, Some more minimal pairs of α-recursively enumerable degrees, to appear.

[48] R.A. Shore, this symposium.

[49] S.G. Simpson, Degree theory on admissible ordinals, Generalized Recursion Theory, North-Holland, 1974, pp. 165-193.

[50] S.G. Simpson, Comprehension and reflection, in preparation.

[51] S.G. Simpson, Countable admissible ordinals and basis theorems, in preparation.

[52] E. Specker, Eine Verschärfung des Unvollständigkeitssatzes der Zahlentheorie, Bull. Acad. Polon. Sci. Ser. Math. Astron. Phys. 5(1957), 1043-1047.

[53] G. Takeuti, On the recursive functions of ordinal numbers, J. Math. Soc. Japan 12(1960), 119-128.

[54] G. Takeuti, Recursive functions and arithmetical functions of ordinal numbers, Logic, Methodology and Philosophy of Science, North-Holland, 1965, pp. 179-196.

[55] C.E.M. Yates, A minimal pair of recursively enumerable degrees, J. Symb. Logic 31(1966), 159-168.

Note: The following definition should be inserted at the bottom of page 359.

Definition 2.6. A set $K \subseteq \alpha$ is said to be α-finite if there exist a primitive recursive relation $R(x,y)$ and parameters p and q less than α, such that

$$K = \{y < p : R(q,y)\}.$$

J.E. Fenstad, R.O. Gandy, G.E. Sacks (Eds.)
GENERALIZED RECURSION THEORY II

WEAKLY INADMISSIBLE RECURSION THEORY

by

Viggo Stoltenberg-Hansen
Oslo

The theory of recursively enumerable degrees for ordinary recursion theory has been generalized quite successfully to α-recursion theory, recursion theory on an admissible ordinal α. The generalization has been extended further, with partial success, in two directions. The first, investigated by Simpson [10] and Stoltenberg-Hansen [11], considers recursion theory on resolvable admissible sets (with urelements), thus studying the need for a recursively well-ordered domain. The second direction (see Friedman [6] for an introduction) investigates the question of how "infinite" the domain must be by considering recursion theory on S_β for an arbitrary limit ordinal β.

In this paper we combine the two generalizations by considering rudimentarily closed structures $\underset{\sim}{M} = \langle M,\in,R\rangle$ which admit what we call an acceptable prewellordering. As in the case of β-recursion theory such structures fall into three disjoint classes determined by how "infinite" the domain is with respect to the recursive power of the structure: admissible, weakly inadmissible and strongly inadmissible.

Not every $\underset{\sim}{M}$-r.e. set is tame when $\underset{\sim}{M}$ is inadmissible, i.e. for inadmissible $\underset{\sim}{M}$ it is not always possible to effectively enumerate the $\underset{\sim}{M}$-finite subsets of an $\underset{\sim}{M}$-r.e. set. Since the notion of "finite" is basic in our generalization we are primarily interested in the tame(ly) r.e. (t.r.e.) sets. Our main result is that for adequate structures $\underset{\sim}{M}$, the structure of the regular t.r.e. $\underset{\sim}{M}$-

degrees is non-trivial (and rich) if and only if $\underset{\sim}{M}$ is admissible or weakly inadmissible (i.e. weakly admissible). Furthermore a regular set theorem for t.r.e. sets holds for adequate weakly admissible $\underset{\sim}{M}$.

The arguments used for weakly admissible structures are recursion theoretic in nature. Thus, although we here restrict ourselves to nice set theoretic structures, we believe that these arguments can be formalized in the axiomatic framework of Moschovakis [9] and Fenstad [3] by weakening their notion of "finite" to cover the inadmissible case. This should be contrasted with the positive and negative results of Friedman ([5],[7]) for strongly inadmissible S_β where $\diamondsuit$ and Fodor's theorem, or effectivized versions thereof, constitute important ingredients.

1. Preliminaries

We will restrict ourselves to transitive rudimentarily closed structures $\underset{\sim}{M} = \langle M,\in,R\rangle$. Clearly the inclusion of urelements, important in applications, would not invalidate any of the results below.

For a treatment of the rudimentary functions we refer to Devlin [2] where a proof of the following lemma can be found.

Lemma 1.1. $\models_{\underset{\sim}{M}}^{\Sigma_n} = \{\langle i,x_1,\ldots,x_m\rangle$: The i:th Σ_n formula ϕ is m-ary and $\underset{\sim}{M} \models \phi(x_1,\ldots,x_m)\}$ is uniformly $\Sigma_n^{\underset{\sim}{M}}$ for transitive rudimentarily closed $\underset{\sim}{M}$.

An immediate consequence is the existence of a universal $\Sigma_1(\underset{\sim}{M})$ formula: $\phi(e,x) \iff \langle (e)_0,x,(e)_1\rangle \in \models_{\underset{\sim}{M}}^{\Sigma_1}$.

The recursion theoretic notions are defined as follows: For

$A \subseteq M$ we say A is $\underset{\sim}{M}$-r.e. if $A \in \Sigma_1(\underset{\sim}{M})$, i.e. A is definable over $\underset{\sim}{M}$ by a Σ_1 formula with parameters from M. A is $\underset{\sim}{M}$-recursive if A and $M - A$ are both $\underset{\sim}{M}$-r.e. If $A \in M$ then A is $\underset{\sim}{M}$-finite. A partial function is $\underset{\sim}{M}$-recursive if its graph is $\underset{\sim}{M}$-r.e. Finally we index the $\underset{\sim}{M}$-r.e. sets by putting $W_e = \{x : \phi(e,x)\}$ where ϕ is a universal $\Sigma_1(\underset{\sim}{M})$ formula.

Definition 1.2. $\underset{\sim}{M}$ admits an acceptable prewellordering if there is an $\underset{\sim}{M}$-recursive prewellordering $\prec$ on M such that

(i) $L^x = \{y \in M : y \prec x\}$ is uniformly $\underset{\sim}{M}$-finite.

(ii) $|\precsim|$ = ordinal of $\prec$ = limit ordinal and for each $\delta < |\precsim|$ $L^\delta \cap |\precsim| = \delta$.

(iii) $\Sigma_1 - cf(|\precsim|)$ = limit ordinal, where $\Sigma_1 - cf(|\precsim|)$ = least γ for which there is $\underset{\sim}{M}$-finite $x \subseteq L^\gamma$ and $\underset{\sim}{M}$-recursive function $f: x \longrightarrow |\precsim|$, unbounded in $|\precsim|$.

(iv) If $x \in L^\tau$ then there is σ such that $x \subseteq L^\sigma$. If $x \in L^\tau$ and $\tau < \Sigma_1 - cf(|\precsim|)$ then there is $\sigma < \Sigma_1 - cf(|\precsim|)$ such that $x \subseteq L^\sigma$.

Note that (iii) asserts that the prewellordering below $\Sigma_1 - cf(|\precsim|)$ is "thin". Clearly S_β (the universe of β-recursion theory) admits an acceptable prewellordering for every limit ordinal β.

Henceforth $\underset{\sim}{M}$ will denote a transitive rudimentarily closed structure which admits an acceptable prewellordering. $\Sigma_1 - cf(|\precsim|)$ will always be denoted by κ.

Lemma 1.3. There is an $\underset{\sim}{M}$-recursive function $\lambda e\sigma\, W_e^\sigma$ (σ varying over $|\precsim|$) such that

(i) $W_e = \cup\{W_e^\sigma : \sigma < |\precsim|\}$

(ii) $\tau < \sigma \Rightarrow W_e^\tau \subseteq W_e^\sigma$.

Proof: Let $\exists z\psi(e,x,z)$ be a universal $\Sigma_1(\underset{\sim}{M})$ formula where ψ is Δ_0. Then

$$W_e^\sigma = v \iff \exists w[w = L^\sigma \ \& \ v = w \cap \{x:(\exists z \in w)\psi(e,x,z)\}] \text{ is } \Sigma_1(\underset{\sim}{M}). \quad \square$$

Lemma 1.4. (Selection operator). There is an $\underset{\sim}{M}$-recursive function q such that $q(e)\downarrow \iff W_e \neq \emptyset$ and $q(e)\downarrow \Rightarrow \emptyset \neq q(e) \subseteq W_e$.

Proof: Let $\exists z\psi(e,x,z)$ be a universal $\Sigma_1(\underset{\sim}{M})$ formula where ψ is Δ_0. Let $r(e) = \text{least } \gamma[(\exists\langle x,z\rangle \in L^\gamma)\psi(e,x,z)]$. Then r is $\underset{\sim}{M}$-recursive since ψ is Δ_0, and $r(e)\downarrow \iff W_e \neq \emptyset$. Define
$q(e) = v \iff \exists\gamma[r(e) = \gamma \ \& \ v = \pi_0'' L^\gamma \cap \{x:(\exists z \in \pi_0'' L^\gamma)\psi(e,x,z)\}]$. $\square$

κ is an important parameter, being an indicator of how admissible $\underset{\sim}{M}$ is. Below κ everything looks quite admissible.

Lemma 1.5. Let $F = \{x \in M:(\exists\beta<\kappa)(x \subseteq L^\beta)\}$.

(i) If B is an $\underset{\sim}{M}$-recursive set and $x \in F$ then $B \cap x \in F$.

(ii) Suppose $f:M \to M$ is $\underset{\sim}{M}$-recursive. Then $g:F \to M$ defined by $g(x) = f''x$ is $\underset{\sim}{M}$-recursive.

Proof: We prove (ii). Let $f(y) = z \iff \exists t\,\phi(y,z,t)$ where ϕ is $\Delta_0(\underset{\sim}{M})$. Then
$g(x) = v \iff \exists w[(\forall y \in x)(\exists z,t \in w)\phi(y,z,t) \ \& \ v = w \cap \{z:(\exists y \in x)(\exists t \in w)\phi(y,z,t)\}]$.
To see that g is defined on F let
$k(y) = \text{some } \gamma[(\exists z,t \in L^\gamma)\phi(y,z,t)]$. For $x \in F$, $k''x$ is bounded by some $\gamma < |\underset{\sim}{M}|$. Thus $g(x)$ is defined. $\square$

Lemma 1.6. (Definition by recursion). Suppose $g:\kappa \times M^{n+1} \to M$ is $\underset{\sim}{M}$-recursive. Then $f(\beta,\vec{x}) = g(\beta,\vec{x},\{\langle\gamma,f(\gamma,\vec{x})\rangle:\gamma < \beta\})$ is $\underset{\sim}{M}$-recursive.

The proof is the usual one (see e.g. Barwise [1]), the point being that we have Σ_1-replacement below κ.

The reducibility notions we consider are the wellknown ones from the admissible case. Thus for sets A and B, A is <u>$\underset{\sim}{M}$-recursive in B</u>, $A \leq_{\underset{\sim}{M}} B$, if there is an $\underset{\sim}{M}$-r.e. set W such that

$a \subseteq A \;\&\; b \cap A = \emptyset \iff \exists c,d(\langle a,b,c,d\rangle \in W \;\&\; c \subseteq B \;\&\; d \cap B = \emptyset)$.

A partial function f is <u>weakly $\underset{\sim}{M}$-recursive in</u> B, $f \leq_{w\underset{\sim}{M}} B$, if $f(\vec{x}) = y \iff \exists c,d(\langle\vec{x},y,c,d\rangle \in W \;\&\; c \subseteq B \;\&\; d \cap B = \emptyset)$ for some $\underset{\sim}{M}$-r.e. set W, and $A \leq_{w\underset{\sim}{M}} B$ if $c_A \leq_{w\underset{\sim}{M}} B$ where c_A is the characteristic function of A. $\underset{\sim}{M}$-degrees are defined in the usual way using $\leq_{\underset{\sim}{M}}$.

As in the admissible case, we need consider two notions of projecta. Let $|\precsim|^* =$ least δ for which there is a partial $\underset{\sim}{M}$-recursive function $q : L^\delta \xrightarrow{\text{onto}} M$, and let $|\precsim|^+ =$ least δ for which there is an $\underset{\sim}{M}$-r.e. set $A \subseteq L^\delta$ such that $A \notin M$. It is easily verified that $|\precsim|^+ \leq |\precsim|^*$. We call $\underset{\sim}{M}$ <u>adequate</u> if $\underset{\sim}{M}$ admits an acceptable prewellordering $\precsim$ for which $|\precsim|^* = |\precsim|^+ =$ limit ordinal. If $M = S_\beta$ then $\underset{\sim}{M}$ is adequate.

It follows from lemma 1.5 that $\underset{\sim}{M}$ is admissible if and only if $\kappa = |\precsim|$. $\underset{\sim}{M}$ is said to be <u>weakly inadmissible</u> if $|\precsim|^* \leq \kappa < |\precsim|$ and <u>strongly inadmissible</u> if $\kappa < |\precsim|^*$. In case $|\precsim|^* \leq \kappa$, $\underset{\sim}{M}$ is weakly admissible.

2. Tamely r.e. sets

As is reflected in the choice of notion of relative computability it is the $\underset{\sim}{M}$-finite sets which are basic in our generalization. It is therefore natural to single out those $\underset{\sim}{M}$-r.e. sets for which there is an $\underset{\sim}{M}$-recursive enumeration of all $\underset{\sim}{M}$-finite subsets of the $\underset{\sim}{M}$-r.e. set. A set A is said to be <u>tamely r.e.</u> (t.r.e.) if $\{a \in M : a \subseteq A\}$ is $\underset{\sim}{M}$-r.e. An $\underset{\sim}{M}$-recursive enumeration $\lambda\,\sigma\,A^\sigma$ of

A is <u>tame</u> if $a \subseteq A \Rightarrow \exists\sigma(a \subseteq A^\sigma)$. Clearly A is t.r.e. if and only if A has a tame enumeration. For admissible $\underset{\sim}{M}$ every enumeration of an $\underset{\sim}{M}$-r.e. set is tame.

Another important property is that of regularity: A set A is <u>regular</u> if $a \cap A \in M$ whenever $a \in M$. For adequate admissible $\underset{\sim}{M}$ every $\underset{\sim}{M}$-r.e. degree has a regular $\underset{\sim}{M}$-r.e. representative. With this in mind it seems reasonable to study the structure of the regular t.r.e. $\underset{\sim}{M}$-degrees.

Let $h : \kappa \to |\precsim|$ be a fixed $\underset{\sim}{M}$-recursive strictly increasing function, unbounded in $|\precsim|$. Such an h exists by lemma 1.6.

<u>Theorem 2.1</u>. Suppose B is a regular t.r.e. set. Then there is a regular t.r.e. set $B^* \subseteq h''\kappa$ such that $B \equiv_{\underset{\sim}{M}} B^*$.

<u>Corollary 2.2</u>. Suppose $\kappa < |\precsim|^+$. Then every regular t.r.e. set is $\underset{\sim}{M}$-recursive in $\emptyset$.

<u>Proof of corollary</u>: Let B be a regular t.r.e. set and let B^* be as in the theorem. Then $B \equiv_{\underset{\sim}{M}} B^* \leq_{\underset{\sim}{M}} h^{-1}[B^*] \in M$. □

Thus, considering adequate $\underset{\sim}{M}$, the structure of the regular t.r.e. $\underset{\sim}{M}$-degrees is trivial for strongly inadmissible $\underset{\sim}{M}$. We shall show it is rich whenever such $\underset{\sim}{M}$ is weakly admissible.

<u>Remark</u>: The hypothesis of regularity in the corollary is necessary. Friedman [4] has exhibited a strongly inadmissible β for which every t.r.e. set is regular and hence of zero degree, while Maass [8] has found a strongly inadmissible β with non-zero t.r.e. degrees.

<u>Proof of theorem</u>: We may assume $B \notin M$. Let $\lambda\sigma\, B^\sigma$ be a disjoint (i.e. $\tau \neq \sigma \Rightarrow B^\tau \cap B^\sigma = \emptyset$) tame enumeration of B, σ varying over κ, such that $(\forall\sigma < \kappa)(B^\sigma \neq \emptyset)$. Put $B^* = \{h(\sigma) : (\exists\tau > \sigma)(B^\tau < B^\sigma)\}$

where $B^\tau < B^\sigma \Longleftrightarrow (\exists x \in B^\tau)(\forall y \in B^\sigma)(x \prec y)$.

It is easily seen that if $A \subseteq \kappa$ is $\underset{\sim}{M}$-r.e. then $h''A$ is t.r.e. It follows that B^* is t.r.e. To show that B^* is regular assume $a \in M$. Let $\sigma_0 < \kappa$ be such that $h(\sigma_0) \cap a = h''\kappa \cap a$, and let $b = h^{-1}[h''\kappa \cap a] \cap \{\sigma : (\exists \tau > \sigma)(B^\tau < B^\sigma)\}$. As $h''b = a \cap B^*$ it suffices to show $b \in M$. Let α be sufficiently large so that $B^{<\sigma_0}$ $(= \cup \{B^\tau : \tau < \sigma_0\}) \subseteq L^\alpha$. By the regularity of B and the tameness of our enumeration, $B \cap L^\alpha \subseteq B^{<\sigma_1}$ for some $\sigma_1 < \kappa$. But then $B^\sigma < B^\tau$ whenever $\tau \geq \sigma_1$ and $\sigma < \sigma_0$. Thus $b = h^{-1}[h''\kappa \cap a] \cap \{\sigma < \sigma_0 : (\exists \tau < \sigma_1)(\tau > \sigma \ \& \ B^\tau < B^\sigma)\} \in M$.

Define $f_1(a) = \text{some } \sigma[a \subseteq L^{h(\sigma)}]$ and $f_2(a) = \text{some } \sigma[B^{<f_1(a)} \subseteq L^{h(\sigma)}]$. Then $r(a) = \text{least } \tau[(L^{h(f_2(a))} - B^{<\tau}) \cap B = \emptyset]$ is total since B is regular and t.r.e., and $r \leq_{w\underset{\sim}{M}} B$. Now $a \cap B^* = \emptyset \Longleftrightarrow (\forall \sigma \in h^{-1}[a])(\forall \tau < r(a))(\tau \leq \sigma \vee B^\sigma \leq B^\tau)$. Hence $B^* \leq_{\underset{\sim}{M}} B$. It remains to show $B \leq_{\underset{\sim}{M}} B^*$. Let $L^{B^\sigma} = L^\beta$ where β is the least ordinal such that $L^\beta \cap B^\sigma \neq \emptyset$. Define $q(a) = \text{least } \sigma[h(\sigma) \notin B^* \ \& \ a \subseteq L^{B^\sigma}]$. Thus $q \leq_{w\underset{\sim}{M}} B^*$. We show q is total. Given $a \in M$ there is, by the regularity and t.r.e.-ness of B, σ_0 such that $\tau \geq \sigma_0 \Rightarrow a \subseteq L^{B^\tau}$. If $h(\sigma_0) \notin B^*$ then we are done. Else there must be $\sigma_1 > \sigma_0$ such that $B^{\sigma_1} < B^{\sigma_0}$. Thus we obtain a sequence $\sigma_0 < \sigma_1 < \ldots$ such that $B^{\sigma_0} > B^{\sigma_1} > \ldots$ which is necessarily finite by the wellfoundedness of $\prec$. Then $h(\sigma_n) \notin B^*$ where σ_n is the last element in the sequence, i.e. $q(a)$ is defined. Clearly $a \cap B = \emptyset \Longleftrightarrow a \cap B^{<q(a)} = \emptyset$ so $B \leq_{\underset{\sim}{M}} B^*$. □

3. The admissible collapse

The main tool used in studying weakly inadmissible structures is the admissible collapse. It is a technique whereby for each weakly inadmissible structure $\underset{\sim}{M}$ one constructs a resolvable admissible structure $\mathfrak{A}$, the admissible collapse of $\underset{\sim}{M}$, such that the regular $\mathfrak{A}$-r.e. degrees can be embedded onto the regular t.r.e. $\underset{\sim}{M}$-degrees.

The idea of an admissible collapse stems from Maass [8], where such a collapse is constructed for the case $M = S_\beta$.

Henceforth let $\underset{\sim}{M}$ be a weakly inadmissible structure and let $h : \kappa = \Sigma_1 - cf(|\precsim|) \to |\precsim|$ be a fixed strictly increasing $\underset{\sim}{M}$-recursive function unbounded in $|\precsim|$. Let $q : L^{|\precsim|^*} \xrightarrow{\text{onto}} M$ be partial $\underset{\sim}{M}$-recursive and let ϕ be a $\Delta_0(\underset{\sim}{M})$ formula such that
$q(x) = y \iff \exists t\, \phi(x,y,t)$. Put
$q^\nu = \{\langle x,y\rangle \in L^\nu \times L^{h(\nu)} : (\exists t \in L^{h(\nu)})\phi(x,y,t)\}$. Then $\lambda\nu q^\nu$ is an $\underset{\sim}{M}$-recursive approximation of q.

Letting $F^\nu = \{a \in L^{h(\nu)} : a \subseteq L^\nu\} \cup \{L^\nu\}$, define

$$N^0 = \emptyset,$$

$$N^{\nu+1} = N^\nu \cup F^\nu \cup \{q^\nu{}''a \cap N^\nu : a \in F^\nu\},$$

$$N^\gamma = \bigcup_{\nu<\gamma} N^\nu \quad \text{if } \cup\gamma = \gamma.$$

Let $N = \bigcup_{\nu<\kappa} N^\nu$.

It follows from lemma 1.6 that $\lambda\nu N^\nu$ is $\underset{\sim}{M}$-recursive and hence that N is $\underset{\sim}{M}$-r.e. Recalling the transitivity condition in 1.2 (iv), it is easily seen that N is transitive.

Define $G^\nu = q^\nu{}''L^\nu \cap N^\nu$.

Lemma 3.1. $\lambda\nu G^\nu$ is $\underset{\sim}{M}$-recursive, $G^\nu \in N^{\nu+1}$, $\nu < \gamma \Rightarrow G^\nu \subseteq G^\gamma$, and $N = \bigcup_{\nu<\kappa} G^\nu$.

Proof: The first three properties are immediate. It follows from the second property, using the transitivity of N, that $\bigcup_{\nu<\kappa} G^\nu \subseteq N$. Suppose $y \in N$. Let $x \in L^{|\precsim|^*}$ be such that $q(x) = y$

and let $\nu < \kappa$ be sufficiently large so that $y \in N^{\nu}$, $x \in L^{\nu}$ and $q^{\nu}(x) = y$. Then $y \in G^{\nu}$ and hence $N \subseteq \bigcup_{\nu<\kappa} G^{\nu}$. □

Lemma 3.2.

(i) If $x \in N$ and $f : x \to |\underset{\sim}{\prec}|$ $(f : x \to \kappa)$ is $\underset{\sim}{M}$-recursive then $f''x$ is bounded (bounded in κ).

(ii) If $x \in N$ and B is $\underset{\sim}{M}$-recursive then $x \cap B \in N$.

(iii) If $f : N \to M$ is $\underset{\sim}{M}$-recursive then $g : N \to M$ defined by $g(x) = f''x$ is $\underset{\sim}{M}$-recursive.

Proof: As a sample we prove (ii). The proof is by induction on ν . Suppose B is $\underset{\sim}{M}$-recursive and let $x \in N^{\nu+1}$. If $x \in F^{\nu}$ then $x \cap B \in N$ by lemma 1.5. Suppose $x = q^{\nu}{}''a \cap N^{\nu}$ where $a \subseteq L^{\nu}$ and $a \in M$. Let $k = q^{\nu} \cap a \times N^{\nu}$ and let $b = \{y \in \mathrm{dom}\ k : k(y) \in B\}$. Then b is an $\underset{\sim}{M}$-recursive subset of L^{ν} so $b \in M$. Hence $b \in F^{\gamma}$ for some $\gamma \geq \nu$. But $x \cap B = k''b = q^{\gamma}{}''b \in N^{\gamma+1}$. □

Let $\exists z\, \phi(x,y,z)$ be a universal $\Sigma_1(\underset{\sim}{M})$ formula where ϕ is Δ_0. Define $\psi(e,y,\beta) \iff e \in \mathrm{dom}\ q^{\beta}\ \&\ (\exists z \in L^{h(\beta)})\phi(q^{\beta}(e),y,z)$. Clearly ψ is $\Delta_1(\underset{\sim}{M})$. Furthermore if W is $\underset{\sim}{M}$-r.e. then there is $e \in N$ such that $y \in W \iff \exists\beta < \kappa\ \psi(e,y,\beta)$.

Definition: $\mathfrak{A} = \langle N, \in \cap N^2, \psi \cap N^3 \rangle$ is the admissible collapse of $\underset{\sim}{M}$.

An easy proof by induction on the definition of $\Delta_0(\mathfrak{A})$ relations shows

Lemma 3.3. Suppose $\phi(\vec{x})$ is a $\Delta_0(\mathfrak{A})$ relation. Then there are $\Sigma_1(\underset{\sim}{M})$ formulas θ_1 and θ_2 such that for all $\vec{x} \in N^n$, $\phi(\vec{x}) \iff \underset{\sim}{M} \models \theta_1(\vec{x})$ and $\neg\phi(\vec{x}) \iff \underset{\sim}{M} \models \theta_2(\vec{x})$.

Theorem 3.4. Suppose $\underset{\sim}{M}$ is weakly inadmissible and $\mathfrak{A}$ its admissible collapse. Then $\mathfrak{A}$ is a transitive resolvable admissible structure such that for each $A \subseteq N$, A is $\mathfrak{A}$-r.e. iff A is $\underset{\sim}{M}$-r.e.

Proof: As already remarked, N is transitive. Extensionality and foundation trivially hold.

Pair: Suppose $x,y \in G^{\nu}$. Then $\{x,y\} \subseteq N^{\nu}$. Let $z,w \in L^{\nu}$ be such that $q^{\nu}(z) = x$ and $q^{\nu}(w) = y$. Choose $\gamma \geq \nu$ such that $\{z,w\} \in F^{\gamma}$. Then $\{x,y\} = q^{\gamma}"\{z,w\} \cap N^{\gamma} \in N^{\gamma+1}$.

Union: First note (using lemma 3.2 (i)) that for each $x \in N$ there is $\delta < \kappa$ such that $x \subseteq G^{\delta}$. Define $f(x) = \text{some } \delta[x \subseteq G^{\delta}]$. Suppose $a \in N$. Then $f"a$ is bounded by some $\delta < \kappa$. Hence $\cup a \subseteq G^{\delta}$. But $\cup a \in M$ and $G^{\delta} \in N$ so $\cup a \in N$ by lemma 3.2 (ii).

Δ_0-separation: Suppose $a \in N$ and $\phi(x)$ is $\Delta_0(\mathfrak{A})$. Choose $\Delta_0(\underset{\sim}{M})$ formulas θ_1 and θ_2 using 3.3 such that for each $x \in N$, $\phi(x) \Leftrightarrow \underset{\sim}{M} \models \exists t\ \theta_1(x,t)$ and $\neg\phi(x) \Leftrightarrow \underset{\sim}{M} \models \exists t\ \theta_2(x,t)$. Let $f(x) = \text{some } \delta[(\exists t \in L^{\delta})(\theta_1(x,t) \vee \theta_2(x,t))]$ and let $\gamma < |\precsim|$ be a bound for $f"a$. Then $a \cap \{x:\phi(x)\} = a \cap \{x:(\exists t \in L^{\gamma})\theta_1(x,t)\} \in N$.

Δ_0-collection: Suppose $\mathfrak{A} \models (\forall x \in a)(\exists y)\phi(x,y)$ where ϕ is $\Delta_0(\mathfrak{A})$ and hence $\Sigma_1(\underset{\sim}{M})$. Let $f(x) = \text{some } \delta[(\exists y \in G^{\delta})\phi(x,y)]$. Then f is $\underset{\sim}{M}$-recursive, $a \subseteq \text{dom } f$ and hence $f"a$ is bounded by some $\gamma < \kappa$. Thus $(\forall x \in a)(\exists y \in G^{\gamma})\phi(x,y)$.

Suppose $A \subseteq N$. If A is $\mathfrak{A}$-r.e. then A is $\underset{\sim}{M}$-r.e. by lemma 3.3 and the fact that N is $\underset{\sim}{M}$-r.e. If A is $\underset{\sim}{M}$-r.e. then there is $e \in N$ such that $x \in A \Longleftrightarrow \underset{\sim}{M} \models (\exists \beta < \kappa)\psi(e,x,\beta) \Longleftrightarrow \mathfrak{A} \models \exists\beta\psi(e,x,\beta)$ so A is $\mathfrak{A}$-r.e. It follows that a partial function $f:N \hookrightarrow N$ is $\underset{\sim}{M}$-recursive iff f is $\mathfrak{A}$-recursive. In particular the function $\lambda\nu G^{\nu}$ is $\mathfrak{A}$-recursive. Thus $\mathfrak{A}$ is resolvable. □

Lemma 3.5. Suppose $A,B \subseteq \kappa$. Then $A \leq_{\mathfrak{A}} B \Leftrightarrow h"A \leq_{\underset{\sim}{M}} h"B$.

Proof: Suppose $A \leq_{\mathfrak{A}} B$ via an $\mathfrak{A}$-r.e. set V. Then

$h''A \leq_{\underset{\sim}{M}} h''B$ via the $\underset{\sim}{M}$-r.e. set $W = \{\langle a,b,h''c,h''(d \cap \kappa)\rangle : a \subseteq h''\kappa$ & $c \subseteq \kappa$ & $\langle h^{-1}[a],h^{-1}[b],c,d\rangle \in V\}$. On the other hand if $h''A \leq_{\underset{\sim}{M}} h''B$ via an $\underset{\sim}{M}$-r.e. set W then $A \leq_{\mathfrak{A}} B$ via the $\mathfrak{A}$-r.e. set $V = \{\langle a,b,h^{-1}[c],h^{-1}[d]\rangle : a,b \in N$ & $a \subseteq \kappa$ & $c \subseteq h''\kappa$ & $\langle h''a,h''(b \cap \kappa),c,d\rangle \in W\}$. □

Theorem 3.6. Suppose $\underset{\sim}{M}$ is weakly inadmissible and $\mathfrak{A}$ its admissible collapse. Then there is an embedding E of the regular $\mathfrak{A}$-r.e. degrees onto the regular t.r.e. $\underset{\sim}{M}$-degrees.

Proof: Applying theorem 2.1 to $\mathfrak{A}$ we see that every regular $\mathfrak{A}$-r.e. degree has a regular $\mathfrak{A}$-r.e. representative which is a subset of κ. For regular $\mathfrak{A}$-r.e. $A \subseteq \kappa$ let $E(\mathfrak{A}\text{-deg}(A)) = \underset{\sim}{M}\text{-deg}(h''A)$. Clearly $h''A$ is regular and t.r.e. for such A. Thus E is an embedding into the regular t.r.e. $\underset{\sim}{M}$-degrees by lemma 3.5. By theorem 2.1 E is in fact onto. □

4. Weakly inadmissible structures

In this section $\underset{\sim}{M}$ will denote a weakly inadmissible structure and $\mathfrak{A} = \langle N,\in,\psi\rangle$ its admissible collapse. We are going to show that whenever $\mathfrak{A}$ is adequate then the structure of the t.r.e. $\underset{\sim}{M}$-degrees is quite satisfactory. It follows immediately from theorem 3.4 that if $\underset{\sim}{M}$ is adequate then $\lambda\sigma L^{\sigma}$ $(\sigma < |\underset{\sim}{\prec}|^{*})$ is an adequate prewell-ordering for $\mathfrak{A}$ in the sense of [11]. Hence if we restrict our attention to adequate structures $\underset{\sim}{M}$, the structure of the regular t.r.e. $\underset{\sim}{M}$-degrees is non-trivial and rich if and only if $\underset{\sim}{M}$ is weakly admissible.

Maass [8] shows that the t.r.e. β-degrees, the regular recursive t.r.e. β-degrees and the recursive β-degrees coincide for weakly inadmissible β. Theorem 4.1 below shows that this is a consequence of the regular set theorem holding for $\mathfrak{A}$.

Theorem 4.1. Suppose every $\mathfrak{A}$-r.e. degree has a regular $\mathfrak{A}$-r.e. representative. Then the following are equivalent for an $\underset{\sim}{M}$-degree $\underset{\sim}{a}$:

(i) $\underset{\sim}{a}$ contains a t.r.e. set.

(ii) $\underset{\sim}{a}$ contains a regular recursive t.r.e. set.

(iii) $\underset{\sim}{a}$ contains a recursive set.

Remark: In [12] it is shown that the regular set theorem holds for every adequate structure $\mathfrak{A}$. Thus the above theorem may be viewed as a regular set theorem for t.r.e. sets and recursive sets for every adequate weakly inadmissible structure $\underset{\sim}{M}$.

Proof: First note that there is a partial $\underset{\sim}{M}$-recursive function $p:N \xrightarrow{onto} M$ such that dom p is $\underset{\sim}{M}$-recursive. For let $q:N \xrightarrow{onto} M$ be partial $\underset{\sim}{M}$-recursive and let $\lambda\sigma A^\sigma$ be an enumeration of dom q. Then define p by $p(<\sigma,x>) = z \Leftrightarrow x \in A^\sigma \;\&\; q(x) = z$.

(i) ⇒ (ii). Suppose A is t.r.e. Let $A_1 = \{x \in M : x \cap A \neq \emptyset\}$. Then $p^{-1}[A_1]$ is $\underset{\sim}{M}$-r.e. and hence $\mathfrak{A}$-r.e. Let $B_1 \subseteq \kappa$ be regular (in $\mathfrak{A}$), $\mathfrak{A}$-r.e. and of the same $\mathfrak{A}$-degree as $p^{-1}[A_1]$ and put $B = \{<\sigma,h(x)> : x \in B_1^\sigma\}$ where $\lambda\sigma B_1^\sigma$ is an enumeration of B_1 in $\mathfrak{A}$. We claim $A \equiv_{\underset{\sim}{M}} B$. Suppose $p^{-1}[A_1] \leq_{\mathfrak{A}} B_1$ via W_1. Then

$$a \cap A = \emptyset \Leftrightarrow a \notin A_1 \Leftrightarrow p^{-1}(a) \cap p^{-1}[A_1] = \emptyset$$

$$\Leftrightarrow \exists b \in N\, (<p^{-1}(a),b> \in W_1 \;\&\; b \cap B_1 = \emptyset)$$

$$\Leftrightarrow \exists b \in N\, (<p^{-1}(a),b> \in W_1 \;\&\; \kappa \times h''b \cap B = \emptyset).$$

Here we view $p^{-1}(a)$ as a non-empty $\mathfrak{A}$-finite subset of the actual $p^{-1}(a)$, chosen effectively by a selection operator. The above reduction then shows $A \leq_{\underset{\sim}{M}} B$. For the converse reduction assume $B_1 \leq_{\mathfrak{A}} p^{-1}[A_1]$ via W_2. Let $r(a) = \text{least}\ \sigma[h^{-1}[\pi_1''a] \cap B_1^\sigma =$

$h^{-1}[\pi_1''a] \cap B_1]$. Then r is total by the regularity of B_1 in $\mathfrak{A}$ and r is calculated $\underset{\sim}{M}$-recursively from B_1 using $\mathfrak{A}$-finite information about B_1. Thus

$$a \cap B = \emptyset \Leftrightarrow \exists\sigma, b \in N[r(a) = \sigma \ \& \ b = h^{-1}[\pi_1''a] \cap B_1^\sigma$$

$$\& \ (\kappa-\sigma) \times h''b \cap a = \emptyset \ \& \ (\forall\tau < \sigma)(\forall x \in b)(x \in B_1^\tau \Rightarrow \langle\tau, h(x)\rangle \notin a)]$$

$$\Leftrightarrow \exists c \in N[\langle a, c\rangle \in V \ \& \ c \cap B_1 = \emptyset]$$

for some $\underset{\sim}{M}$-r.e. set V. Furthermore for $c \in N$,

$$c \cap B_1 = \emptyset \Leftrightarrow \exists d \in N[\langle c, d\rangle \in W_2 \ \& \ d \cap p^{-1}[A_1] = \emptyset]$$

$$\Leftrightarrow \exists d \in N[\langle c, d\rangle \in W_2 \ \& \ p''(d \cap \mathrm{dom}\, p) \cap A_1 = \emptyset]$$

$$\Leftrightarrow \exists d \in N[\langle c, d\rangle \in W_2 \ \& \ \bigcup p''(d \cap \mathrm{dom}\, p) \cap A = \emptyset].$$

Combining these reductions, using the fact that $\mathrm{dom}\, p$ is $\underset{\sim}{M}$-recursive, we have $B \le_{\underset{\sim}{M}} A$.

Clearly B is $\underset{\sim}{M}$-recursive. Note that

$$a \subseteq B \Leftrightarrow \exists\sigma < \kappa[a \subseteq \kappa \times h''\kappa \ \& \ h^{-1}[\pi_1''a] \subseteq B_1^\sigma$$

$$\& \ (\forall\tau < \sigma)(\forall x \in h^{-1}[\pi_1''a])(\langle\tau, h(x)\rangle \in a \Rightarrow x \in B_1^\tau)],$$

so B is t.r.e. Finally suppose $a \in M$. Using the regularity of B_1 choose $\sigma < \kappa$ such that $h^{-1}[\pi_1''a] \cap B_1^\sigma = h^{-1}[\pi_1''a] \cap B_1$. Then

$$a \cap B = a \cap [(\kappa-\sigma) \times (\pi_1''a \cap h''B_1^\sigma) \cup \bigcup_{\tau<\sigma} \{\langle\tau, h(x)\rangle : x \in B_1^\tau\}].$$

Thus $a \cap B \in M$ so B is regular.

<u>(iii) ⇒ (i)</u>. Suppose A is $\underset{\sim}{M}$-recursive. Let $A_1 = \{\langle 0, x\rangle : x \cap A \ne \emptyset\} \cup \{\langle 1, x\rangle : x \cap (M-A) \ne \emptyset\}$. Choose regular $\mathfrak{A}$-r.e. set $B_1 \subseteq \kappa$ such that $p^{-1}[A_1] \equiv_{\mathfrak{A}} B_1$ and let $B = h''B_1$. B is clearly t.r.e. We show $A \equiv_{\underset{\sim}{M}} B$. First suppose $p^{-1}[A_1] \le_{\mathfrak{A}} B_1$ via W_1. Viewing $p^{-1}(x)$ as in the previous case we have

$$a \subseteq A \Leftrightarrow \langle 1, a\rangle \notin A_1 \Leftrightarrow p^{-1}(\langle 1, a\rangle) \cap p^{-1}[A_1] = \emptyset$$

$$\Leftrightarrow \exists b \in N(\langle p^{-1}(\langle 1, a\rangle), b\rangle \in W_1 \ \& \ b \cap B_1 = \emptyset)$$

$$\Leftrightarrow \exists b \in N(\langle p^{-1}(\langle 1, a\rangle), b\rangle \in W_1 \ \& \ h''b \cap B = \emptyset).$$

The analogous reduction holds for negative neighbourhood conditions of A. Thus $A \leq_{\underset{\sim}{M}} B$. For the converse reduction assume $B_1 \leq_{\mathfrak{a}} p^{-1}[A_1]$ via W_2. Then

$$
\begin{aligned}
a \cap B = \emptyset &\Leftrightarrow h^{-1}[a] \cap B_1 = \emptyset \\
&\Leftrightarrow \exists b \in N[\langle h^{-1}[a], b\rangle \in W_2 \quad \& \quad p''(b \cap \operatorname{dom} p) \cap A_1 = \emptyset] \\
&\Leftrightarrow \exists b \in N[\langle h^{-1}[a], b\rangle \in W_2 \quad \& \quad \cup \pi_1''(p''(b \cap \operatorname{dom} p) \cap \{0\} \times M) \cap A = \emptyset \\
&\qquad \& \quad \cup \pi_1''(p''(b \cap \operatorname{dom} p) \cap \{1\} \times M) \subseteq A] . \qquad \square
\end{aligned}
$$

Now we can transfer results for admissible structures from [11] to weakly inadmissible structures, using theorems 3.6 and 4.1. As a sample we give

Theorem 4.2. Suppose $\underset{\sim}{M}$ is an adequate weakly admissible structure. Then

(i) Every non-zero t.r.e. $\underset{\sim}{M}$-degree is the join of two strictly smaller (and hence incomparable) t.r.e. $\underset{\sim}{M}$-degrees.

(ii) There is a largest t.r.e. $\underset{\sim}{M}$-degree.

(iii) If $\underset{\sim}{a}$ is a t.r.e. $\underset{\sim}{M}$-degree strictly between $\underset{\sim}{0}$ and the largest t.r.e. $\underset{\sim}{M}$-degree then there is a t.r.e. $\underset{\sim}{M}$-degree incomparable with $\underset{\sim}{a}$.

References

[1] K.J. Barwise, Admissible Sets and Structures, Springer Verlag, 1975.

[2] K.J. Devlin, Aspects of Constructibility, Springer Lecture Notes in Mathematics no. 354, 1973.

[3] J.E. Fenstad, Computation theories: An axiomatic approach to recursion on general structures, in: Logic Conference Kiel 1974, Springer Lecture Notes in Mathematics no. 499, 1975, 143-168.

[4] S. Friedman, β-recursion theory, to appear.

[5] ———, Post's problem without admissibility, to appear.

[6] S. Friedman, An introduction to β-recursion theory, this volume.

[7] ————, Negative solutions to Post's problem, I, this volume.

[8] W. Maass, Inadmissibility, tame r.e. sets and the admissible collapse, to appear.

[9] Y.N. Moschovakis, Axioms for computation theories - first draft, in: R.O. Gandy and C.E.M. Yates (eds.), Logic Colloquium '69, North-Holland, 1971, 199-255.

[10] S.G. Simpson, Post's problem for admissible sets, in: J.E. Fenstad and P.G. Hinman (eds.), Generalized Recursion Theory, North-Holland, 1974, 437-441.

[11] V. Stoltenberg-Hansen, Finite injury arguments in infinite computation theories, Preprint series in mathematics, No. 12, 1977, Oslo.

[12] ————, A regular set theorem for infinite computation theories, Preprint series in mathematics, No. 15, 1977, Oslo.

J.E. Fenstad, R.O. Gandy, G.E. Sacks (Eds.)
GENERALIZED RECURSION THEORY II

The 1-Section of a Non-Normal Type - 2 Object

Stanley S. Wainer.

Leeds University.

When analyzing recursions in a continuous functional F one naturally adopts a formally constructive attitude towards application $F(g) = x$ by first simply regarding F as a rule which elementwise transforms a given sequence $\langle g_i \rangle_{i<\omega}$ of finite approximations to g into a sequence $\langle x_i \rangle_{i<\omega}$ with limit x. (This is the basic idea behind Hyland's filter - and limit - space approaches to the continuous functionals in [4]). The rule associated with F is encodable as a real h_F, trivially recursive in F. However the value x of F(g) cannot, in general, be computed from h_F alone - we need in addition a "modulus of continuity" for the sequence $\langle x_i \rangle_{i<\omega}$, i.e. a number m such that $x_i = x_m$ for every $i > m$. An associate of F (Kleene[5]) is a real a_F which encodes h_F together with a suitable "modulus-of-continuity function" which determines for each g a number m such that F is constant on all extensions of $\bar{g}(m)$. Thus for irreducible type-2 continuous F's, of which there are by now several interesting examples (Bergstra [1], Gandy and Hyland [2], Normann [6]), we have $h_F <_T F <_T a_F$. The aim of this paper is to bridge the gaps between h_F and F and between F and a_F by (i) finding a structural characterization of 1-sc(F) in terms of h_F alone, and (ii) finding a weaker (intensional) notion of "associate" which is more closely related t the recursion theory of F.

Clearly the 1-section of a type-2 F is generated entirely "from within" in the sense that F is only applied to functions already recursive in F (so if G coincides with F on 1-sc(F) then 1-sc(G) = 1 - sc(F)). Thus we may as well broaden our scope here by considering those type-2 functionals F which are <u>continuous on their 1-sections</u> (i.e. those F's such that, given any g recursive in F and any recursive-in-F sequence $\langle g_i \rangle_{i<\omega}$ with $\lim g_i = g$,

$\lim F(g_i) = F(g)$). Grilliot [3] proved that these are exactly the non-normal functionals (i.e. those in which ${}^2E$ is not recursive), and our results here are obtained as refinements of his work.

From now on F will be an arbitrary non-normal type-2 object. With F is associated the real h_F defined, for finite sequences s, by

$$h_F(s) = F(\lambda x . s(x))$$

where s(x) is the x-th component of s if $x <$ length (s) and $s(x) = 0$ otherwise. Clearly $h_F \in 1\text{-sc}(F)$. Now in order to compute F(g) for any $g \in 1\text{-sc}(F)$ it suffices to be given any recursive-in-F sequence $\langle g_i \rangle$ of finite sequences approximating to g. For then, by the non-normality of F,

$$F(g) = \lim h_F(g_i) = h_F(g_{i_o})$$

for any i_o satisfying $\forall i \geqslant i_o (h_F(g_i) = h_F(g_{i_o}))$. F therefore coincides, on its 1-section, with a functional partial recursive in the jump of h_F, and so $1\text{-sc}(F) \subseteq \Delta_2^0(h_F)$ whereas in contrast, $1\text{-en}(F) = \Pi_1^1(h_F)$. From this we see that every $g \in 1\text{-sc}(F)$ can be canonically represented by a certain sequence $\langle g_i \rangle$ primitive recursive in h_F, and hence that F(g) can be computed from h_F and the (finite) set $D_g = \{i \mid \exists j > i (h_F(g_j) \neq h_F(g_i))\}$ which is clearly r.e. in h_F uniformly in an index for g. We will show that each such D_g is itself recursive in F, again uniformly in g's index. From this will follow our characterization of non-normal 1-sections, as certain inductively-defined coded limit-spaces L(h), generated by $\Pi_1^1(h)$ initial segments of the r.e.-in-h degrees, thus answering a question of Normann [6].

The Inductive Limit Spaces L(h).

For a general discussion of limit spaces and their applications to the theory of continuous functionals see Hyland [4]. In our present, more restricted, context suffice it to say that a limit space is just a collection of reals g determined by certain sequences of finite approximations $\langle g_i \rangle_{i<\omega}$. We write "$\lim g_i = g$" or "$g_i \to g$" to mean $\forall x \exists k \forall i \geq k \; (g_i(x) = g(x))$. Notice that if $f_i \to f$ then the functions recursive in f form a limit space in which each $\{e\}^f$ is determined by the primitive-recursive-in- $\langle f_i \rangle$ sequence $\langle \{e\}_i^{f_i} \rangle_{i<\omega}$ where as usual, $\{e\}_i^{f_i}$ is the sequence of length i such that for each $x < i$, $\{e\}_i^{f_i}(x) = U(\mu y < i \; T(\bar{f}_i(y), e, x))$ if there is such a y, and 0 otherwise.

We begin here with a completely arbitrary function h, which we think of as an elementwise map from sequences $\langle s_i \rangle_{i<\omega}$ to sequences $\langle x_i \rangle_{i<\omega}$ where $x_i = h(s_i)$. The aim is to build up a "good" recursion - theoretic limit space relative to this map, wherein the "index" of each real g encodes a canonical (primitive) - recursive - in - h sequence determining g. Our earlier considerations, and those in [4], suggest that if $g = \lim g_i$ is in the space, then so should be the set $D_g = \{i \mid \exists j > i (h(g_j) \neq h(g_i))\}$. Thus L(h) will be generated from h by closing off under relative recursion and applications of this scheme. (Note however that the arbitrariness of h means that the sequences $\langle h(g_i) \rangle_{i<\omega}$ need not converge, so the sets D_g may sometimes = N). There are now two options open to us in defining L(h): either we write down Kleene - type schemes for partial functions and then extract the total objects, or we generate directly a set of codes for the total functions we are interested in. We will take the second, more hierarchical, approach as it seems rather more convenient for our purposes. In order to generate the required codes we first need some technical machinery for constructing canonical sequences.

<u>Definition</u>. Let $[x]^h$, $x < \omega$, be a standard enumeration of all functtions primitive recursive in h. Then a <u>limit index</u> for g is a number e such that for each i, $[e]^h(i)$ is a finite sequence g_i, and $g = \lim g_i$.

<u>Lemma 1</u> (Canonical Sequences). (i) There is a primitive recursive

d such that if z is a limit index for f and $\{e\}^f$ is total then $d(e,z)$ is a limit index for (the characteristic function of) the set $D = \{n \mid \exists m > n(h(\{e\}_m^{f_m}) \neq h(\{e\}_n^{f_n}))\}$.

(ii) There is a primitive recursive function l such that if z is a limit index for f and for each $x, e_x = \{e\}^f(x)$ is a limit index for (the characteristic function of) a set E_x, then $l(e,z)$ is a limit index for the set $E = \{\langle n,x\rangle \mid n \in E_x\}$.

Proof. (i) Let s_i be the finite sequence such that for $n < i$, $s_i(n) = 1$ if $\exists m(n < m < i \ \& \ h(\{e\}_m^{f_m}) \neq h(\{e\}_n^{f_n}))$ and 0 otherwise. Since $f_m = [z]^h(m)$ for each m, $\lambda_i . s_i = [d]^h$ for some d primitive recursively computable from e and z. Clearly $D = \lim s_i$.

(ii) Let s_i be the finite sequence such that for $\langle n,x\rangle < i$, $s_i(\langle n,x\rangle)$ is computed as follows: First find $e_{x,i} = \{e\}_i^{f_i}(x)$. Then find the greatest $j < i$ (if there is one) such that $[e_{x,i}]^h(j)$ can be computed within i steps. If there is such a j and $[e_{x,i}]^h(j)$ is a sequence, give out its n-th component as the value of $s_i(\langle n,x\rangle)$. Otherwise put $s_i(\langle n,x\rangle) = 0$. Clearly $\lambda i . s_i = [l]^h$ for some l primitive recursively computable from e and z, and it's easily checked that $E = \lim s_i$.

We will show in Lemma 2 that the sequences built up by means of Lemma 1 have the crucial property that if $\langle s_i\rangle$ is any such sequence and $s_i \to g$ then a "modulus" for $\langle s_i\rangle$ can be computed from g. But first, the definition of L(h):

Definition. The system of codes C^h and sets D_a^h for $a \in C^h$, are defined inductively according to the following three clauses (we henceforth omit the superscript h). For each code a, $(a)_1$ will be a limit index for D_a, and we shall denote the correponding sequence $\langle [(a)_1]^h(i)\rangle_{i<\omega}$ of approximations to D_a by $\langle D_{a,i}\rangle_{i<\omega}$.

(1) $\langle 0,k\rangle \in C$ where k is a fixed limit index for the sequence $\langle \bar{h}(i)\rangle_{i<\omega}$, and $D_{\langle 0,k\rangle} = h$.

(2) If $a \in C$ and $\{e\}^{D_a}$ is total then $b = \langle 1, d(e,(a)_1), e, a\rangle \in C$ and and $D_b = \{n \mid \exists m > n(h(\{e\}_m^{D_{a,m}}) \neq h(\{e\}_n^{D_{a,n}}))\}$.

(3) If $a \in C$ and $\phi = \{e\}^{D_a}$ is a total function such that $\phi(0) = a$ and $\phi(x) \in C$ for every x, then $c = \langle 2, l(e_1,(a)_1), e, a\rangle \in C$, where e_1 is such that $\{e_1\}^{D_a}(x) = (\{e\}^{D_a}(x))_1$, and $D_c = \{\langle n,x\rangle \mid n \in D_{\phi(x)}\}$.

Finally we set $L(h) = \{f \mid f \leqslant_T D_a$ for some $a \in C\}$.

The following is an extension of Shoenfield's Modulus Lemma to $L(h)$.

Lemma 2.

(i) There is a recursive function M such that if $a \in C$ then for every x and every $i \geqslant \{M(a)\}^{h,D_a}(x)$, $D_{a,i}(x) = D_a(x)$.

(ii) There is a recursive function N such that if $a \in C$ and $\{e\}^{D_a}$ is total then $\{N(e,a)\}^{h,D_a}$ is total and for each n and every $i \geqslant \{N(e,a)\}^{h,D_a}(n)$, $\{e\}_i^{D_{a,i}}(x) = \{e\}^{D_a}(x)$ for all $x \leqslant n$.

Proof.

(i) M is defined by induction over $a \in C$, using the Recursion Theorem:

If $a = \langle 0,k\rangle$ then it is only necessary to choose $M(a)$ so that $\{M(a)\}^{h,D_a}(x) = x+1$.

If $a = \langle 1, d(e,(b)_1), e, b\rangle \in C$ then by Lemma 1, $D_{a,i} = \{x \mid \exists m$ s.t. $(x < m < i \;\&\; h(\{e\}_m^{D_{b,m}}) \neq h(\{e\}_x^{D_{b,x}}))\}$. Thus we let $i_x = \mu i(D_{a,i}(x) = D_a(x))$, then for every $i \geqslant i_x$, $D_{a,i}(x) = D_a(x)$. But $\lambda i,x.\, D_{a,i}(x)$ is primitive recursive in h with index computable from a, and so i_x is recursive in h, D_a with index $M(a)$ given by some fixed primitive recursive function of a.

If $a = \langle 2, l(e_1,(b)_1), e, b\rangle \in C$ then $D_a = \{\langle n,x\rangle \mid n \in D_{\phi(x)}\}$ where $\phi = \{e\}^{D_b}$. Recall the definition of $\langle D_{a,i}\rangle_{i<\omega}$ as given in the proof of Lemma 1(ii). Given $\langle n,x\rangle$ we want to compute i_0 such that $D_{a,i}(\langle n,x\rangle) = D_a(\langle n,x\rangle)$ for every $i \geqslant i_0$. First, using the induction hypothesis and part (ii) below, we can compute from $M(b)$, e, h and $D_b = \{m \mid \langle m,0\rangle \in D_a\}$, a number i_1 such that the limit $\phi(x)$ of the sequence $\langle \{e\}_i^{D_b,i}(x)\rangle_{i<\omega}$ is "decided" by stage i_1. Then, again by the induction hypothesis, we can compute from $M(\phi(x))$, h and $D_{\phi(x)} = \{m \mid \langle m,x\rangle \in D_a\}$, a number i_2 such that $D_{\phi(x),i}(n) = [(\phi(x))_1]^h(i)(n)$ is constant from stage $i=i_2$ onward. Finally we compute i_3 such that $[(\phi(x))_1]^h(i_2)$ is computable from h within i_3 steps. We therefore need to set $i_0 = \max(i_1,i_2,i_3)$, and notice that i_0 is recursive in h,D_a with index $M(a)$ given by a fixed primitive recursive function of a and m, where m is an index of M.

(ii) Given n we want to compute i_n such that for every $i \geq i_n$, $\{e\}_i^{D_a,i}(x) = \{e\}^{D_a}(x)$ for each $x \leq n$. This is done as follows: first find the least y_n such that $\forall x \leq n \, \exists y < y_n \; T(\overline{D_a}(y),e,x)$. Then compute $z_n = \max_{y<y_n} \{M(a)\}^{h,D_a}(y)$, so that for every $i \geq z_n$, $D_{a,i}(y) = D_a(y)$ for each $y < y_n$. It then suffices to put $i_n = \max(y_n,z_n)$. Clearly i_n is recursive in h,D_a with index $N(e,a)$ primitive recursively computable from e and $M(a)$.

Although the spaces $L(h)$ may be of some independent interest, we are henceforth concerned only with the case where $h = h_F$ for some non-normal type-2 F. (Thus each D_b defined by clause (2) will be finite).

The Characterisation of Non-Normal 1-sections.

Theorem 1. For any non-normal type-2 object F,

$$1\text{-sc}(F) = L(h_F).$$

This is proved by means of Lemmas 3 and 4 below, but first we need a convenient definition of 1-sc(F). We will assume from now on that F is such that, for every real g, $g(0) = (F(g))_0$. Clearly no generality is lost by this, since we can always deform F to $F' = \lambda g . \langle g(0), F(g)\rangle$ so that $1\text{-sc}(F) = 1\text{-sc}(F')$ and $L(h_F) = L(h_{F'})$. Now by [7] the 1-section of F consists of just those reals which are primitive recursive in the f_a's, $a \in \mathcal{O}^F$, generated inductively according to the scheme: $f_1(x) = 0$, $f_{2^a}(x) = F([x]^{f_a})$, and $f_{3^a5^e}(x) = f_{\psi(x_0)}(x_1)$ where $\psi = [e]^{f_a}$.

Lemma 3. There are primitive recursive functions p and q such that if $a \in \mathcal{O}^F$ then f_a is recursive in $D^{h_F}_{p(a)}$ with index q(a). Hence $1\text{-sc}(F) \subseteq L(h_F)$.

Proof. p and q are defined simultaneously by induction over $a \in \mathcal{O}^F$, using the Recursion Theorem:

If $a = 1$ then put $p(a) = \langle 0,k \rangle$ and choose q(a) to be an index of the zero function.

If $a = 2^b \in \mathcal{O}^F$ then by the induction hypothesis, f_b is recursive in $D_{p(b)}$ with index q(b) and so for each x, $f_a(x) = F([x]^{f_b}) = F(\{e_x\}^{D_{p(b)}})$ where $e_x = tr(x, q(b))$ for some primitive recursive function tr. Let $\phi(0) = \langle 0,k \rangle$ and $\phi(x+1) = \langle 1, d(e_x, (p(b))_1), e_x, p(b) \rangle$. Then $\phi = \{z\}^h$ for some z primitive recursively computable from p(b) and q(b), and every $\phi(x)$ is a code $\in C$.

Now set $p(a) = \langle 2, l(z_1,k), z, \langle 0,k \rangle \rangle$. It remains to show that f_a is recursive in $D_{p(a)} = \{\langle n,x \rangle \mid n \in D_{\phi(x)}\}$. But for each x, $f_a(x) = \lim h_F(\{e_x\}_i^{D_{p(b)},i}) = h_F(\{e_x\}_m^{D_{p(b)},m})$ where m is the least element of $\overline{D_{\phi(x+1)}} = \{n \mid \langle n, x+1 \rangle \notin D_{p(a)}\}$. Therefore, since the sequence $\langle \{e_x\}_i^{D_{p(b)},i} \rangle_{i<\omega}$ is primitive recursive in h_F and $h_F = D_{\phi(o)}$, f_a is recursive in $D_{p(a)}$ with index q(a) primitive recursively computable from p(b) and q(b). Note that ϕ is recursive and each $D_{\phi(x)}$ is (uniformly) r.e. in h_F, and so $D_{p(a)}$ is r.e. in h_F.

Finally, if $a = 3^c5^e \in \mathcal{O}^F$ then $f_a(x) = f_{\psi(x_0)}(x_1)$ where $\psi = [e]^{f_c}$, and by the induction hypothesis, each $f_{\psi(y)}$ is recursive in $D_{p(\psi(y))}$ with index $q(\psi(y))$. Thus from c,e, and an index of p, we can first primitive recursively compute p(a) so that $D_{p(a)} = \{\langle n,y \rangle \mid n \in D_{p(\psi(y))}\}$, and then compute q(a) using an index of q, such that f_a is recursive in $D_{p(a)}$ with index $q(a)$.

Lemma 4. There is a primitive recursive function r such that if $c \in C^{h_F}$ then $D_c = \{r(c)\}^F$. Hence $L(h_F) \subseteq 1\text{-sc}(F)$.

Proof. Again, r will be defined by induction over $c \in C$ using the Recursion Theorem.

The cases $c = \langle 0,k \rangle$ (defining h_F) and $c = \langle 2, l(e_1,(a)_1), e, a \rangle$ (effective joins) are straightforward.

Suppose $c = \langle 1, d(e,(a)_1), e, a \rangle \in C$. Inductively we can assume $D_a = \{r(a)\}^F$ and so the function $g = \{e\}^{D_a}$ is recursive in F with index primitive recursively computable from e and r(a). For each i let g_i be the finite sequence $\{e\}_i^{D_{a,i}}$ so $\lambda i.g_i$ is primitive recursive in h_F and $g_i \to g$. We must show that the set $D_c = \{n \mid \exists m > n(h_F(g_m) \neq h_F(g_n))\}$ is recursive in F, uniformly in c (D_c is of course finite). For each n compute the function g_n^* from h_F, D_a as follows: given x, first compute $i_x = \{N(e,a)\}^{h,Da}(x)$ by Lemma 2(ii). Then see if there is an m such that $n < m < i_x$ and $h_F(gm) \neq h_F(g_n)$. If there is one let m_o be the least such and put $g_n^*(x) = g_{m_o}(x)$. If there is no such m by stage x set $g_n^*(x) = g(x)$. Notice that if x is the first stage at which m_o appears then the same m_o will be used at every later stage and for every $y < x$, $g_{m_o}(y) = g(y) = g_n^*(y)$. Thus $g_n^* = \lambda x.g_{m_o}(x)$ and hence $F(g_n^*) = h_F(g_{m_o})$ if such an m_o is ever encountered, and $g_n^* = g$ otherwise. Therefore $n \in D_c$ if and only if either $h_F(g_n) \neq F(g)$, or $h_F(g_n) = F(g)$ but $h_F(g_n) \neq F(g_n^*)$, and so D_c is recursive in F with index r(c) primitive recursively computable from e, a and r(a).

Remark. The above computation of D_c typifies recursions in a non-normal F, and shows that in certain contexts, F can compute a "single" quantifier. This idea was first brought out by Grilliot [3], and later applied by Bergstra [1].

Theorem 2. A class of functions K is the 1-section of a non-normal type-2 object if and only if $K = L(h)$ for some h satisfying the two conditions:

(i) $\forall f \in L(h)\ \exists n\ \forall s(s \text{ extends } \bar{f}(n) \rightarrow h(s) = h(\bar{f}(n)))$.

(ii) $\forall s\ (h(s) = h(s*0))$.

Proof. If $K = 1\text{-sc}(F)$ with F non-normal, then h_F automatically satisfies (i), (ii) and $K = L(h_F)$ by Theorem 1. Conversely if $K = L(h)$ with h satisfying (i) and (ii) then by (i) we can define a functional F by $F(f) = \lim h(\bar{f}(i))$ if the limit exists, 0 otherwise; and by (ii) $h_F(s) = F(s*0*0*0*....) = h(s)$. The proof of Lemma 3 then shows that $1\text{-sc}(F) \subseteq L(h)$, and so F is non-normal and $K = L(h) = L(h_F) = 1\text{-sc}(F)$.

Remark. Condition (ii) can be slightly weakened to

(ii)′ $\forall s\ \exists x\ (h(s) = h(s*x))$.

Also notice that the class $L(h)$ is highly sensitive to variations in h, even within the same primitive recursive degree. For example, if h is everywhere zero the $L(h)$ consists of just the recursive functions, whereas if $h = h_G$ for Normann's functional G [6] then $L(h)$ contains an ω_1-high tower of r.e. degrees.

Theorem 3. For every non-normal type-2 object F there is a $\Pi^1_1(h_F)$ set B of r.e.-in-h_F indices such that $1\text{-sc}(F) = \{g \mid g \leq_T W_e^{h_F} \text{ for some } e \in B\}$. i.e. $1\text{-sc}(F)$ is generated by its r.e.-in-h_F elements.

Proof. If $g \in 1\text{-sc}(F)$ then $g \leq_T f_a$ for some $a \in \mathcal{O}^F$ of the form $a = 2^b$. But, as was noted in the proof of Lemma 3, each such f_a is recursive in some $D_{p(a)}$ which is r.e. in h_F. Thus we need only let B consist of those numbers e such that $W_e^{h_F} = D_c$ for some code $c \in C$. That B is Π^1_1 in h_F follows immediately from the inductive definition of $\{\langle c,x,0\rangle \mid c \in C \ \&\ x \notin D_c\} \cup \{\langle c,x,1\rangle \mid c \in C \ \&\ x \in D_c\}$.

Corollary. By Sacks' Density Theorem (suitably relativized) every topless 1-section of a type-2 object contains dense chains of degrees. Thus no topless well-founded initial segment of degrees can form a 1-section.

Remark. In [6] Normann constructed a continuous G with a non-collapsing hierarchy, by iterating Bergstra's [1] "small jump" operator F_a^b along a recursive ordering of r.e. sets whose maximal well-founded initial segment is Π_1^1 but not Δ_1^1. Theorem 3 suggests that for each non-normal F it might be possible to construct a continuous G_F, along similar lines to Normann's example, such that $1\text{-sc}(F) = 1\text{-sc}(G_F)$.

Weak Associates.

Just as an associate a_F extensionally encodes the behaviour of a continuous functional F on all reals, so a "weak associate" α_F will intensionally describe the continuous behaviour of a non-normal functional F on the reals $\{e\}^F$, but only with respect to certain canonical sequences approximating $\{e\}^F$.

Given $\{e\}^F \in 1\text{-sc}(F)$ we can, by Lemma 3, compute from e a code $c \in C^{h_F}$ and an index e_1 such that $\{e\}^F = \{e_1\}^{D_c}$. From e_1 and c we can then compute a limit index j(e) such that for each i, $[j(e)]^{h_F}(i) = \{e_1\}_i^{D_{c,i}}$ and hence $\lim [j(e)]^{h_F}(i) = \{e\}^F$. The function j is primitive recursive and we call $<[j(e)]^{h_F}(i)>_{i<\omega}$ the canonical sequence for $\{e\}^F$. A "modulus" for the sequence $<h_F([j(e)]^{h_F}(i))>_{i<\omega}$ approximating $F(\{e\}^F)$ is then any number n in $\overline{D_{c'}}$, where $c' = <1, d(e_1(c)_1), e_1, c>$. But $D_{c'} = \{r(c')\}^F$ by Lemma 4, and so we have:

Lemma 5. There is a function m_F partial recursive in F such that whenever $\{e\}^F$ is total then $\lambda n.m_F(e,n)$ is the characteristic function of $\{n \mid \forall i \geq n(h_F([j(e)]^{h_F}(i)) = F(\{e\}^F))\}$.

Definition. A weak associate for a non-normal type-2 functional F is a partial function α such that whenever $\{e\}^F$ is total then

(i) $\lambda n.\alpha(e,n)$ is total and $\exists n(\alpha(e,n) > 0)$, and

(ii) if $\alpha(e,n) > 0$ then $\forall i \geq n(h_F([j(e)]^{h_F}(i)) = F(\{e\}^F) = \alpha(e,n)-1)$.

The final result re-casts Theorem 1 in terms of weak associates, and suggests that from a purely recursion-theoretic point of view, the intensional notion of "associate" is perhaps the more appropriate one.

Theorem 4. There is a partial recursive functional Φ such that for each non-normal type-2 object F, $\alpha_F = \lambda e,n.\Phi(F,e,n)$ is a weak associate for F and $1\text{-sc}(F) = 1\text{-sc}(\alpha_F)$.

Proof. Our work throughout has been completely uniform in F. In particular there is a fixed index u such that for every non-normal F, $\{u\}^F$ is the m_F of Lemma 5. Thus we need only define Φ so that $\Phi(F,e,n) = 1 + h_F([j(e)]^{h_F}(n))$ if $m_F(e.n) \simeq 1$, 0 if $m_F(e.n) \simeq 0$. Then every total function recursive in α_F will be recursive in F and conversely, since $F(\{e\}^F) = \alpha_F(e,\mu n(\alpha_F(e,n) > 0)) - 1$ for total $\{e\}^F$, an application of the Recursion Theorem will yield for each e an e′ such that if $\{e\}^F$ is total then $\{e\}^F = \{e'\}^{\alpha_F}$.

References.

[1] J.A. Bergstra 1976, Computability and continuity in finite types, Dissertation, Utrecht.

[2] R.O. Gandy and J.M.E. Hyland 1977, Computable and recursively countable functions of higher type, in: Logic Colloquium 76, North-Holland, Amsterdam, pp. 407-438.

[3] T. J. Grilliot 1971, On effectively discontinous type-2 objects J.S.L. 36, 245-248.

[4] J.M.E. Hyland 1977, Filter spaces and continuous functionals, to appear.

[5] S.C. Kleene 1959, Countable functionals, in Constructivity in Mathematics, North-Holland, Amsterdam, 81-100.

[6] D. Normann 1976, A continous type-2 functional with a non-collapsing hierarchy, J.S.L. to appear.

[7] S.S. Wainer 1974, A hierarchy for the 1-section of any type two object, J.S.L. 39, 88-94.